Glimpses of Indian Organic Farming

NIPA® GENX ELECTRONIC RESOURCES & SOLUTIONS P. LTD.
New Delhi-110 034

About the Authors

Dr. Mude Ramesh Naik, is currently working as Scientist (SS) in Research Systems Management Division, at ICAR-National Academy of Agricultural Research Management (NAARM), Hyderabad, Telangana State. He belongs to 97th Foundation Course for Agricultural Research Services (FOCARS) batch of Agricultural Research Service (ARS). He is specialized in Agronomy. He worked as a Scientist at ICAR-Central Research Institute for Jute & Allied Fibres (ICAR-CRIJAF), Barrackpore, Kolkata from 2013 to 2019. Since 2019, he has been serving in ICAR-NAARM as Scientist (SS). He has published research papers (22) in national and international journals, authored books (7), book chapters (5), policy papers (2), technical bulletins (7), booklets/leaflets (11), popular articles (16), workshop/conference organized (3), training manual edited (2). Also he has organized 22, 9 and 5 need based skill development training programmes for farmers, scientists/faculties and students, respectively. He is also an active member in 13 Societies and received best scientist awards and oral presentation awards from different professional societies. He is also acting as editor in various research journals and popular magazines.

Areas of Interest: Nutrient Management, Resource conservation technologies, organic farming and weed management.

Dr. A.V. Ramanjaneyulu has obtained B.Sc. (Ag.) in 1998 and M.Sc. (Ag.)-Agronomy in 2000 from Acharya N.G. Ranga Agricultural University, Andhra Pradesh and Ph.D (Agro.) in 2006 from IARI, New Delhi. Currently he is working as Principal Scientist (Agro.) & Head, AICRP on Agroforestry at P.J.T. Agricultural University, Hyderabad, India. He is a recipient of Dr P.S. Deshmukh Young Agronomist award from ISA, New Delhi, ICAR-Best Scientist award (PJTAU), Meritorious Research scientist award (ANGRAU), Best editor award of IJBSM, Kolkata etc., He has published 70 research articles, 76 popular articles, >40 book chapters, 04 edited text books, 01 policy paper, several technical bulletins and is acting as Managing editor (Associate) of IJBSM and IJEP and Managing editor of Chronicle of Bioresource Management and reviewed articles for several national and international journals.

Areas of Interest: Organic and natural farming, Dryland Agriculture, Micro-irrigation, Agroforestry

Glimpses of Indian Organic Farming

M. Ramesh Naik
Scientist (SS)
Research Systems Management Division
ICAR-National Academy of Agricultural Research Management (NAARM)
Hyderabad, Telangana

A.V. Ramanjaneyulu
Principal Scientist (Agro.) & Head
AICRP on Agroforestry
P.J.T. Agricultural University
Hyderabad, Telangana

NIPA® GENX ELECTRONIC RESOURCES & SOLUTIONS P. LTD.
New Delhi-110 034

NIPA® GENX ELECTRONIC RESOURCES & SOLUTIONS P. LTD.

101,103, Vikas Surya Plaza, CU Block
L.S.C. Market, Pitam Pura, New Delhi-110 034
Ph : +91-11-43860225, Mob.: +91 9717133558, 9540816132
E-mail: newindiapublishingagency@gmail.com
Website: www.nipaersources.com

Print ISBN: 978-93-58877-76-2
ebook ISBN: 978-93-58875-46-1

Composed and Designed by NIPA®.

भा.कृ.अनु.प
ICAR

Dr. Shanti Kumar Sharma
Ph.D. (Agronomy), IARI
Assistant Director General
Human Resource Management

डॉ. शान्ति कुमार शर्मा
सहायक महानिदेशक (मानव संसाधन प्रबंधन)

भारतीय कृषि अनुसंधान परिषद
INDIAN COUNCIL OF AGRICULTURAL RESEARCH
कृषि शिक्षा संभाग/
AGRICULTURAL EDUCATION DIVISION
मानव संसाधन प्रबंधन एकक/ HRMUNIT
कृषि अनुसंधान भवन-II, पूसा, नई दिल्ली - 110012
Krishi Anusandhan Bhavan-II, Pusa, New Delhi - 110012

Message

Food production is connected with the balance of human activities & technologies with five mahabhootas of nature (Earth, Water, Energy, Air and Space). In this regard, organic farming has emerged as an alternative production system to fine tune this balance in modern era. Organic farming, rooted in the principles of ecological balance and sustainability, is not just an agricultural practice but a holistic approach to nurturing the environment, enhancing biodiversity and ensuring the well-being of all living beings. In today's world, where the challenges of climate change, soil degradation and food insecurity are ever-pressing, the significance of organic farming cannot be overstated. This book beautifully encapsulates the essence of India's rich agricultural heritage and the innovative strides we are making towards a more sustainable future.

The chapters within this book cover a wide array of topics from soil health management and crop residue utilization to organic pest and disease control, providing valuable knowledge and practical solutions for farmers, researchers and policymakers. The empirical evidence, field observations and traditional wisdom presented here serve as a robust foundation for fostering innovation and informed decision-making in sustainable agriculture.

I am confident that "Glimpses of Indian Organic Farming" will serve as an invaluable resource for those dedicated to creating a resilient equitable and sustainable agricultural landscape. May this book inspire and guide our collective efforts towards achieving ecological harmony and ensuring food security for future generations.

Dated: 20th December, 2024

Shanti Kumar Sharma

Telefax: 011-25842179, Phone: 011-25842381, 011-25842262/Extn. 1319 Email: shanti sharmal@icar.gov.in # Website: https://hrm.icar.gov.in/

Preface 1

Organic agriculture is a cultivation methodology that emphasizes using natural substances and processes, avoiding synthetic agrochemicals and GMOs to maintain ecological balance and biodiversity. Techniques such as crop rotation, composting, and biological pest control are employed to maintain soil fertility, manage pests, and promote biodiversity. The goal is to produce healthy, nutritious food while preserving soil fertility, biodiversity, and environmental health, thereby minimizing negative impacts on the environment and promoting the well-being of plants, animals, and humans. In the modern agricultural landscape, organic farming stands as a symbol of sustainability, health, and resilience, crucial for feeding a growing global population while safeguarding our planet.

India, with its ancient agricultural heritage and remarkable biodiversity, plays a significant role in the global quest for sustainable food systems. The book "Glimpses of Indian Organic Farming" showcases the contributions of scientists dedicated to advancing organic agriculture in India. Through research, innovative practices, and interdisciplinary collaborations, these experts have deepened our understanding of agro-ecological systems and developed practical solutions promoting environmental sustainability, human health, and social equity. Each chapter covers diverse topics, including organic farm design, research, extension and policy needs, guidelines for conducting research on organic farming, organic seed production, soil health management, and non-chemical pest and disease management approaches, offering valuable insights and recommendations for all stakeholders involved directly or indirectly in organic farming.

This book highlights the importance of science in evidence-based decision-making and fostering innovation in sustainable agriculture. By bridging the gap between theory and practice, it aims to ignite a transformative movement toward holistic and regenerative farming systems that nurture the land and the well-being of communities, the environment, and the planet. It serves as a valuable resource for researchers, educators, policymakers, and practitioners striving to create a more resilient, equitable, and sustainable future for agriculture in India and beyond.

Dr. Mude Ramesh Naik
Scientist, RSM Division
ICAR-NAARM, Hyderabad, India

Preface 2

The journey of organic farming in India is a testament to the nation's commitment to sustainable agricultural practices and environmental stewardship. "Glimpses of Indian Organic Farming" presents a comprehensive exploration of this transformative approach, highlighting the efforts of scientists, farmers, and policymakers dedicated to promoting ecological balance and biodiversity. This book delves into the principles and practices that define organic agriculture, offering a holistic perspective on how traditional wisdom and modern innovations converge to create resilient and sustainable farming systems.

Organic farming is more than an alternative agricultural methodology; it is a philosophy that values the health of the soil, the integrity of ecosystems, and the well-being of communities. This preface underscores the importance of understanding the interconnectedness of all components within the agricultural landscape. The book covered some of the unique chapters which are different from that of other published documents. Through chapters that cover topics such as nexus between agroforestry and organic farming, organic livestock production, liquid biofertilizers, efficient non-chemical approaches of biotic stress management, readers will gain insights into the multifaceted benefits of organic farming. These practices not only enhance soil fertility and crop productivity but also mitigate the impacts of climate change and ensure food security.

"Glimpses of Indian Organic Farming" is a tribute to the tireless efforts and pioneering spirit of those who have dedicated their lives to advancing organic agriculture in India. By showcasing empirical evidence, field observations, and success stories, this book aims to inspire and inform stakeholders across the agricultural spectrum. It is our hope that this work will serve as a catalyst for further innovation and collaboration, guiding us towards a future where sustainable farming practices are the norm, ensuring a healthy planet for generations to come.

Dr. A.V. Ramanjaneyulu
Principal Scientist (Agro.) & Head,
AICRP on Agroforestry,
P.J.T. Agricultural University, Hyderabad, India

Contents

Acronyms

ABDP	Association of Biodynamic Plant Breeders
APEDA	Agricultural and Processed Food Products Export Development Authority
AIFOF	All India Federation of Organic Farming
AMF	Arbuscular Mycorrhizal Fungi
BBF	Broad Bed and Furrow
BDF	Biodynamic Farming
BFOA	Breeding for Organic Agriculture
BGA	Blue Green Algae
Bt	*Bacillus thuringiensis*
BNF	Biological Nitrogen Fixation
CEC	Cation Exchange Capacity
CV	Cytoplasmic Virus
ETL	Economic Threshold Level
CWDE	Cell Wall Degrading Enzymes
CSI	Chitin Synthesis Inhibitors
COC	Copper Oxychloride
DAS	Days After Sowing
EB	Evolutionary Breeding
EC	Electrical Conductivity
EVM	Ethno-Veterinary Medicine
EPN	Entomo Pathogenic Nematodes
FYM	Farm Yard Manure
FAO	Food And Agriculture Organization
FFF	Food-Feed-Fallow
FSSAI	Food Safety and Standards Authority of India
GMOs	Genetically Modified Organisms
GHG	Greenhouse Gas
HDPE	High Density Polyethylene
ICAR	Indian Council of Agricultural Research
ITK	Indigenous Technical Knowledge

IOFS	Integrated Organic Farming System
IOAM	Indian Organic Agriculture Movement
ISOFAR	International Society of Organic Agriculture Research
IPM	Integrated Pest Management
IFOAM	International Federation of Organic Agriculture Movements
IFS	Integrated Farming System
IOFS	Integrated Organic Farming System
ICM	Integrated Crop Management
IGP	Indo-Gangetic Plains
IAA	Indole-3-Acetic Acid
KTI	Kuntz Trypsin Inhibitor
MT	Metric Tonnes
mt	Million Tonnes
MBC	Microbial Biomass Carbon
NMSA	National Mission for Sustainable Agriculture
NPOP	National Program for Organic Production
NF	Natural Farming
NPV	Nuclear Polyhedrosis Virus
NFTs	Nitrogen Fixing Trees
NC	Neem cake
NADH	Nicotinamide Adenine Dinucleotide
OA	Organic Agriculture
OF	Organic Farming
OPB	Organic Plant Breeding
OCIA	Organic Crop Improvement Association
OSA	Organic Seed Alliance
OLFs	Organic Liquid Formulations
PGPR	Plant Growth-Promoting Rhizobacteria
PM	Poultry Manure
PSB	Phosphate Solubilizing Bacteria
PPB	Participatory Plant Breeding
QPM	Quality Planting Material
SOC	Soil Organic Carbon
SOM	Soil Organic Matter
SCO	Seed Certification Officer
SDGs	Sustainable Development Goals

SMS	Straw Management System
USDA	United States Department of Agriculture
VAM	Vesicular Arbuscular Mycorrhiza
VOC	Volatile Organic Compounds
VC	Vermicompost
WHO	World Health Organization
ZBNF	Zero Budget Natural Farming

1

Conversion and Development of An Organic Farm

M.V. Ramana, T.L. Neelima and A.V. Ramanjaneyulu

Professor Jayashankar Telangana Agricultural University, Rajendranagar Hyderabad-500 030, Telangana, India

Abstract

Organic farming necessitates a shift from conventional to non-chemical agricultural methods and practices. For this, it needs creation of facilities for the production of organic inputs, adoption of crop rotation, green manuring and other best management practices for the organic production of crops and livestock. Further, the 'towards organic approach' in intensive agricultural areas and 'scientific organic farming' in niche areas have to be implemented for development of an organic farm. The synthesized information on how to convert an inorganic farm into an organic farm followed by its sustainable maintenance is the need of the hour. This chapter holistically deals with designing and maintaing the whole ecosystem organic.

Keywords: *Buffer zone, contamination, crop rotation, GM crops, organic farm*

1. Introduction

Nowadays, the growers, policymakers and consumers are talking about organic farming (OF) across the globe. For practitioners, it is still a confusion regarding what and how to practice OF. There is a wrong notion that simple addition of organic manures and the use of chemical pesticides for growing different crops is considered as organic farming. In sense real, organic farms are the cultivated fields without addition of chemical fertilizers, pesticides, livestock additives, etc., and only organic and (or) non-chemical methods are practiced, and the concept or practice is known as organic farming. A large number of farms have been converted from conventional into chemical farms during the last more than 50 years by using synthetic chemicals. In view of the increasing demand for organic products and spiraling soil and human health problems, there is need to transform them gradually into organic farms. So that, it can help enhance soil organic carbon status and fertility which in turn can support crop growth and produce healthy foods besides protecting the environment. However, the extent of chemical used might be varying from one farm to another depending on the crops cultivated, nutrient requirement, pest/ disease severity, availability and accessibility to chemical use, socio-economic condition of the farmers, geographical location, etc., Thus, such fields can be divided into

1. Areas with high external input consumption
   a) Less diverse, monocropping, cash crops, high profits and less risky
   b) More diverse, various enterprises, risky and less profits
2. Mixed farms with a medium level of external input use (Farmers are familiar with organic practices and can learn easily from others)
3. Areas with less to meager external input addition, rainfed, hilly and tribal areas

2. Conversion of an Inorganic Farm into an Organic Farm

The transformation of a conventional farm into an organic farm is not possible within a short time. It takes a minimum period of two years for annual crops and three years for perennial crops excluding pastures and meadows based on the soil, season, type of crop, adoption of organic practices, etc. The conversion to an organic farm needs a thorough understanding of the existing farms in terms of infrastructure, the extent of chemical used and economic condition of the growers (FAO, 2015). However, the conversion period may be extended or reduced by the certification bodies based on the history of land use and crop management. This period is known as the '*conversion or transition*' period. During this phase, organic practices should be followed liberally, scrupulously but flawlessly. Scientific evidence clearly established that conversion of high-intensive agriculture areas to organic systems lead to a reduction in crop yields considerably (up to 25-30%), especially during the initial 3-4 years; before the soil system regains and crop yields come to a comparable level. The development of an organic farm needs a holistic approach to foster agro-ecosystem health, maintenance of habitat and biodiversity. The four principles of health, ecology, fairness and care (Eyhorn *et al.*, 2003) have to be taken care off during conversion and thereafter.

2.1. Farms with high external input use, less diverse, monocropping cash crops, high profits and less risky

Farmers possessing such farms might be cultivating one or two commercial crops with the addition of a high quantity of synthetic fertilizers and pesticides, but, a meager emphasis on organics, over a longer period of time. They might be earning high profits thus less risk and have low climate resilience. Such farmers should adopt the following practices to convert inorganic farms to organic farms.

- Liberal application of organic manures, composts, practicing green manuring/green leaf manuring over and above the recommended doses should be followed. Synthetic chemicals in all forms should be disallowed.
- Grow low to medium-nutrient-requiring leguminous crops (chickpeas on heavy soils only, pigeonpea, mungbean, urdbean, soybean, beans, peanuts, lentil, lupins, alfalfa, clover) and millets (pearl millet, finger millet, foxtail millet, proso millet, kodo millet, barnyard millet, little millet, pseudo millets during initial years.
- Intercropping can also be practiced with the crop combinations during the rainy season as insurance against crop failure. So that, the yield penalty will be less thus farmers will not incur negative returns.
- Crop rotation with legume crops to improve soil fertility Grow trap crops and ornamental crops to repel the pests.
- Avoid crop residue burning but adopt on-site crop residue incorporation with suitable machinery for efficient on-farm recycling.
- Application of vermicompost prepared using cattle wastes and crop residues.
- Grow cover crops during the summer season to prevent loss of organic carbon driven by high temperature. It also reduces soil and water erosion.
- Give importance to livestock and their health.

2.2. Farms with high external input use, more diverse, various enterprises, risky and less profits

The owners of these farms might be raising a variety of crops and other enterprises with less reliance on artificial substances, earning less income but meeting the household needs besides some marketable surplus, thus farming is less risky. Their climate resilience capacity might be medium. The following practices should be adopted by them.

- Relinquish use of chemical fertilizers and pesticides completely
- Avoid residue burning rather incorporate *in-situ*
- Efficient use of on-farm resources and on-farm composting
- Soil health management should be given utmost importance by planting N fixing legumes and border planting of leguminous trees/fodders, crop rotation and intercropping
- Cost-effective soil, water and wind erosion control measures like contour farming, diversion drains, contour bunding, windbreaks, etc., should be adopted
- Keep constant vigilance on the maintenance and health of animals
- Integrated farming systems involving crop and livestock
- Avoid indiscriminate cutting of trees for firewood

2.3. Mixed farms

The managers of mixed farms might be maintaining various enterprises and having knowledge of organic methods and practices and earning less profits. They may be having better climate resilience. They are knowledgeable enough to seek and adopt suggestions from fellow farmers and experts if required. Such farmers should implement the following activities.

- Cover cropping and crop rotation to manage weeds, pests and diseases
- Recycle farm waste into manure followed by application to fields
- Use healthy seeds
- Monitor the pests
- Improve diversification in the farm to improve farm productivity and profitability
- Provide habitat for beneficial insects

2.4. Areas with less to meager external input addition, rainfed, hilly and tribal areas

The farmers in rainfed areas use nil to less amount of chemicals due to lack of moisture following partial or complete failure of monsoon, economic backwardness etc., Further, peasants in north-east, hilly and tribal areas are usually not habituated to apply chemicals due to high inherent organic carbon status, inaccessibility to such inputs following poor infrastructure facilities, weak socio-economic condition etc., These farmers can follow below mentioned measures to convert to/continue organic farming

- Use organically produced seeds or planting material only
- On-farm *in-situ* residue recycling
- Adopt crop and enterprise diversification for effective recycling of resources
- Mulching, crop rotation, cover cropping, etc.
- Maintain green leaf manure trees and legume fodders
- Protect biodiversity
- Keep away polluted, contaminated and industrial waters
- Maintain buffer zones to avoid drift from conventional fields
- Take maximum care during harvesting, threshing, storage and marketing to avoid mixing with chemically produced produce

3. Assessment of Existing Facilities at The Farm and Creation of New Facilities

An organic farm should have the following inventory of facilities and resources (Table 1). If not, these have to be created.

Table 1. Organic farm inventory of facilities and resources.

S.No.	Particulars
I. Animal related	
1	Cattle/sheep/goat shed(s)
2	Cow urine collection and maintenance tanks
3	Arrangement for cattle feed and storing mulching material
4	Border planting with few trees (eg. *Azadirachta indica*, *Tamarindus indica*, *Pongamia pinnata*, *Sesbania sesban* etc.,) for providing shade to composting unit and cattle sheds
II. Manure and plant protection related	
1	Compost pit(s) to aggregate cattle dung, sheep and goat manure and crop residues
2	Vermicompost/NADEP compost unit(s)
3	Arrangement for preparation of concoctions/formulations (beejamrit, jeevamrit, brahmastra, panchagavya, neem seed kernel extract etc.,)
4	Azolla unit
III. Quality and contamination related	
1	Buffer zones dividing adjacent conventional fields and organic farms
2	Soil and water analysis report
3	Soil and water conservation structure in accordance with land capability classification
IV. Irrigation water	
1	Well/farm pond with pumping unit
2	Diversion drains/trenches

V. Mechanization	
1	Suitable machinery for land preparation sowing, manure application, intercultivation, harvesting, threshing, winnowing, crop residue incorporation etc.,
2	Arrangement (threshing floor/tarpaulin sheets) for drying and safe storage (storage structures/godowns)
VI. Access to the market	

After analysing the existing infrastructure, the same may be discussed with the scientists and experienced organic growers to create required facilities depending on the need.

4. Getting Familiar with Agronomic Interventions for Organic Farming

4.1. How to create and maintain habitat and biodiversity?

The habitat and biodiversity in organic farms should be created and maintained. For this purpose, native plant vegetation with diverse plant species have to be maintained throughout the organic farm. So that, it enhances beneficial organisms like pollinators (bees and bats), improves groundwater recharge, provides shelter for birds and mammals. Further, it also filters pollution controls wind and water erosion.

Organic growers should also create conditions to improve the soil microbiome diversity by adding organic materials and growing deep rooting cover crops as a part of crop rotation. Border plantation with neem (*Azadirachta indica*), tamarind (*Tamarindus indica*) pongamia (*Pongamia pinnata)*, drumstick (*Moringa oleifera*), curry leaf (*Murraya koenigi*), glyricidia (*Glyricidia sepium*/*maculata*), ber (*Ziziphus mauritiana*), aonla (*Emblica officinalis*), subabul (*Leucaena leucocephala*) in the farm depending on the edapho-climatic conditions. They may harbour birds, predators and beneficial insects which in turn control the crop pests. Farmers can use the branches and leaf of glyricidia, pongamia, neem etc., as green leaf manure for incorporation into rice fields well ahead of transplanting. Some of the local plants having medicinal properties like *Adhathoda vasica*, *Vitex negundo*, *Calotropis procera*/*gigantia*, *Datura* spp., *Ipomoea carnea* etc. should be planted as they are useful for pest and disease control in crop plants and animals and also for preserving ecological diversity in organic farming (Bhanu *et al.,* 2017) Marigold can be grown on the rice field bunds as an ecological engineering measure to reduce the incidence of brown plant hopper. Marigold and castor when grown in rainfed or irrigated dry crops, act as trap crop against *Helicoverpa* sp. and *Spodoptera* sp., respectively.

4.2. How to prevent contamination

A contiguous area may be converted into an organic farm to avoid contamination from adjacent conventional fields. Further, genetically modified (GM) crops like Bt cotton, Bt mustard, Bt potato, golden rice, etc., must be avoided under organic farming. The land meant for growing organic crops should be free from prohibited substances. Polluted, contaminated, sewage water or industrial waters should not be allowed for irrigation purpose in organic farms.

4.2.1. Buffer zones

A designated land area to protect the organic field from conventional fields and safeguard the soil from any unwanted materials is known as a 'buffer zone'. They should also shield organic farms from chemical drift, arrest run-off from adjacent areas and cross-pollination from GM crops. Further, they should also attract pollinators, provide habitats for the natural enemies and encourage wild life and biodiversity. The size of the buffer zone depends on risk involved due to harmful substances, airborne and waterborne contaminants, pollution levels in soil and water, topography and location of the organic farm. Though there is no specific size, but, upto 50 feet (15 m) can be a safe width between organically managed crops and potential sources of contamination (Fig.1). They play a greater role in keeping the soil healthy and unsoiled. The produce from buffer zone may be sold in the regular market, but, not as organic.

The buffer zones can be maintained by planting perennial vegetation like alfalfa or other forage grasses or tall growing green manure crops or green leaf manure trees (subabul, pongamia, tamarind etc.,), maintaining wind breaks/shelter belts (*Casuariana, Pongamia, Azadirechta, Acacia planifrons, A. auriculiformis, Tamarindus, Albizzia, Peltophorum, Melia azadirechta, Eugenia, Mangifera, Ailanthus, Sesbania, Moringa, Eucalyptus sp., Artocarpus and Grevillea in dry and arid regions* and *Anacardium, Ailanthus, Acacia, Casuariana, Pongamia, Thespesia, Calophyllum, Mani, Gravillia, Cassia sp.* in coastal areas) (Fig.2) (https://agriinfo.in). Buffer zones may be avoided if there is already an unattended pasture/permanent trees or shrubs/lawn/solid fence, hedgerow or tall annual crops between non-organic and organic growing areas.

Fig. 1. Buffer zone between conventional and organic farm

4.2.2. Contaminants soils

The presence of organic (wood, bones, motor oil etc.,) and inorganic (Pb, Zn, Cu, hydrocarbons, pesticide residues etc.,) pollutants either naturally and with human intervention, respectively make drastic changes in the soil physico-chemical and biological properties. It is also a great threat to both wildlife and human beings. Hence, it is always advisable to avoid contaminated or polluted sites for organic farming to prevent the potential risk of hazardous residues coming into foods. However, in any case, when there is no choice to avoid them, physical removal (wherever possible) or phytoremediation or bioremediation techniques have to be employed to cleanse such soils to make them fit for organic farming (Vidali, 2001; Ramanjaneyulu and Gajendra, 2006). However, it is a costly affair and takes long time.

Fig. 2. Wind break with *Sesbania sesban*

4.3. Gaining technical knowledge on tradition, innovation and science

The local and undocumented knowledge of local population at a given location is known as Indigenous Technical Knowledge (ITK). It indicates the historical experience of people (Haverkort, 1995). It is stored in their minds but not published anywhere. It is eccentric to a particular society or culture. They differ from one geographical region to the other. They were developed, modified, refined and practised for various purposes by farmers over the years in different zones. These are highly valuable, hence, needs to be preserved and documented by the scientists and NGOs, for the benefit of present and future generations. It further helps to reduce dependence on external synthetic inputs, cut down the cultivation costs (Sundramari and Ranganathan, 2003) and also get patents. Further, organic growers should combine the ITKs with modern scientific methods for sustaining the organic systems besides enhancing the productivity and profitability.

4.4. Organic wild collections

Wild collections include some of the medicinal and aromatic plants, aquatic plants or products present in forests, sea coasts or natural habitats. In view of high demand domestically and internationally for their use in various industries, it fetches good income to the locals. However, there are twin challenges. First one is, many local people particularly tribals have been surviving on the forest products for their livelihoods, over the centuries. Second one is, if harvested injudiciously, it will exhaust the resources and affect the ecosystem. So, to strike a balance between these two, government has imposed certain restrictions as detailed below. It helps to provide employment to the specific people on one hand and doesn't disturb the ecological balance on the other hand.

- Collect only from designated forest area
- Maintain certain defined distance from conventional areas, pollution or contamination
- Should possess permission from respective authorities
- Not to disturb biodiversity but sustain the natural ecosystem
- Not to deplete entire resources
- Not to be treated with inputs except with those allowed in organic production

4.5. Composting and liberal application of organic manures

An organic farm should have provision for a compost pit of convenient size. The livestock excreta and other crop waste should be regularly dumped in pits to make the compost through either general composting, NADEP composting or vermicomposting. These organic manures are eco-friendly and their application provide macro and micro nutrients for the crop growth. Further, they help in amelioration of problematic soils and enhances soil physical and biological health. Increased soil organic matter (SOM) due to regular application of organic manures will reduce soil erosion and degradation. It further enhances soil microbial biodiversity. It reduces soil erodibility due to improved soil structure, soil aeration and aggregate stability thus improves water storage and supplying capacity. It helps the crop to grow better even during dry spells.

As organic manures are produced on-farm by the farmers, production cost will be less thus total cost of crop cultivation. It also ensures self-sufficiency and quality besides saving of transport costs. Use of organic manures is being encouraged by the government under Paramparagat Krishi Vikas Yojana (PKVY) of National Mission for Sustainable Agriculture (NMSA). Further, farmers can also make required quantity of concentrated organic manures i.e. oilseed cakes from crops like groundnut, cotton, castor, safflower, rapeseed, sesame, produced on-farm. They will help in meeting partial nutrient requirement of crops grown under organic farming.

4.6. Land preparation

Deep ploughing of fields during summer months (April-May) upon receipt of summer showers, help in controlling the weed menace, breaking the hardpans, improving the rain water infiltration (Ramana and Neelima, 2021), disrupting the life cycle of pest and disease organisms and destroying the pest and disease propagules. Further, it enhances the yield of rainy season crops by 50% (Chaturvedi *et al.,* 2020). However, in tropical and sub-tropical climate where spike in temperatures will be upto 40-45°C during summer, loss of soil organic carbon (SOC) and high soil respiration are the twin problems. Otherwise, SOC is very important for sustaining the soil fertility thus healthy soil. It acts as a guide to apply required quantity of nutrients through organic sources in organic farming. Hence, repeated or intense deep ploughing should not be done in organic fields. But, regulated deep tillage depending on the need only has to be done to conserve the organic carbon and moisture.

Minimum tillage is preferred in organic farms as it causes minimum disturbance, maintains organic matter and crumby soil structure. It also reduces nutrient losses due to slow decomposition of organic matter. Further, ploughing should not be done in the presence of high moisture as it leads to compaction, reduced nutrient availability and aeration. This inturn restricts crop growth.

Primary tillage has to be done with normal tractor drawn cultivator instead of deep tillage implements, to loosen the soil and create favorable conditions for sowing and crop growth. Secondary tillage has to be done with rotavators to break or crush the clods and smoothen the surface. In case of high weed menace, stale seed bed technique has to be adopted but avoid heavy machinery.

4.7. Choice of crops

Not all the crops can be grown during conversion period. The choice of crops depends on the soil fertility status, rainfall, climate, household needs and ready market. It is always advisable to raise only low to medium nutrient requiring crops (legumes, millets, oilseeds, certain vegetables, fruit crops, aromatic plants etc.,) having low pest and disease load or cultivars of such crops with host plant resistance to endemic pests or diseases. So that, these crops yield reasonably better thus less loss to the farmer. However, after conversion period when the soil fertility is improved significantly, cereals (rice, maize, barley, wheat etc.,) and commercial crops (cotton, sugarcane, tobacco etc.,) can be cultivated with recommended organic package of practices. Further, marketing facility should be given top priority in the vicinity of organic production zones. Besides meeting the household needs, the concept of *produce locally, shop locally and eat locally* should be promoted to create demand for organic foods even in rural areas also. Further, they should also establish the backward linkages with super markets or malls in nearby towns or cities to sell out marketable surplus, or else, with local people based on trust.

4.8. Planting material and treatment

The planting material or seeds of different cultivars have to be selected based on local adaptability and resistance or tolerance to endemic pests or diseases. They should be produced organically to avoid any kind of contamination. Before sowing, they must be treated with appropriate material like biofertilizers *viz.*, *Rhizobium, Azotobacter, Azospirillum*, *Azolla,* BGA to ensure N fixation and with PSB or biophos or VAM for better P availability and with *Pseudomonas*, *Trichoderma, beejamrith etc.,* to ensure crop growth with minimal disease/ pest menace. Genetically modified crops like Bt cotton, Bt Brinjal, Bt mustard etc., should be totally avoided under organic farming.

4.9. Soil and water conservation methods

In slopy areas, the risk of contamination is one of the major problems. So, an organic farmer has to take precautions to avoid soil, water and nutrient flow from conventional fields to organic fields. Further, lack of moisture due to drought or absence of irrigation facilities is an another constraint for successful crop production in rainfed agriculture. Hence, following soil and water conservation measures have to be taken up (Yellamanda Reddy, 2017).

- Formation of diversion drains/trenches to divert rain water from conventional fields. Practice contour farming i.e. performing all the agronomic operations across the slope irrespective of the crop and season.
- Adopt ridge and furrow (R & F), compartmental bunding, tied ridging, broad bed and furrow (BBF), check basin method of land configuration to conserve moisture in low rainfall events. Further, R & F and BBF will also be useful for draining excess water thus avoid excess moisture stress on crops.
- Contour bunding (<6% slope) and graded bunding (>6% slope) depending on the slope.
- In India, nearly 683 MT of crop residues are generated every year, of which significant share is used for fodder and fuel. Further, 84-141 MT biomass is burnt by the farmers to take up sequential crops. But, still, 178 MT is available as surplus in the country (TIFAC, 2018) and this can be utilised for mulching purpose in various crops under organic farming. Mulches are one of the cheaper source of nutrition produced available on-farm with the respective farmers. Mulching with crop residues will help conserve moisture, regulate soil temperature, controls weed growth and alkalinity and improves soil fertility (Fig.3) and It can reduce the water requirement of crops in moisture stress conditions and also help to improve overall crop growth, development and yield.

- As a part of ex-situ moisture conservation, farm ponds of convenient size depending on the catchment area, slope, soil type and rainfall, can be dug and lined with 500 micron HDPE geomembrane sheet, with in the farm. The run-off water in case of high intensity or cyclonic rains can be harvested in the farm pond and the same can be utilized for scheduling life irrigation to rainfed crops (Ramanjaneyulu *et al.,* 2021a). However, care has to be taken to harvest rainfall received in the catchment area with in that particular organic farm only, to avoid contamination from adjacent fields.
- Vegetative barriers with Vetiver grass (*Vetiveria zizanioides* or *Chrysopogon zizanioides*) have to be established in wider strips across the slope to control the soil and water erosion and stabilize the soil. Further green manure crops or any other cover crops can be grown in the inter rows to further stop erosion (Ramanjaneyulu *et al.,* 2021).

Fig. 3. Mulching for soil moisture conservation in rainfed castor

Fig. 4. Mulching for soil moisture conservation in rainfed sunflower

4.10. Green manures

The leguminous green manure crops like dhaincha, sunhemp, mungbean, urdbean, cowpea etc., have to be broadcasted in the field and incorporated *insitu* at 50% flowering stage. This can be followed in either irrigated or rainfed ecosystem. Similarly, the inter and intra row space in the orchards can also be utilized for this purpose. They can be incorporated using tractor drawn rotavator, disc harrow, rotary mulcher, rotary auger plough etc., It will be decomposed in the presence of optimum moisture content in the soil. Furthermore, branches and leaves of tree species like neem, *pongamia*, *glyricidia* etc., can also be incorporated 10-15 days before transplanting in rice. Such practice adds organic matter, improves soil fertility, water holding capacity, soil structure, control the weeds, soil salinity and nematode menace. Various green manure species produce 8-20 t ha^{-1} green biomass and fixes 30-150 kg N ha^{-1} as reviewed by Ramanjaneyulu *et al.* (2021b).

According to Dabholkar, a mathematician turned agriculturist, seeds from five categories of crops like cereals, pulses, oilseeds, legumes and spices totalling 20-25 kg as detailed below, should be mixed and broadcasted in the field. Subsequently, the entire biomass should be incorporated insitu into the soil at 45-60 DAS. This method helps to replenish the soil fertility depleted due to intensive agricultural practices. It can also be left as mulch over the soil surface, which helps to reduce loss of organic carbon in tropical climate. This process further helps to improve soil physical, chemical and biological properties. Optimum soil moisture should be maintained for better growth and decomposition of these crops. (http://www.ecosecretz.com)

- Cereals and millets: 5 kg (2 kg each of jowar and bajra+500 g korra +500 g sama)
- Pulses: 5 kg (1-2 kg each of mungbean, urdbean, pigeonpea and chickpea)
- Oilseeds: 5 kg (500 g sesame, 1-2 kg each of groundnut, sunflower and castor)
- Legume crops: 5 kg (2 kg each of dhaincha and sunhemp, 500 g each of mothbean and 1kg of horsegram)
- Spices: 5 kg (0.5-1.0 kg each of mustard, coriander, methi and ajwain)

4.11. Growing fodders and ensiling

A separate fodder block has to be established near by a permanent irrigation source. To make the whole farm or ecosystem organic, fodders must be grown organically, so that, cow based products will also be organic in nature. Further, to meet the off-season fodder demand of cattle, there should be arrangement for making silage (ensiling) without using chemicals. The cow dung and urine slurry can also be further channelised with irrigation water for irrigating the fodders or other crops in organic farm. This eliminates the need for chemical fertilizer application.

4.12. Azolla production

An organic farm should also have an azolla production unit (Fig.5). Azolla is an aquatic fern present in tropical and temperate regions. It possesses cavities on the dorsal side of leaves in which BGA (blue green algae i.e. anabaena algae) lives and fixes atmospheric N symbiotically. It can be intercropped with rice or applied to soil before transplanting. It can be used for all types of vegetables and plantation crops. It acts as a green manure crop and its incorporation help in meeting the partial N requirement of crops grown organically. Its application improves yield by 15-20%. It covers water surface thus reduces the weed growth by reducing light availability in rice growing environment. Thus, it reduces the use of chemical pesticides and fertilizers and environmental pollution. Furthermore, it can be used as fodder for dairy cattle, feeding pigs, chicken, ducks and fish.

Fig. 5. Azolla production

4.13. Intercropping

If two or more crops are grown together without distinct row proportion it is known as mixed cropping. If the crops are grown in definite rows, it is called as row intercropping. It helps to accrue profits to the farmers in case of good rainy season or atleast provides some income from one crop in case of drought year. When the crops are raised in distinct rows but widely spaced crops are alternated with closely spaced crops in wide strips, it is termed as strip intercropping. In all these methods, legume crops must be grown as one of the component crops.

Advantages of row or strip intercropping include covering the ground at a faster rate, weed control, soil and water erosion control on slopy lands, improved soil fertility, higher biomass, productivity and profits, efficient land and other resources use, increase in cropping intensity, higher land equivalent ratio etc., Further, it also helps to maintain on-farm biodiversity (Scherr and McNeely, 2008). This practice is crucial in harbouring pollinators, bees, predators and encouraging beneficial microorganism in soil in organic fields (Nicholls and Altieri, 2013; Maitra and Ray, 2019). It also helps in enhancing availability of N through fixation and P through solubility and also micronutrients like Fe and Zn thus nutrient economy. Besides, it reduces the incidence of fungi viz., *Fusarium spp.* (Huifang Lv *et al.,* 2018) and *Phytophthora spp.* (Zhang *et al.,* 2020). Furthermore, this practice also helps to diversify production and meet the household needs. But, compatible component crops have to be selected keeping in view local climate, market demand, availability of resources and advise of the experts. Thus, intercropping system should be an integral part of organic agriculture.

4.14. Crop rotation

Crop rotation is growing different types of crops in succession on the same piece of land. It avoids monocropping i.e., growing same crop year after year on a given land. It refers to Roman's principle of food-feed-fallow (FFF), in which the land is divided into three parts and different crops are rotated. The crop rotation has to be done with legumes, cover crops and grasses. This method is helpful for the farmer in many ways as detailed below.

- Improves organic matter and soil fertility through leaf fall and biological N fixation (BNF), respectively and also by recycling nutrients (Mohler and Johnson, 2009)
- Reduce the menace of weeds, pests and diseases through alteration of micro climate and harboring natural enemies
- The crops with differential root systems and forage area helps in improving the soil structure, exploits the water and nutrients from different soil profile layers rather than single layer, break the hard pan if any, and improves porosity and infiltration
- Produces diversity of products
- Climate change mitigation
- Reduction in erosion

4.15. Cover crops

The crops which can quickly cover the soil rather than for the purpose of being harvested are known as 'cover crops' (CC). It may be an off-season crop planted after harvesting the main crop. They are nurse crops and they support the survival of main crop. They are generally grown in winter/summer to cover the land in the absence of a main crop. They may be grasses (barley, oat, rye grass, buckwheat, sorghum-sudan hybrids), legumes (mungbean, urdbean, horsegram, cowpea, all clovers, *Medicago* spp., soybean, peas, beans, field pea, hairy vetch), non-legumes (spinach, flax) and brassicas (radish, turnip).

These crops should possess certain characters like easy to establish and multiply, faster ground coverage, produce more biomass, resistance to biotic and abiotic stresses (Poeplau and Don, 2015), ease of termination (at least two weeks before main crop) and ecological and social acceptance (Arbuckle and Roesch-McNally, 2015; Reeves, 1994; Baas *et al.*, 2015). Further, their cultivation should be less expensive. These crops should not be allowed to go to seed, otherwise, volunteer plants will become major weeds.

CC act as live mulch and enhance biodiversity. They help in reducing NO_3-N leaching, extract nutrient and water due to extensive root system, reducing run-off, soil loss and increased soluble P movement (Sharpley and Smith,

diamond back moth. The incidence of sorghum ear head fly is extremely low when sorghum is intercropped with red gram and incidence of root rot in cotton is meagre due to intercropping with dew gram (*Phaseolus aconitifolius*). Gangwar *et al.,* (2011) observed that coriander/garlic/ fennel intercropping was found to minimise incidence of top borer in sugarcane.

Fig. 5. Intercropping (Onion+carrot)

4.16.2. Other approaches

Grow cultivars with resistance/tolerance to major abiotic and biotic stresses local to that particular area. Arrange 'T' shaped bird perches @ 50 ha^{-1} at one feet above crop canopy. These perches could be animate or inanimate type depending on the availability. They provide good habitat for birds which inturn will provide biological control of crop pests (Lingaraju *et al.,* 2014) thus reduce the reliance on chemicals. Installation of pest specific pheromone traps @ 10 ha^{-1} for monitoring and 20 ha^{-1} for mass trapping will be useful for mating disruption thereby controlling pests (Ahmad and Kamarudin, 2011). They lure, trap and kill them. These are eco-friendly and biorational way of controlling the pests in organic farming (Sohrawardy *et al.,* 2021)

Biological control is environmental friendly technology for controlling the pest population and is cheaper than other strategies and non-toxic to the public, plants and ecosystem (Sharma *et al.,* 2013). Some of the examples for biological control include *Trichogramma japonicum/chilonis* against lepidopterous pests in rice, sugarcane, *Bracon brevicornis against* several lepidopterous pests, *Stenobracon deesae for* graminaceous stemborers, *Oomyzus sokolowskii, a* pupal parasitoid of diamondback moth (*Plutella xylostella*), *Trichogrammatoidea bactrae, an* indigenous egg parasitoid of several lepidopterous hosts like *Bactra venosana, Chilo spp., Helicoverpa armigera, Pelopidas mathias*. Above methods help in managing the pests and disease by encouraging pollinators, natural enemies, parasites and predators and promote crop growth through creation of favourable environment.

4.17. Integrated organic farming system (IOFS)

Integrating crops, fodders, vegetables, trees, livestock, poultry, small ruminants etc., on the lines of an Integrated Farming System (IFS) helps to utilize all the on-farm resources efficiently in a sustainable manner. Crop residue or fodder can be fed to cattle under the shade of trees. In turn, livestock animals provide solid and liquid manure for crops. The excess residue if any can be used for composting/vermicomposting which inturn will be used for manuring crops and improving the soil fertility. Diverse crops with cereals, pulses, vegetables will provide a balanced diet containing carbohydrate, proteins, fats, minerals etc., to the households. Other enterprises like bee keeping, fishery can also have combined in this model depending on the resources and market. Over the years, well developed IFS systems can be converted into IOFS (Fig. 6).

Fig. 6. Integrated farming system

5. Conclusion

The non-chemical practices have to be followed continuously in an integrated manner to maintain the farm fully organic and optimizing the input supply for better crop growth and yield from all the enterprises. Collective community action is required to make the whole ecosystem an organic. The products/ substances permitted by National Progamme on Organic Production (NPOP) only have to be used in organic farms. Material used for packaging should be eco-friendly, reusable and biodegradable and it should not contaminate the food products.

6. Way Forward

Once started, farmers should not discontinue after few seasons or years or atleast should continue in some part of the land with all precautions, which otherwise will be a futile exercise. Efforts should be made for organic certification with the help of accredited certification bodies. It helps them to sell organic products at premier price to make the farming a profitable and sustainable one.

7. References

Acuña, J.C.M. and Villamil, M.B. 2014. Short-term effects of cover crops and compaction on soil properties and soybean production in Illinois. Agronomy Journal 06: 860-870.

Ahmad, S.N. and Norman Kamarudin, N. 2011. Pheromone trapping in controlling key insect pests: progress and prospects. Oil Palm Bulletin 62: 12-24.

Arbuckle, J.G., and Roesch-McNally, G. 2015. Cover crop adoption in Iowa: The role of perceived practice characteristics. Journal of Soil and Water Conservation 70: 418-429.

Baas, D., Casteel, S.., Clay, D.E., Clay, S., Gross, P., Hoorman, J., Hoormann, R., Kaspar, T., Kladivko, E., Ruark, M., Millar, N., Murray, H., Noggle, S. and Wyse, D. 2015. Integrating cover crops in soybean rotaions: Challenges and recommendations for the north central region. Midwest cover crop Council, North Central Soybean Research. https://mccc.msu.edu.

Bhanu, C., Ravisankar, N. and Panwar, A.S. 2017. Pest, disease identification and management in organic crop production. In: Ravisankar N, Panwar A S, Kamta Prasad, Vipin Kumar and Bhaskar S (eds). Organic farming crop production guide. Network project on organic farming. ICAR-Indian Institute of Farming Systems Research, Modipuram, Meerut-250 110, Uttar Pradesh, India, p 81.

Chaturvedi, S., Shrivastava, A.K., Rathore, F. and Bunker, N. 2020. Deep ploughing in summer will increase kharif yield. Bhartiya Krishi Anusandhan Patrika 35(1-2): 114-116.

Eyhorn, F., Heeb, M. and Weidmann, G. 2003. Training manual for organic agriculture in the tropics. IFOAM, pp 243.

FAO, 2015. Training manual for Organic Agriculture. Energy and Tenure (NRC) of Food and Agriculture organization (FAO) of the UN, pp 11-27, 39-49 and 81-85.

Gangwar, B., Gangwar, K.S. and Tomar, O.K. 2011. Need of cropping systems based approach of agro-technology for higher productivity, profitability and soil health. Indian Farming 61(2): 3-6.

Hadiya, H.R., Patel, D.R., Pathak, D.M. and Patel, S.G. 2020. Impact of sowing periods on incidence of sucking pest on summer mungbean. Journal of Entomology and Zoology Studies 8(1): 815-818.

Haverkort, B. 1995. Agricultural development with a focus on local resources: ILEIAS view on Indigenous Knowledge. In: Warren DM, Slikkerveer LJ and Brokensha D (eds). The cultural dimensions of development: Indigenous knowledge systems. Intermediate Technology Publications Ltd., London, pp 454-457.

http://agropedia.iitk.ac.in.

http://www.ecosecretz.com.

https://agriinfo.in.

Huifang Lv., Haishun Cao, Nawaz, M.A., Sohail, H., Huang, Y., Cheng, F., Kong, Q. and Bie, Z. 2018. Wheat intercropping enhances resistance of watermelon to fusarium wilt. Frontiers in Plant Sciences 9: 696.

Lingaraju, G.H., Hasansab, A., Nadaf, Anand Katti, Vishwe Gowda, Sunil Puranik and Nagaraja, H. 2014. Bird perches in pest management. Agrobios Newsletter 12(11): 63.

Maitra, S., Ray, D.P. 2019. Enrichment of biodiversity, influence in microbial population dynamics of soil and nutrient utilization in cereal-legume intercropping systems: A review. International Journal of Bioresources Science 6: 11-19.

Mohler, C.L. and Johnson, S.E. 2009. Crop rotation on organic farming: A planning manual. Natural Resource, Agriculture, and Engineering Services (NRAES), Cooperative Extension, Ithaca, NY.

Nicholls, C.I. and Altieri, M.A. 2013. Plant biodiversity enhances bees and other insect pollinators in agroecosystems. A review. Agronomy for Sustainable Development 33, 257-274.

Poeplau, C. and Don, A. 2015. Carbon sequestration in agricultural soils via cultivation of cover crops-A meta-analysis. Agriculture Ecosystem and Environment 200: 33-41.

Ramana, M.V. and Neelima, T.L. 2021. Organic farm – Design, conversion and system planning. In: Module I: Organic Agriculture. Manual on 'Online short-term course for skill and entrepreneur's development in Organic Farming' (Eds.)., Jagadeeshwar, R. and Gouri, P.V.S.M. pp 8-23.

Ramanjaneyulu, A.V. and Gajendra, G. 2006. Phytoremediation - A review. Agricultural Reviews 27(3): 216-222.

Ramanjaneyulu, A.V., Sainath, N., Swetha, D., Reddy, R.U. and Jagadeeshwar, R. 2021 b. Green manure crops: A review. Biological Forum – An International Journal 13(2): 445-455.

Ramanjaneyulu, A.V., Venkata Ramana, M., Srinivas, A., Ramulu, V., Neelima, T.L. and Madhavi, A. 2021 a. Enhancing rainfed crop and water productivity through life saving irrigation with farm pond water. In: Singh HP, Jagadeeshwar R, Ramulu V, Singh Babita and Choudhary H (eds). Book of abstracts. Global conference on "Innovative approaches for enhancing water productivity in Agriculture including Horticulture" held during 16th-19th September at PJTSAU, Hyderabad, p 76.

Ramanjaneyulu, A.V., Sainath, N. and Venkata Ramana, M. 2021. Multifarious uses of vetiver grass. Chronicle of Bioresource Management 5(2): 026-032.

Reese, C., Clay, D.E., Clay, S.A., Bich, A.D., Kennedy, A.C., Hansen, S.A. and Moriles, J. 2014. Winter cover crops impact on corn production in semiarid regions. Agronomy Journal 106: 1479-1488.

Reeves, D.W. 1994. Cover crops and rotations. In: J.L. Hatfield and B.A. Steward, editors, Advances in soil science: Crops residue management. CRC Press, New York. p 125-172.

Scherr, S.J. and McNeely, J.A. 2008. Biodiversity conservation and agricultural sustainability: Towards a new paradigm of 'eco-agriculture' landscapes. Philosophical Transactions of the Royal Society of B 363: 477-494.

Sharma, A., Diwevidi, V.D., Singh, S., Pawar, K.K., Jerman, M., Singh, L.B., Singh, S. and Srivastawa, D. 2013. Biological control and its important in Agriculture. International Journal of Biotechnology and Bioengineering Research 4(3): 175-180.

Sharpley, A.N. and Smith, S.J. 1991. Effect of cover crops on surface water quality. In: W.L. Hargrove, editor, Cover crops for clean water. Soil Water Conservation Society. Ankeny, IA, p 1-91.

Shelton, A.M. and Perez, A.M.B. 2006. Concepts and applications of trap cropping in pest management. Annual Reviews in Entomology 51: 285-308.

Sohrawardy, H., Debnath, J., Fatema, Rahat, A.A.M. and Islam, T. 2021. Eco-friendly management of insect pests in Agriculture by pheromones: A review. Ecology Journal 3(2): 203-215.

Sundramari, M. and Rangnathan, T.T. 2003. Indigenous agricultural practices for sustainable farming. Agrobios (India), Jodhpur, India, pp 168.

TIFAC. 2018. Estimation of surplus crop residue in India for biofuel production. Joint report of Technology Information Forecasting and Assessment Council of India (TIFAC) and Indian Agricultural Research Institute (IARI), New Delhi, October 2018, pp 464.

Vidali, 2001. Bioremediation - An overview. Pure and Applied Chemistry 73 (7): 1163-1172.

Wagger, M.G. and D.B. Mengel. 1988. The role of nonleguminous cover crops in the efficient use of water and nitrogen. In: Hargrove WL, Ellis BG, Cavalier T, Johnson RC, and Reginato RJ (eds). Cropping strategies for efficient use of water and nitrogen. ASA Spec. Publ. 51. ASA, CSSA, and SSSA, Madison, WI, p 115–127.

Yellamanda Reddy, T. 2017. Soil conservation. In: Dryland Agriculture – Principles and Practices. Kalyani Publishers, New Delhi – 110 002, pp 154-190.

Zhang, H., Yang, Y., Mei, X., Li, Y., Wu, J., Li, Y., Wang, H., Huang, H., Yang, M., He, X., Zhu, S. and Liu, Y. 2020. Phenolic acids released in maize rhizosphere during maize-soybean intercropping inhibit phytophthora blight of soybean. Frontiers in Plant Sciences 11: 886.

2

Breeding Aspects in Organic Farming

C.V. Sameer Kumar[1], N. Aswini and M. Ramesh Naik[2]

[1] Professor Jayashankar Telangana Agricultural University, Hyderabad-500 030 Telangana, India

[2]ICAR-National Academy of Agricultural Research Management, Hyderabad-500 030, Telangana, India

Abstract

All hybrids and varieties perform better under conventional farming in response to excess use of chemical fertilizers and pesticides, which are harmful to humans, animals and environment. Hence, there is a need to develop varieties/hybrids exclusively for organic farming. Breeding aspects of organic farming include breeding tools/methods which are used to breed varieties/hybrids including testing and multiplication under organic conditions duly following the principles of organic farming as defined by IFOAM. These cultivars should have high nutrient use efficiency, resistance to diseases and pests, competitive ability against weeds, higher yield and stability across environments along with enhanced nutritional quality. This book chapter briefly discusses the aspects of breeding that are involved in generating cultivars suitable for organic farming.

Keywords: *Conventional farming, organic farming, organic plant breeding*

1. Introduction

Currently, organic farmers rely heavily on conventionally developed and produced varieties. Nevertheless, in order to optimize their production under organic agriculture, it is necessary to develop cultivars that are more suitable for organic cultivation. According to some studies, the naturalness of organic agriculture should go beyond simply by avoiding chemical inputs and substituting natural and organic components during the transition from conventional to organic farming.

This book chapter included precise assertions of what was meant by "naturalness" refers to the relationship between (agri) culture and nature sustainability, arable cropping, biotechnology, agro-ecosystem, bio-ethics as well as food and nutrition. There are many statements made concerning the naturalness of organic farming as given by Van Bueren (2002).

- The naturalness of organic agriculture in comparison to conventional agriculture is well acknowledged
- The goal of controlling the industrial process is not to break free from nature. Control that honours nature and natural processes is what it is. As a result, organic farming is more in tune with nature than conventional farming
- Culture and nature are intended to be integrated via organic agriculture
- Genetic modification is viewed as unnatural and at contrast with key principles of organic farming, such as respect for life and the distinctions between species
- Although biotechnology can have good benefits on the environment, it is not a "green" technology in ecological sense
- Sustainability is outlined in terms of the agro-ecosystem as a whole, in which no technologies or substances that are irrelevant to the system are used
- The methods used in organic cultivation must respect the environment and all living things
- Unique characteristics and integrity of plant species are protected
- Research on organic farming demonstrated a progressive change from growing in soil to growing in greenhouses. With the exception of hydroponics, it is difficult to draw a clear line based on the notion of naturalness
- Consumers value organic foods.
- The demands of organic agriculture are not completely met by the modern cultivars of field-grown crops. Therefore, more attention is needed on the development of cultivars adapted to conditions prevailing in organic farming and complying with the specific nature of organic agriculture.

2. Need for Organic Plant Breeding

The main goal of organic plant breeding is to keep the good health of the environment and all living creatures and produce nutritious crops in sustainable way.

Till today, organic cultivation mainly depends on plant varieties developed through traditional plant breeding. However, traditional breeding had great focus on the development of varieties in response to the intensive agriculture where more yield is achieved through excess application of inorganic fertilizers, herbicides and insecticides (Table 1). On the other hand, organic farming needs a separate crop improvement strategy as it needs varieties that suit well to organic conditions (Mader *et al.,* 2002).

Table 1. Difference between organic and traditional plant breeding

S. No.	Features	Organic plant breeding	Traditional plant breeding
1	Breeding method used	Resist using radiation and transgenics	All crop improvement tools
2	High preference for	Quality aspects	Yield improvement
3	Weeds suppressiveness	High	Low
4	Uniformity of produce	Low	High
5	Adaptation	High	Low
6	Use of chemicals	Not permitted	Permitted
7	Use of radiations	Not permitted	Permitted
8	Biodiversity	High	Low
9	Genetic base	Broad	Narrow
10	Production cost	Low	High
11	Eco-friendly	Yes	No
12	Principles of organic farming	Strictly followed	Not followed

Further, organic system prefers varieties that have resistance to diseases, higher nutrient use efficiency and pests and reduce the weed growth and possess unique qualities like taste and colour. The diversity of agro-ecological aspects of organic farming indicates the need for an ideotype plant architectural and growth dynamical features which contribute directly and indirectly to yield stability over locations and years and lowers the risks, yield loss as well as quality (Van Bueren, 2002).

The above requirements for organic varieties depict the importance of breeding and selecting varieties for organic cultivation. The organic ideotype per crop and per market can help in selection of best available varieties among existing ones that can propagate organically in short and medium term. In the long-term breeder can integrate organic traits in future crop improvements in addition developing suitable cultivars organic conditions.

Thus, to maintain soil health, prevent environmental pollution, enhance biodiversity, mitigate greenhouse effect and global warming, no gene pollution, organic plant breeding play crucial role to develop varieties suitable for organic condition.

3. Characteristic Features of Organic Plant Breeding

- It does not use any synthetic chemicals at any stage which cause adverse effect on soil, human and animal health.
- Organic plant breeding gives both commodities and non-commodities output for public
- Quality of crop produce is given priority rather than yield.
- Promotes numerous bee activity for pollination in crops.
- Importance is given to the species diversity and biodiversity as it involves diversified crop rotation and mechanical weeding.
- The seed that has been produced by following organic principles at least for one generation called as organic seed.
- The varieties developed from organic crop breeding methods known as organic plant varieties.
- Organic produce is of two types, such as organic certified A and organic certified B. When crops are grown after 3 years without use of restricted chemicals, the produce is said to belong to organic certified A. If crops are produced in the first and second year without use of banned chemicals, the produce is considered as organic certified B
- Development of varieties with high degree of resistance to biotic and abiotic stresses (Alliance, 2014).

4. Recommended Breeding Tools for Organic Plant Breeding

4.1. Suitable for organic plant breeding

4.1.1. Creation of variation

- Varietal crossing
- Bridge cross to transfer gene(s)
- Back cross breeding

4.1.2. Selection methods

- Pure line selection
- Mass selection
- Pedigree breeding
- Bulk method of breeding
- Ear to row method
- Back crossing and selection
- Test crossing and selection

- Multi-line method
- Indirect selection
- DNA diagnostic tools

4.1.3. Maintenance and propagation

- Sexual propagation
- Asexual propagation using tubers, buds, bulbil Layers, cut and graft shoots, rhizomes

4.2. Not Suitable for organic plant breeding but provisionally allowed

4.2.1. Creation of variation

- Embryo/ Ovary culture
- In-vitro pollination

4.2.2. Selection methods

- In-vitro selection

4.2.3. Maintenance and propagation

- Anther/ pollen culture
- Meristem culture
- Micro-propagation
- Somatic embryogenesis

4.3. Not suitable or to be banned immediately

- Cytoplasmic hybrids without fertility restorer gene
- Protoplast fusion
- Radiator mentor pollen
- Induced mutation
- Genetic modification
- Genome editing

5. Varieties for Organic Farming can be Bred by Following Ways

5.1. Conventional breeding (Category I)

In this, varietal testing is carried out under conventional system whereas multiplication of varieties under organic conditions.

5.2. Conventional breeding (Category II)

In this, both testing and multiplication of varieties are carried out under organic conditions.

1991), reduce rain drop impact and compaction, enhancing soil aggregation, mycorrhizal activity, fixing N and scavenging of N, P and K to main crops and regeneration of degraded soils (Acuña and Villamil, 2014; Reese *et al.*, 2014). N_2O emissions may be increased or decreased depending on the type of legume, non-legume, brassicas or grassy cover crops, N application rates, wetness, root exudates and microbiome in the soil. When non-leguminous cover crops are grown, though immobilization of N occurs during decomposition, increased soil N availability after that can decrease the succeeding crop's N requirement (Wagger and Mengel, 1988).

4.16. Eco-friendly pest and disease management

Chemical based insecticides, fungicides, acaricides, nematicides, livestock additives etc., in no case should be used in organic crop and livestock production. Hence, integrated pest and disease management strategies devoid of synthetics, have to be employed as detailed below.

4.16.1. Agronomic approaches

- Regulated summer deep ploughing helps in exposing hibernating pupae and troublesome weed and disease propagules to hot scorching sun
- Adoption of optimum sowing window helps to create favourable environment for better crop growth and helps in escape of pest or disease attack. For eg. Early sowing (i.e. within 7-10 days of the onset of monsoon) of jowar can help to escape shoot fly incidence, during rainy season. While for winter season, sowing during september and october reduces shoot fly incidence (http://agropedia.iitk.ac.in). In case of summer mungbean, mid late sowing (3rd week of February) was found to be the best due to low sucking pest incidence than early sowings (1st week of February) (Hadiya *et al.,* 2020).
- Marigold and castor act as traps crops when planted on bunds or around the main crops (Shelton and Perez, 2006) can reduce the load of *Helicoverpa* and *Spodoptera,* respectively. Raising few rows of tall growing crops like jowar, bajra, maize etc., obstructs the movement of sucking pests by acting as physical barriers
- Crop rotation with legumes helps in controlling pests, disease, weeds besides improving soil fertility. Intercropping with suitable component crop(s) has the potential to reduce the pest or disease incidence. For eg., maize+bittergourd reduces fruit fly menace, cabbage+carrot for managing the diamond back moth, brinjal+coriander/fennel against fruit and shoot borer. Certain crops (tomato, garlic, clover, coriander, marigold, mustard, onion, lucern) when intercropped in cabbage repel the

5.3. Breeding for organic farming (Category III)

This primarily targets the final product's suitability for an organic environment, therefore breeding programme for varietal development, testing and multiplication are performed under organic conditions.

6. Approaches for Organic Plant Breeding (OBP)

Various approaches proposed by Van Bueren and Struik (2004) are given below.

6.1. Naturalness

This concept given by Verhoog *et.al.,* (2003), has 3 categories.

6.1.1. Non-chemical approach

One of the best approaches to organic agriculture, wherein there is no use of any inorganic fertilizers, chemical pesticides and synthetic chemicals and totally depends on the organic materials. It creates a better environment for natural selection of promising cultivars for organic farming.

6.1.2. Agro-ecological approach

It mainly considers the environmental safety. It is against the use of chemicals. The varieties with resistance to insect and disease, ability to perform with organic substances and higher yield with great quality have to be selected (Van Bueren and Struik 2004).

6.1.3. Integrity approach

Organic plant breeding targets mainly to develop a genotype which is higher yielder, species-specific expression potential with quality traits and eco-friendly. Hence, organic plant breeding (OPB) balances the environment without affecting the living creature and through providing the quality crop varieties.

6.2. Emotive dimension

This dimension removes fears from that chemicals have adverse effect on human health, as organic breeding does not use any chemicals. There should be variety suited for organic condition with higher yield. Farmers have to follow all the cultivation aspects in organic condition only. Varieties developed should be adaptable to variable environment with high yield under organic conditions. In this dimension, farmers play a vital role in maintenance and development of genetic variability.

6.3. Normative dimension

Chemicals are not used to control pests and diseases instead only organic substances are used. The breeding techniques should be in accordance with organic conditions, variety developed in organic condition should be adaptable to organic environment with higher yield.

7. Quality Aspects of Organic Plant Breeding

The organic breeding focuses on following quality aspects.

7.1. Ecological aspect

Farmers need varieties with features which are highly suitable to organic conditions. The required cultivar characteristics include acclimatization to organic soil fertility system, with lower organic inputs, more competitiveness with weeds, better root system better resistance to insects and diseases, greater yield and stability, ensuring sustainable use of the variety.

7.2. Ethical aspect

The techniques and strategies of organic breeding should be in accordance with the principles of organic farming.

7.3. Edible aspect

Encouraging edible aspect of the organic variety consists of quality parameters, like better taste, structure and vitality. In those organic varieties, colour, taste, form, aroma, keeping quality and other nutritional values must be retained and enhanced. The different criteria for desired varietal character for Organic Plant Breeding are listed in Table 2.

Table 2. Criteria for desired varietal characters for organic plant breeding

S. No.	Variety characteristics	Criteria
1	Suitability to soil fertility system	Suitability to organic inputs Greater buffering Higher water and nutrient use efficiency Deep and dense root system Capacity to interact with beneficial soil microbes
2	Weed suppressiveness	Early soil cover and more light uptake Allelopathy on weeds
3	Plant health	Mono-factorial and poly-factorial Horizontal resistance Field tolerance Plant architecture for resistance Compatibility for crop varietal mixture
4	Seed vigor	Resistance to pathogens during seed production and storage High rate of seed germination

5	Quality	Early maturity Good processing quality Better taste Long-term storability
6	Stable yield	High stable yield with low inputs across the environments

8. International Organizations Related to Organic Plant Breeding

- European Consortium for Organic Plant Breeding (ECO-PB), Freiburg, Germany.
- Association of Biodynamic Plant Breeders (ABDP), Milwaukee, Wisconsin.
- International Society of Organic Agriculture Research (ISOFAR), Rennes, France.
- European Association for Plant Breeding Research (EUCARPIA), Wageningen (the Netherlands).
- International Federation for Organic Agriculture Movements (IFOAM), Bonn, Germany.

9. Advantages of Organic Plant Breeding

- The cultivation of organic varieties helps in lowering the cost of cultivation as there is no use of chemicals in cultivation.
- Organic varieties are eco-friendly in nature which increases the natural enemies of destructive insects and leads to efficient natural control.
- More importance is given to quality rather than yield.
- Practice of organic cultivars leads to sustainable agriculture.
- It encourages the utilization of organic sources of nutrition and enhances the physicochemical properties of soil.

10. Disadvantages of Organic Plant Breeding

- Mutations play a vital role in creating novel variations and enhancing diversity and helping in developing distinct novel varieties. Banning the induced mutations in OPB makes a greater reduction in crop improvement.
- When it is impossible to transfer genes across species and genera, we prefer somatic hybridization. But this technique is prohibited in OPB, hence desirable gene(s) transfer is not possible.
- When a desirable target trait is not available in cross-compatible species, it can be transferred from desirable through transgenic technology and

develop suitable cultivar known as GMOs but, GMOs are banned in OPB.

- Various tissue culture tools such as anther/ovary/embryo culture etc. generate a novel variation lead development of new varieties or crop species which can withstand various stresses and with superior quality traits. Ban on these techniques hinders the development of superior organic varieties.
- Inter-specific hybrids are made fertile by chromosome doubling using colchicine chemical. Banning of usage of colchicine will have greater negative effect on inter-specific hybridization.
- Restrictions on the exchange of breeding materials between conventional and organic methods have had a significant negative impact on agricultural progress.

11. Challenges in Organic Plant Breeding

- The world's ability to feed itself through organic farming is a critical concern that has arisen due to the rapidly expanding human population. As a result, the sustainability of this farming model is sometimes questioned due to the lower yields of organic agriculture (Kumar *et al.*, 2019).
- Organic plant breeding should not be seen as a method that simply forbids specific practices, but rather as a way to expand the possibilities for improving organic varieties in areas that have hitherto received little attention.
- Determining specific selection criteria relevant to each crop, establishing selection methods, creating suitable legal and socio-economic conditions to encourage organic breeding programmes is challenging.
- The area under organic cultivation is very less, it is the major barrier to creating targeted breeding approaches in organic systems, particularly in some important crops like cereals, millets, pulses and oilseeds.
- The conventional seed economy is pushed to its limits by the organic sector to come up with innovative and improved methods for propagating and breeding organic plants.
- As a first step toward organic plant breeding, it will also be necessary to maintain varieties appropriate for organic agricultural systems.

12. Conclusion

The varieties that work well in a conventional farming system may or may not work well in an organic farming system. As per the required standards of naturalness, genetically engineered crops were exempted from use in organic agriculture. Breeding for Organic Agriculture (BFOA), substantially facilitates cultivar development through abrupt selection and other techniques for the improvement of the crop of interest. Hence, in order to meet the current demands of healthy food, conserve natural resources and preserve the agro-ecosystems, practicing organic farming is preferable and so, organic plant breeding may play a crucial role in developing the varieties and hybrids suitable for organic conditions.

13. Way Forward

Wide multi-location testing is required to well exploit G × E to identify main locations showing higher yield. Implementation of shuttle breeding to develop varieties suited for organic conditions. Screening of larger number of germplasms conserved in gene banks to identify valuable genetic material for organic breeding is the need of the hour. Also, evaluation of prediction possibility of genotypic performance in important areas with the help of high throughput genotyping-phenotypic evidence obtained from multi-location tests is required. Estimation of crop improvement of perennial plants suitability and sustainability for organic condition is to be kept in mind. To comprehend the extent of genetic variation of desirable traits in organic condition, it is needed to undertake quantitative and association genetics research also use of multiline concepts for biotic and abiotic stress resistance should be explored.

14. References

Alliance, O.S. 2014. Introduction to On-farm Organic Plant Breeding. Organic Seed Alliance, Port Townsend, WA, p 39.

Kumar, Pardeep, K., Lokendra, S and Sushil, K. 2019. Organic plant breeding: Opportunities and Challenges. Biomolecule Reports-An International e-newsletter. p 8.

Mader, P., Fliessbach, A., Dubois, D., Gunst, L., Fried, P. and Niggli, U. 2002. Soil fertility and biodiversity in organic farming. Science 296 (5573): 1694-1697.

Van Bueren, E. T. 2002. Organic plant breeding and propagation: concepts and strategies. [external PhD, WU, Wageningen University]. https://edepot.wur.nl/121365.

Van Bueren, E.L. and Struik, P.C. 2004. The consequences of the concept of naturalness for organic plant breeding and propagation. NJAS-Wageningen Journal of Life Sciences 52(1): 85-95.

Verhoog, H., Matze, M., Van Bueren, E.L. and Baars, T. 2003. The role of the concept of the natural (naturalness) in organic farming. Journal of Agricultural and Environmental Ethics 16(1): 29-49.

3

Seed Production Aspects in Organic Farming

R.T. Maruthi[1], K. Gangadhara[2] and A. Anil Kumar[3]

[1]ICAR-Sugarcane Breeding Institute, Coimbatore-641 007, Tamil Nadu, India

[2]ICAR-Central Tobacco Research Institute, Rajahmundry-533 105, Andhra Pradesh India

[3]ICAR-Indian Institute of Oilseeds Research, Hyderabad-500 030, Telangana, India

Abstract

Increased awareness of the importance of quality and nutrient-rich food among the majority of the population created a need for the development of organic production systems and suitable varieties/hybrids for cultivation. The quality of soil, water and air is affected by the overuse of agrochemicals in agriculture. The use of organic varieties helps in reducing the use of various agrochemicals and high-quality pesticides. Organic varieties can address environmental issues and climate change. The world organic seed market is valued at USD 4.2 Bn in 2022 and it is expected to grow at an annual rate of around 8.7% between 2022 and 2031. The demand for organic seeds in the market has created new opportunities for organic farmers. By following standard organic crop production practices and seed certification guidelines, they can produce quality organic seed or planting material on their own farm.

Keywords: *Organic seed, organic breeding, production practices, seed certification*

1. Introduction

Worldwide, organic agriculture is a rapidly growing sector. The demand for organic food markets is increasing recently with the fast-growing movement of community support for organic agriculture. A total of 96.4 m-ha were grown organically with 2% of the worldwide agricultural land. The region with the most organic agricultural land is Oceania (53.2 m-ha) followed by Europe (18.5 m-ha), Latin America (9.5 m-ha) Asia (8.8 m-ha), North America (3.6 m-ha) and Africa (2.7 m-ha). Australia stands first in area with 53.0 m-ha followed by India (4.7 m-ha) and Argentina (4.1 m-ha) (FiBL & IFOAM,

2024). The global organic farming market will grow from $169.04 billion in 2022 to $287.83 billion in 2027 at a Compound Annual Growth Rate (CAGR) of 11.3% (Organic Farming Global Market Report, 2023).

1.1. Organic farming: Global context

Organic agriculture defined as a whole system approach based upon a set of processes resulting in a sustainable ecosystem, safe food, good nutrition, animal welfare and social justice (IFOAM, 2002). Building a harmonic environment by cycling of natural resources, balancing ecology and conserving biodiversity are important component of organic agriculture as per the National Organic Program, 2015. During 1972, The International Federation of Organic Agriculture Movements was founded as a member-based organization working to improve true sustainability in agriculture, value chains and consumption (IFOAM-Organics International, 2020). IFOAM has developed standards and certifications of organic agriculture. In 1973, The Research Institute of Organic Agriculture (FiBL) was established to conduct practical research, offer advice, providing training and information (Schmidtke, 2020). FiBL collaborates with IFOAM-Organics International and other partners to collect data on worldwide organic agriculture (Willer *et al.*, 2021) to review the status and planning of activities of organic agriculture. The Organic Crop Improvement Association (OCIA) International is a nonprofit, member owned and largest leader in the organic certification. The Organic Seed Alliance (OSA) is nonprofit organic seed institution in the USA works on framing national policies, ethical seed solutions, educating farming community to strengthen organic seed systems (Loriz, 2020).

1.2. Organic farming: Indian context

In India, IOAM (Indian Organic Agriculture Movement), a member of IFOAM, adopted the IFOAM international standards under Indian conditions. At present in India, the following six authorized accreditation agencies (Agricultural & Processed Food Product Export Development Authority, Coffee Board, Spices Board, Tea Board, Coconut Development Board, and Cocoa & Cashew nut Board) has been approved by the Ministry of Commerce, Government of India. In addition, there are four Certification Agencies accredited by APEDA such as IMO Control Pvt. Ltd., Bangalore, Skal International (The Netherlands) Bangalore, SGS (Societe Generale de Surveillance, Switzerland) India Pvt. Ltd. Gurgaon and ESCOCERT (Ecological Certification, France) International, Germany. The Government of India has implemented the National Programme for Organic Production (NPOP), which is accreditation programme for Certification Bodies, formulating standards for organic production and promotion of organic farming. In 2002, India's first ever local Organic

Certification Body, Indian Organic Certification Agency (INDOCERT), was established with an objective to offer a reliable and affordable organic inspection and certification services to farmers, processors, input suppliers and traders.

2. Plant Breeding for Sustainable Organic Agriculture

Organic plant breeding is a holistic approach that respects natural crossing barriers and is based on fertile plants that can establish a viable relationship with the living soil and respecting the intrinsic value and integrity of plant (Lammerts van Bueren and Osman, 2001). Breeding for Organic activities usually integrate breeding goals of the organic sector into their running breeding programme. A wide range of locally adaptable crop cultivars need to validation for exploiting full potential of organic farming in diverse areas of agriculture. There is need to rescue local varieties and develop their own system of selection and distribution. These empirical systems have been based on the selection of individuals better adapted to the local environment and that are more resistant to pests and diseases. Considering the need for a wide gene pool to improve and multiply genetic resources for food and agriculture. Seeds and breeds from the formal and informal sectors should be included harmoniously in local and national breeding programmes.

Thus far most of seed and vegetative propagation material used in organic farming currently are grown organically but not bred organically. This may lack specific adaptation traits relevant to organic cultivars. Because of currently available cultivars were developed mainly for input responsiveness and other selection criteria over the years (Kotschi *et al.*, 2022). There is mismatch between supply and demand of organic seed. As an alternative to conventional plant breeding, organic plant breeding has emerged as novel option for organic seed market (Horneburg, 2011). Organic breeding programmes will need to be certified only when allowed techniques have been used for development of variety. Several of currently employed techniques would be banned from the development of organic varieties *viz.*, genetic engineering, cytoplasm male sterile hybrids without restorer genes, protoplast fusion, radiated mentor pollen and induced mutations, culture, ovary culture and in-vitro pollination. The primary objective for organic plant breeding is development of improved varieties under low input conditions prevailing in organic agriculture. Plant breeding plays a major role in development of high yielding varieties/hybrids suitable for organic agriculture. Biological diversity is the strength for resilience and adaptability of cropping systems (Thrup, 1998). Organic plant breeding is defined mainly by the breeding technologies used (Messmer, 2015). However, the principles of organic agriculture based on ecological criteria of naturalness, which includes here complementary approaches (Verhoog *et al.*, 2003) consists of the non-chemical, agroecological and the integrity values respecting the integrity of living organisms. Organic farming considers mainly sustainability, health and safety concerns reflecting four major principles of health, ecology, fairness and care as defined by the International Federation of Organic Agriculture Movements (IFOAM, 2005).

2.1. Future strategies for organic plant breeding

The progress can be achieved for development of cultivars suitable to organic conditions in case of plant breeding programs combine the selection of the progeny in optimal and organic or low input environments (Braun *et al.*, 1996). The detailed germplasm characterization and evaluation of gene bank should be undertaken to study the genetic diversity to meet the needs of organic agriculture sector. Selection for varieties for stable yields under low input conditions and build up soil fertility needs specific requirements of ideotype characteristics viz., rapid root growth, higher N fixing ability and nutrient use efficiency (Brancourt-Hulmet *et al.*, 2005; Dawson *et al.*, 2008; Fess *et al.*, 2011; Hirel *et al.*, 2011), weed suppressing ability (Ostergard *et al.*, 2008; Bertholdsson, 2011), resistance to pest and diseases, higher nutritional quality and good processing advantage. Shuttle breeding may help to exchanging the segregating generations across the different environments to achieve wider adaptation to biotic and abiotic stresses. To exploit full genetic potential is advisable to carry out selection solely under organic environments (Murphy *et al.*, 2007). Also, participatory plant breeding (PPB) and evolutionary breeding (EB), have been proposed as suitable breeding methods to target Organic farming to facilitate selection for local adaptation and considering the needs of farmers and industry (Desclaux, 2005; Ceccarelli, 2014). To exploit the G×E under multilocations testing, identify the key locations within regions to conduct yield evaluation trails (Desclaux *et al.*, 2008; Bueren *et al.*, 2008). Largescale screening of germplasm deposited in the gene bank to identify genetic resources/donors for organic plant breeding (Konvalina *et al.*, 2010) using high throughput genotyping platforms and accurate phenomics data from multi locations testing. Deployment of multi-lines/cultivar mixtures to manage host plant resistance in agroecosystems prone to various kinds of stresses (Kiaer *et al.*, 2009). Priority based breeding for quality trait like micronutrients and storage quality using new breeding technologies (Revilla *et al.*, 2012).

3. Crop Production Practices for Organic Seed Production

Farmers practicing organic farming often face problem of timely availability of good quality organic seeds. Such uncertainties can be overcome by producing quality organic seed in his own farm. Organic seed growers have to follow below elucidated crop production practices for the production of genetically pure high quality organic seed.

3.1. Field standards

Land selected for seed production should be organically managed, if the whole farmland is not converted a buffer zone or a natural barrier should be maintained to ensure a clear separation between organic and conventional production. Since the seed crop has to stay for longer duration in the field and also need to ensure high seed quality and purity. The field selected should be fertile with adequate irrigation and good drainage facilities. The field should be devoid of high weed incidence and free from volunteer plants. Crop rotation should be followed to avoid volunteer plants which hampers genetic purity through open pollination. It also helps in reducing pest and disease build-up in the field. Land preparation equipment's or tools should be cleaned before use. Seedbed preparation is most crucial step that optimize seed germination and survival rate. The process of preparation will vary with the species selected and the equipment used. Small seeded crops require finer seedbed than bold seeded crops. Homogeneity of seedbeds assures good soil-seed contact, also favours uniform emergence and seedling development.

3.2. Nutrient management

Nutrient recycling is the basic principle of nutrient management in organic farming. Adequate quantities of biodegradable material of microbial, plant or animal origin produced on organic farms should be returned to the soil to maintain soil fertility and the biological activity within it. Biodegradable material produced on organic farm like green manure, crop residues, straw, farmyard and poultry manure can be used. A large number of fast-growing legume and non-legume crops can be grown as green manure crop. Sunnhemp (*Crotalaria juncea*), dhaincha (*Sesbania spp.*), berseem (*Trifolium alexandrinum*) and green gram (*Vigna radiata*) are the more commonly used green manure crops that can fix atmospheric nitrogen to the extent of 60-100 kg ha^{-1} through symbiotic association with rhizobium bacteria. Sunnhemp can produce biomass of 8900-13000 kg ha^{-1} in about 8-12 week and accumulates about 170-200 kg nitrogen per hectare (Balkcom and Reeves, 2005; Perin *et al.*, 2006). Dhaincha can accumulate 80-86 kg N ha^{-1} in 50 days of growth period (Sharma and Ghosh, 2000). Green manuring enhances soil fertility by improving physical, chemical and biological properties. Legume green manures improve bacterial and fungal populations and enzyme activities in the soil. Increased soil microbial activity intern improves many soil physical properties and indirectly offers many ecosystem services.

Non synthetic mineral fertilisers and brought in fertilisers of biological origin should be regarded as supplementary and not a replacement for nutrient recycling. Appropriate management practises should be followed to ensure minimum nutrient losses. Soil pH is one of the important factors, it affects both the cation exchange capacity (CEC) of the soil and nutrient availability. Therefore, adequate pH levels should be maintained in the soil to optimize fertilizer use efficiency and crop yield. Generally, soil pH 6.0-7.5 is acceptable for most plants as most nutrients become available in this pH range. In highly acidic soil, aluminum and manganese can become more available and more toxic to plant while calcium, phosphorus, and magnesium are less available to the plant. In highly alkaline soil, phosphorus and most micronutrients become less available. When soil pH is deviated from ideal range it needs to be adjusted to produce optimum plant growth. Application of limestone and wood ashes can increase the soil pH. Similarly, Soil pH can be reduced most effectively by adding elemental sulphur.

3.3. Selection of crop and planting method

The variety or hybrid of any crop of choice adopted to the soil and climatic conditions can be selected for seed production. Since, the use of synthetic pesticides and fertilizers are not allowed, the variety selected must be resistant to pests and diseases. Genetically engineered transgenic varieties are not allowed. Adaptation is key to achieving resilience in agricultural system. Hence, more importance to be given to varieties adopted to specific farm conditions and climates. Organic farmers are required to use organic seed or planting materials. If it is commercially unavailable chemically untreated conventional materials shall be used for first year and for subsequent years' organic materials shall be used.

Planting method varies with the crop type. Seeds are generally planted directly in field using seed drill or transplanting from a greenhouse-grown seedling. Optimum row spacing and plant density must to followed for maximum plant development. Proper spacing ensures good vegetative growth and provide pollinators an unrestricted access to inflorescence to ensure proper seed set. It also ensures good sunlight penetration into the crop canopy and improved air circulation among plants which restricts disease and pest incidence. It is therefore important to know the nature of the crop species, and space accordingly both between and within rows.

The seed crop has to be planted at a safe distance from other varieties of same crop located in the adjacent area in order to maintain the genetic purity of the seeds. It is known as isolation distance and is fixed based on the pollination behaviour, pollinating agent, flying capacity of the pollen and nature of variety or hybrid. Isolation distance is minimum for self-pollinated crops and maximum for cross pollinated crops. Maintaining appropriate isolation distance avoids cross pollination and crossing of genes. Isolation can be maintained by distance, by timing of flowering and by growing barrier crop. Isolation by distance is the most reliable method, which involves providing adequate distance between a variety and any sources of contaminating pollen to ensure that the variety remains true to type.

3.4. Rouging

It is defined as selective removal of undesirable or off-type plants from seed crop on the basis of visual field inspection in order to improve one or more quality attributes of the seed lot to be harvested (Laverack and Turner, 1995). Removal of diseased plants affected by seed borne disease, noxious weeds (wild oat in wheat, and *Argemone Mexicana* in Brassica species), off-type plants (other distinct variety of same crop) and volunteer plants (previous season crop plant) from seed production plot is regarded as rouging. Generally rouging is to be carried in three phases; (1) Pre flowering, (2) During flowering and (3) Post flowering or before harvest.

3.5. Weed management

Weed management in seed production is more important. Crop-weed competition can result in reduced seed yields and quality, and in some cases increased disease and pest risk. Weed seeds can also contaminate harvested crop seeds and create problem in seed certification. Weed management in seed crop is hard because the crop has to stay in field for twice as long as the conventional crop. Long season exposes the field to more wind-disseminated weed seeds, more weed seed germination windows, and more time for weeds to mature and set seed. Hence, organic seed growers should develop a multiple-tactic systems approach to manage weed which results in fewer weed seeds in the soil and fewer opportunities for their germination, growth, and seed production. As a first step understand the weed cycle then design the cropping system tools for effective weed control. Preventive cultural techniques like crop rotation, green manuring, mulching and mechanical control helps in limiting weed growth. Crop rotation is very effective in reducing weed population density and biomass by breaking weed cycle (Liebman and Dyck, 1993; Marks *et al.*, 2018a, b; Weisberger, 2019). Plant residues like rice straw, oat straw and maize harvest residue etc., as biodegradable mulch materials were proved most effective in weed controlling (Radics and Szne Bognar, 2004; Anzalone *et al.*, 2010, Campiglia *et al.*, 2010). Fast growing green manure crops can compete strongly with weeds (for space, light, etc.), preventing them from proliferating and seeding. Some of the plants are allelopathic, so can further inhibit weed plant growth and also the germination of weed seeds, and in addition their residues/ground cover can prevent the emergence of weeds (Dhima *et al.*, 2009; Das *et al.*, 2016). In no case, herbicides should be used.

3.6. Pest and disease management

Pest and diseases can have a significant effect on seed yield and quality. Generally, management strategies are either preventive (measures taken to avoid or reduce the possibility of pest or disease occurrence) or curative (treatments that eliminate or reduce the effects of a particular pest or disease after it has become established). Since organic farmers have got very few effective curative practices available to them, their primary focus of disease management efforts has to be on preventive cultural practices. Slight modification in the cultural practices will have an impact on the whole ecosystem. Some of the practices include use of pathogen-free seed, selecting fields of low inoculum potential and locations with good air movement, adopting wide row spacing, orienting the crop rows to maximize air movement between rows and reduce humidity built-up in the canopy.

Seeds are a major means of survival of some plant pathogens and of introducing new pathogens to a field or region. Pathogens usually remain viable for longer in seed than in vegetative parts of the plant or in the soil. Hence, selecting good quality seed preferably use of disease and pest resistant variety is recommended in organic farming. Even when insect resistant cultivars are not available, some varieties may be less attractive to pest species or tolerate more damage than others such varieties can be used. Leaf size, shape, coloration, leaf hairs, and natural chemicals (both attractants and repellents) can affect the outcome of insect crop colonization (Haldhar *et al.*, 2015a, b). Through proper crop rotation less mobile foliar pests with one generation a year may be significantly suppressed by creating no host crop to pest insects (Weisz *et al.*, 1994). Brassica and related crops as green manures proven to be most effective for management of multiple soil borne insect pests and diseases due to their biofumigation potential. Brassica species can produce volatile sulfur compounds through a process known as biofumigation, and also affect soil microbial ecology. Variety of plant pathogens can be controlled, including species of *Rhizoctonia, Verticillium, Sderotinia, Phythophthora, Pythium, Aphanomyces* and *Macrophomina* (Larkin and Griffin, 2007; Larkin, 2013).

Conservation of natural enemies and creating on farm favourable environment to enhance the survival, fecundity, longevity, and behavior of natural enemies to increase their effectiveness. Parallelly if there is a shortage of natural enemies biological control agents such as insect predators, parasitoids, insect pathogens can be applied or released externally for colonization. They have a greater role to play in controlling the insect pests in an insecticide free organic farming environment. Biocontrol agents like *Bacillus thuringiensis, Metarhizium anisopliae, Beauveria bassiana, Trichogramma spp.* and Nuclear polyhedrosis virus (NPV) are effective in controling Lepidopteran and coleopteran pests (Stiling and Cornelissen, 2005; Kumaranag *et al.*, 2013; Mohan *et al.*, 2013). Insect predators like Lady bird beetles and *Chrysoperla* spp. effective against aphids, whiteflies and mealy bugs (Shalaby, 2012; Ahirwar *et al.*, 2015).

The crude extracts as well as commercial formulations from plants like neem, pongamia, garlic, margosa and tobacco have been shown to be effective against number of insect pest (Guleria and Tiku, 2009; Khater, 2012). The use of botanicals and other commercially available products for the control insect pests and diseases were used as last options in the organic agriculture. There are a number of biological and natural disease management products available in the market that are approved for use on organic farms. To use these products to control pest and diseases, communicate with certifying agency to make sure the materials are permitted for use on organic farms. Strict regulation of the products that are allowed for insect pest and disease management in organic cultivation is monitored by NPOP (National Programme for Organic Production).

3.7. Harvesting and drying

Among the crop management aspects influencing initial seed quality, harvest and drying practices are of considerable significance. Seeds of high quality can be obtained by adopting proper harvesting and drying techniques, careful handling and processing to minimize mechanical injuries and unwanted seed mixing with other crop/variety seed, and ensuring minimum deterioration before packing and storage. However, the method of harvesting and post-harvest handling largely depends on the type of seed being produced. Generally, orthodox seeds (like cereals and legumes) attain the ability to withstand desiccation during the end of grain filling period (Ellis *et al.*, 1987; Ellis and Pieta Filho, 1992). At that stage moisture content of seed will be too high to harvest and thresh without mechanical injuries. Seeds succumb to extensive threshing damage if crop is harvested at an early stage of maturity (Andersen and Andersen, 1972). Hence, seeds are normally allowed to dry on the plant until they reach harvestable maturity, i.e., the moisture content at which seeds can be safely harvested and threshed mechanically (Ellis *et al.*, 1987). Generally, for majority of grain crops a moisture range of 16-18% was considered as ideal for harvest. Some crops have seed shattering problem, in such cases stalks of the plant need to be cut when they are still green and field dried to allow uniform seed maturity.

Sun or shade drying and mechanical drying by using forced natural or heated air methods were generally used to reduce moisture content. Sun drying is very common, easy and cheap method of drying, but overheating and erratic drying were two major drawbacks of sun drying (Probert and Hay, 2000). Direct drying under the sun and on surfaces that would heat up much (>40 °C) should be avoided. Seeds should be spread in thin layers and turned over frequently to promote uniform drying. Drying should be done at a lower temperature, FAO (2013) recommended drying seeds at 10-25 % relative humidity and 5-20 °C to maximise the benefit of improved seed longevity. Sun drying is advisable only in the morning and evening hours. Drying in mid noon causes damage to seed quality. Nowadays mechanical drying machines are available. They are more efficient, drying is uniform and done within a short span of time. But the cost of the equipment and fuel requirement is very expensive.

3.8. Cleaning, packaging and storage

After harvest the seeds are to be processed by various methods in order to maintain physical purity and also to increase the shelf life. Stem bits and chaff collected along with the seeds may harbour insect pests which can damage stored seeds. So, to prevent such damage cleaning is essential. Winnowing is an age-old method to remove chaff from the seeds by tossing them in the air. It helps to remove stem bits, old petals, husks and other parts of the flower and debris mixed with the seeds. There are also mechanical winnowers available with high cleaning efficiency.

4. Certification Procedure for Organic Seeds

As per the NPOP guidelines farmers following organic cultivation methods should use organically certified seeds or planting materials in their farms. However, certified organic seeds or planting materials are not commonly available in the markets. To overcome this situation, certified organic farmers could take up the production of quality certified organic seeds in own farm following seed certification guidelines. Every farmer should have timely access to healthy seeds which are genetically pure, with high seed vigour and good germination percentage. Quality seed produced by following efficient certification procedures play a major role in increasing productivity as well as overall food production of our country.

Farmers willing to produce certified organic seeds have to register their land with any of the accredited organic certification bodies and also the seed crop with the Department of Seed Certification in the respective State. Farmers should follow all the organic certification standards according to the National Programme for Organic Production (NPOP) in addition to the seed certification standards for the production of certified organic seeds. Seeds certified by the Seed Certification Agency are called certified seeds, which passes through both the field and seed standards specified by the certification body. Seed standards are uniform throughout the country, whereas the seed certification procedures and fee vary from state to state.

Following are the steps involved in certification of organic seed

4.1. Application for seed production

Any person who wants to take up certified seed production should submit a sowing report to the Assistant Director of Seed Certification to register the crop and season by paying prescribed fee. The sowing report contain information about the seed producer, location of the farm, type of the crop and acreage of cultivation, name of variety and source of seed (tag to be submitted). Separate sowing reports are required for different crop varieties, and even for same variety if seed production fields are separated by more than 50meters, sowing dates differ by more than 7 days. After thorough scrutiny Assistant Director of Seed Certification registers the seed farm and assign a Seed certification number for each sowing report.

4.2. Field inspection

Field inspections are conducted by Seed Certification Officer (SCO). His role is to check for the factors that may affect the genetic purity and physical health of the seeds, to sensitise and guid the growers to follow appropriate seed certification practices. Number of field inspections vary from crop to crop and stage of the crop. Generally field inspections would be carried out during the following growth stages of the crop.

Growth stage of crop	Observations to be made
Pre flowering stage	Location and area of seed farm, source of seed, isolation distance, identification and removal of off-types
Flowering stage	Ensuring appropriate isolation distance maintained and proper rouging is done
Post flowering and Pre harvest stage	Confirming that objections made in the previous visit are taken care, guiding grower when and how to harvest the crop
Harvest stage	Verification of crop maturity and guiding grower processing and handling techniques

4.3. Seed processing

After the seeds are harvested, they should only be taken to approved seed processing units for processing. Each seed lot must be accompanied by the processing report. Processing includes cleaning, drying, grading, treating and other operations to improve the seed quality. The organic seeds are generally treated with materials from organic sources like botanicals, biofertilizer formulations and biocontrol agents. The seed Certification Officer will inspect the processing plant to verify the possibility of mechanical mixtures. Once the seed lots are processed, the seeds must be properly weighed, bagged, sealed and labeled. Seed lot numbers are assigned at this stage. The labels contain information on the kind, variety, seed class, lot number, sampling date, quantity in kg, total numbers of bags together with the signature of the Seed Certification Officer.

4.4. Seed sampling and analysis

The seed sample must be sent to the seed testing laboratory for analysis through the Assistant Director of Seed Certification. Seed lots that meet prescribed seed standards such as purity, free of inert matter, moisture percentage and germination capacity alone will be allotted the certification label. Seed certification standards varies from crop to crop.

4.5. Tagging

Approved seed lots should be tagged with certification tag within two months from the date of the receipt of seed analysis report. On receipt of the seed tags, it is verified by the Seed Certification Officer. All the prescribed details are entered in the tag without any omission. The green colour producer tag should also be attached to the seed lot along with the certification tag. Producer tag contains following information

- Label number
- Crop
- Variety
- Lot No.
- Date of Test (Date/Month/Year)
- Validity period (Date/Month/Year)
- Germination % (Minimum)
- Physical purity (Minimum)
- Genetic purity (Minimum)
- Net weight
- Organic certification logo
- Organic seed producer label
- Name and address of the producer

5. Conclusion

Seed is the primary input for agriculture and it has enormous impact on how we farm and what we eat. More specifically organic farmers rely on seed adapted to their specific farm conditions and climates more than other farmers because they don't use synthetic pesticides and fertilizers. Organic crop production demand crop varieties with genetic constitution that can defend plants from pests and diseases, and withstand weeds and diverse weather conditions. Farmers have to focus on locally adaptable crop cultivars to exploit full potential of organic farming. Further, quality seed and its timely availability are also equally important. Farmers can produce quality seed on their own farm by follow crop production standards set by National Programme for Organic Production (NPOP). In addition, if the seed growers follow the seed certification standards set by Seed Certification Agency for the production of certified organic seeds, not only it ensures high genetic purity but also fetches prime price for the seed in the market.

6. Way forward

Organic breeding could be the next-generation ecological approach in sustainable organic farming and organic seed production. The need for 100% organic seed spurred breeding cultivars for optimized organic systems. Certified organic seed should be seed produced from the organic variety bred under organic systems and commercialized and regulated. Through adherence to standardized organic crop production practices and seed certification guidelines, farmers can actively participate in the production of high-quality organic seed and planting material. This decentralized approach empowers farmers to contribute to the organic farming movement at a grassroots level while promoting fairness and sustainability throughout the agricultural supply chain.

7. References

Ahirwar, R., Devi, P. and Gupta, R. 2015. Seasonal incidence of major insect- pests and their biocontrol agents of soybean crop (Glycine max L. Merrill). Scientific Research and Essays 10(12): 402-406.

Andersen, S. and Andersen, K. 1972. Longevity of seeds of cereals and flax. Acta Agriculturae Scandinavica 22: 3-10.

Anzalone, A., Cirujeda, A., Aibar, J., Pardo, G. and Zaragoza, C. 2010. Effect of biodegradable mulch materials on weed control in processing tomatoes. Weed Technology 24(3): 369-377.

Balkcom, K.S. and Reeves, D.W. 2005. Sunn-hemp utilized as a legume cover crop for corn production. Agronomy Journal 97: 26-31.

Bertholdsson, N-O. 2011. Use of multivariate statistics to separate allelopathic and competitive factors influencing weed suppression ability in winter wheat. Weed Research 51: 273-283.

Brancourt-Hulmel, M., Heumez, E., Pluchard, P., Beghin, D., Depatureaux, C., Giraud, A. and Le Gouis, J. 2005. Indirect versus direct selection of winter wheat for low-input or high-input levels. Crop Science 45: 1427-1431.

Braun, H.J., Rajaram, S. and van Ginkel, M. 1996. CIMMYT's approach to breeding for wide adaptation. Euphytica 92: 175-183.

Bueren, E.T.L., Ostergard, H., Goldringer, I. and Scholten, O. 2008. Plant breeding for organic and sustainable, low-input agriculture: dealing with genotype-environment interactions. Euphytica 163: 321-322.

Campiglia, E., Mancinelli, R., Radicetti, E. and Caporali, F. 2010. Effect of cover crops and mulches on weed control and nitrogen fertilization in tomato (Lycopersicon esculentum Mill.), Crop Protection 29(4): 354-363.

Ceccarelli, S. 2014. GM crops, organic agriculture and breeding for sustainability. Sustainability 6: 4273-4286.

Das, A., Kumar, M., Ramkrushna, G.I., Patel, D.P., Layek Jayanta., Naropongla., Panwar, A.S. and Ngachan S.V. 2016. Weed management in maize under rainfed organic farming system. Indian Journal of Weed Science 48(2): 168-172.

Dawson, J.C., Huggins, D.R. and Jones, S.S. 2008. Characterizing nitrogen use efficiency in natural and agricultural ecosystems to improve the performance of cereal crops in low-input and organic agricultural systems. Field Crop Research 107: 89-101.

Desclaux, D. 2005. Participatory plant breeding methods for organic cereals: review and perspectives. In: Lammerts Van Bueren ET, Goldringer I and Ostergard H (eds) Proceedings of the COST SUSVAR/ECO-PB Workshop on Organic Plant Breeding Strategies and the use of Molecular Markers. Eco-PB Congress (January 17-19, 2005), Louis Bolk Institute, Driebergen, pp 17-23.

Desclaux, D., Nolot, J.M., Chiffoleau, Y., Goze, E. and Leclerc, C. 2008. Changes in the concept of genotype × environment interactions to fit agriculture diver-sification and decentralized participatory plant breeding: pluridisciplinary point of view. Euphytica 163: 533-546.

Dhima, K.V., Vasilakoglou, I.B., Gatsis, Th.D., Panou-Philotheou, E. and Eleftherohorinos, I.G. 2009. Effects of aromatic plants incorporated as green manure on weed and maize development. Field Crops Research 110(3): 235-241.

Ellis, R.H. and Pieta Filho, C. 1992. The development of seed quality in spring and winter cultivars of barley and wheat. Seed Science Research 2: 9-15.

Ellis, R.H., Hong, T.D. and Roberts, E.H. 1987. The development of desiccation-tolerance and maximum seed quality during seed maturation in six grain legumes. Annals of Botany 59: 23-29.

FAO. 2013. Genebank standards for plant genetic resources for food and agriculture. FAO, Rome.

Fess, T.L., Kotcon, J.B. and Benedito, A.V. 2011. Crop breeding for low input agriculture: a sustainable response to feed a growing world population. Sustainability 3: 1742-1772.

FiBL and IFOAM. 2024. The world of organic agriculture. http:// www. organicworld.in.

Guleria, S. and Tiku, A.K. 2009. Botanicals in pest management: current status and future perspectives. Integrated Pest Management: Innovation-Development Process 1: 317-329.

Haldhar, S.M., Choudhary, B.R., Bhargava, R. and Gurjar, K. 2015b. Host plant resistance (HPR) traits of ridge gourd (Luffa acutangula (Roxb.) L. against melon fruit fly, (Bactrocera cucurbitae (Coquillett)) in hot arid region of India. Scientia Horticulture 194: 168-174.

Haldhar, S.M., Choudhary, B.R., Bhargava, R. and Meena, S.R. 2015a. Antixenotic and allelochemical resistance traits of watermelon against Bactrocera cucurbitae in a hot arid region of India. Florida Entomologist 98: 827-834.

Hirel, B., Tetu, T., Lea, J.P. and Dubois, F. 2011. Improving nitrogen use efficiency in crops for sustainable agriculture. Sustainability 3: 1452-1485.

Horneburg, B. 2011. Organic plant breeding: Achievements, opportunities, and challenges. In: Proceedings of the Keynote Lecture at the Organic World Congress (26 September -5 October, 2011), Namyangju, Korea, pp 13.

IFOAM. 2002. Basic standards for organic production and processing. International Federation of Organic Agricultural Movements. Tholey-Theley, Germany. pp 85.

IFOAM. 2005. The IFOAM Norms for Organic Production and Processing (Ver. 2005), IFOAM.

IFOAM-Organics International. 2020. About IFOAM-Organics International. IFOAM-Organics International. https://www.ifoam.

Khater, H.F. 2012. Prospects of botanical biopesticides in insect pest management.Pharmacologia 3(12): 641-656.

Kiaer, L.P., Skovgaard, I.M. and Ostergard, H. 2009. Grain yield increase in cereal variety mixtures: a meta-analysisof field trials. Field Crop Research 114: 361-373.

Konvalina, P, Capouchova, I., Stehno, Z. and Moudry, J. 2010. Morphological and biological characteristics of the land races of the spring soft wheat grown in the organic farming system. Journal of Central European Agriculture 11: 235-244.

Kotschi, J., Schrimpf, B., Waters-Bayer, A. and Horneburg, N. 2022. Financing organic plant breeding-new economic models for seed as a common. Sustainability 14:10023.

Kumaranag, K.M., Kedar, S.C. and Patil, M.D. 2013. Insect pest management in organic farming systems. Popular Kheti 1(4): 3.

Lammerts van Bueren, E.T. and Osman, A. 2001. Stimulating GMO-free breeding for organic agriculture: a view from Europe. LEISA/ILEIA Newsletter 17(4): 12-14.

Larkin, R. P. 2013. Green manures and plant disease management. CABI Reviews, pp. 1-10

Larkin, R.P. and Griffin, T.S. 2007. Control of soilborne potato diseases using brassica green manures. Crop Protection 26 (7): 1067-1077.

Laverack, G.K. and Turner, M.R. 1995. Roguing seed crops for genetic purity: a review. Plant Varieties and Seeds 8(1): 29-45.

Liebman, M. and Dyck, E. 1993. Crop rotation and intercropping strategies for weed management. Ecological Applications 3(1): 92-122.

Loriz, C. 2020. Organic Seed Alliance-Our Story. Organic Seed Alliance. https://seedalliance.org.

Marks, M., Jastrzebska, M. and Kostrzewska, M.K. 2018a. Long-term Experiments in Agricultural Studies in Poland. UWM Publications p. 1e280 (in Polish).

Marks, M., Rychcik, B., Treder, K. and Tyburski, J. 2018b. A 50-year study on growing plants in crop rotation and continuous cropping systems e a source of knowledge and a monument of farmin culture. In: Marks M, Jastrzebska M and Kostrzewska MK (eds) Long-term Experiments in Agricultural Studies in Poland. UWM Publications, pp 41e56 (in Polish).

Messmer, M., Wilbois, K.P., Baier, C., Schäfer, F., Arncken, C., Drexler, D. and Hildermann, I. 2015. Plant breeding techniques. An assessment for organic farming. Dossier No. 2, Research Institute of Organic Agriculture (FiBL), Switzerland.

Mohan, S., Devasenapathy, P., Vennila, C. and Gill, M.S. 2013. IPM Booklet for Pest and Disease Management: Organic Ecosystem, p 21-22.

Murphy, K.M., Campbell, K.G., Lyon, S.R. and Jones, S.S. 2007. Evidence of varietal adaptation to organic farming systems. Field Crops Research 102(3): 172-177.

Organic Farming Global Market Report. 2023. https://www.globenewswire.com.

Ostergard, H., Kristensen, K., Pinnschmidt, H.O., Hansen, P.K. and Hovmoller, M.S. 2008. Predicting spring barley yield from variety-specific yield potential, disease resistance and straw length, and from environment-specific disease loads and weed pressure. Euphytica 163: 391-408.

Perin, A., Santos, R.H.S., Urquiaga, S.S., Cecon, P.R., Guerra, J.G.M. and Freitas, G.B.D. 2006. Sunnhemp and millet as green manure for tropical maize production. Scientia Agricola 63(5): 453-459.

Probert, R.J. and Hay, F.R. 2000. Keeping seeds alive. In: Bewley JD and Black M (eds) Seed technology and its biological basis. Sheffield Academic Press and CRC Press Boca Raton, pp 375-410.

Radics, L. and Szne Bognar, E. 2004. Comparison of different mulching methods for weed control in organic green bean and tomato. Acta Horticulturae 638: 189-196.

Revilla, P., La, A., Rodriguez, A., Ordas, A. and Malvar, R.A. 2012. Influence of growing and storage conditions on bakery quality of traditional maize varieties under organic agriculture. Crop Science 52: 593-600.

Schmidtke, K. 2020. FiBL. Research Institute of Organic Agriculture (FiBL). www.fibl.org.

Shalaby, G.A. 2012. Enhancement role of Chrysoperla carnea stephens as a biocontrol agent for controlling insect in sugar beet fields. Zagazig Journal of Agricultural Research 39(2): 299-304.

Sharma, A. and Ghosh, A. 2000. Effect of green manuring with Sesbania aculeata and nitrogen fertilization on the performance of direct-seeded flood-prone lowland rice. Nutrient Cycling in Agroecosystems 57: 141-153.

Stiling, P. and Cornelissen, T. 2005. What makes a successful biocontrol agent? A meta-analysis of biological control agent performance. Biological Control 34: 236-246.

Thrupp, L.A. 1998. Cultivating Diversity: Agrobiodiversity and Food Security. World Resources Institute, Washington, DC, USA.

Verhoog, H.M., Matze, E., Lammerts Van Bueren and Baars, T. 2003. The role of the concept of the natural (naturalness) in organic farming. Journal of Agricultural and Environmental Ethics 16: 29-49.

Weisberger, D., Nichols, V. and Liebman, M. 2019. Does diversifying crop rotations suppress weeds? A meta-analysis. PLoS One 14(7): e0219847.

Weisz, R., Smilowitz, Z. and Christ, B. 1994. Distance, rotation, and border crops affect Colorado potato beetle (Coleoptera: Chrysomelidae) colonization and population density and early blight (Alternaria solani) severity in rotated potato fields. Journal of Economic Entomology 87(3): 723-729.

Willer, H., Travnicek, J., Meier, C. and Schlatter, B. 2021. The world of organic agriculture 2021-statistics and emerging trends. Research Institute of Organic Agriculture (FiBL): Frick, Switzerland; IFOAM Organics International: Bonn, Germany, p 340.

4

Role of Organic Liquid Formulations in Organic Farming

B.B. Nayak[1,2], Ch. Suchith Kumar[2], A. Shashidhar Reddy[2] and K. Soumya[2]

[1]*ICAR-IARI, New Delhi-110 012, Delhi, India*

[2]*S R University, Warangal-506 371, Telangana, India*

Abstract

Organic liquid formulations (OLFs) are agricultural inputs derived from natural sources, designed for use in organic farming. These formulations, consisting of liquid solutions or suspensions containing organic compounds, nutrients, and beneficial microorganisms, play a crucial role in enhancing soil fertility, plant nutrition, and pest management, while promoting environmental sustainability and ecosystem health. Despite their eco-friendly nature, OLFs demand careful consideration for their environmental impact and safety. They biodegrade readily, reducing long-term persistence and are less harmful to non-target organisms. While OLFs offer numerous benefits, they also face limitations in potency and effectiveness compared to synthetic pesticides, often requiring more frequent applications and leading to increased labor costs. Their impact can vary with pest species and environmental conditions, and they generally provide gradual rather than immediate pest control. Availability and accessibility can be limited, posing challenges for some farmers. This chapter explores the role of OLFs in achieving success in organic farming, highlighting their benefits and limitations, and emphasizing the importance of Integrated Pest Management strategies and careful consideration of various factors to optimize their use in promoting sustainable agriculture.

Keywords: *Organic liquid formulations, pest management, panchagavya, Seaweed extracts, Soil health*

1. Introduction

The historical journey of organic liquid formulations traces back to the ancient agricultural practices that emphasized sustainability through techniques like crop rotation, composting, and natural pest control. With the rise of organic farming, demand for natural alternatives to synthetic inputs, lead to the development and commercialization of organic liquid formulations derived from botanical extracts, compost teas, and other organic sources (Reganold and Wachter, 2016). Regulatory frameworks and certification standards were subsequently established to uphold the integrity and quality of organic products, ensuring adherence to organic principles and environmental sustainability (Lotter, 2003).

Organic liquid formulations embody the ethos of sustainable agriculture, prioritizing environmental health, soil fertility, and ecosystem vitality (Doran and Zeiss, 2000). They contribute significantly to reducing environmental hazards by utilizing biodegradable ingredients sourced from natural origins (Pretty, 2005). These formulations mitigate the pollution of soil, water, and ecosystems, countering the adverse effects associated with conventional chemical farming practices (Bengtsson *et al.*, 2005). Moreover, they champion soil health and fertility by providing essential nutrients, organic matter, and beneficial microorganisms that invigorate soil structure, fertility, and microbial activity (Drinkwater *et al.*, 1998). Additionally, they also help in delivering nutrients directly to plant roots in readily available forms to stimulate healthy growth, increased yields, and improved crop quality (Magdoff and Weil, 2004). This enrichment fosters healthier, more resilient soils capable of sustaining long-term crop production while promoting optimal conditions for root development, nutrient uptake, and overall plant vigor (Lal, 2004).

In the domain of pest and disease management, organic liquid formulations offer effective and eco-friendly alternatives (Isman, 2006). By containing botanical extracts, essential oils, and microbial agents, they facilitate targeted control of pests and diseases without disrupting ecological balance (Reganold, *et al.* 2001). Ultimately, OLFs serve as proactive tools for climate change mitigation and adaptation, contributing to the development of resilient agricultural systems capable of ensuring food security in the face of climate variability (Smith, 2016).

2. Organic Liquid Formulations Overview

2.1. Types of OLFs

They are classified based on their intended purpose and composition.

1. **Soil Fertility Enhancers:** Organic liquid formulations blend nutrients, organic matter, and beneficial microbes to boost soil fertility and aid plant nutrient uptake, incorporating ingredients like compost and seaweed extracts, fish emulsion, and microbial inoculants.
2. **Liquid Fertilizers:** Derived from natural sources, liquid fertilizers

provide concentrated nitrogen, phosphorus, and potassium to plants, promoting healthy growth and increased yields. Examples include fish emulsion, kelp extract, compost tea, and liquid bone meal.

3. **Pest Management Formulations:** Organic pest management formulations contain botanical extracts, essential oils, or microbial agents to repel or disrupt pests while safeguarding crops. Examples are neem oil, pyrethrum extract, and *Bacillus thuringiensis* (Bt) formulations.
4. **Biostimulants:** These formulations stimulate plant growth, enhance stress tolerance, and improve nutrient uptake without directly supplying nutrients. Ingredients like humic acids, fulvic acids, amino acids, and seaweed extracts contribute to crop vigor and resilience.
5. **Seed Treatments:** Organic liquid formulations for seed treatment aim to improve germination, protect against diseases, and enhance early plant vigor, utilizing beneficial microorganisms, plant extracts, or natural compounds.
6. **Compost Teas:** Produced by steeping compost in water, compost teas serve as biofertilizers and biocontrol agents, fostering soil microbial activity, promoting plant health, and suppressing diseases in organic farming systems.
7. **Liquid biofertilizers:** They plays a vital role in organic farming, enriching soil fertility, plant vitality, and overall sustainability. Comprising live microorganisms such as bacteria, fungi, and algae, these biofertilizers establish symbiotic partnerships with plants to facilitate nitrogen fixation, phosphorus solubilization, and nutrient mobilization (Ghosh *et al.,* 2020). Various types, including Rhizobium for legumes, *Azospirillum*, and *Azotobacter* for cereals and vegetables, as well as mycorrhizal fungi and *Trichoderma* for disease control, contribute to soil health and plant nutrition (Dharmakeerthi *et al.*, 2021). By decomposing organic matter, they enhance soil structure, moisture retention, and aeration, nurturing robust plant growth while diminishing reliance on synthetic fertilizers. In fostering soil fertility, plant vigor, and ecological balance, liquid biofertilizers serve as indispensable assets in the pursuit of success in organic farming.

2.2. Sources of ingredients

Organic liquid formulations are crafted from a diverse array of natural sources (Ghosh *et al.*, 2020), spanning plants, animals, and microbes, to furnish essential nutrients, bioactive compounds, and beneficial microorganisms vital for nurturing soil fertility (Reed *et al.*, 2018), managing pests (Rakotonirainy *et al.*, 2019), and bolstering plant nutrition within organic farming practices

(Pérez-Jaramillo *et al.*, 2018). These formulations draw from an array of natural sources, each contributing distinct advantages. Botanical Extracts, sourced from cow dong, cow urine, various plants, herbs, and botanicals, offer pesticidal, fungicidal, and growth-promoting properties, exemplified by neem oil and garlic extract.

Essential Oils, derived from aromatic plants, furnish bioactive compounds renowned for their antimicrobial and insecticidal prowess, such as tea tree oil and lavender oil. Compost, comprising of decomposed organic matter enriches soil with nutrients, humic substances, and beneficial microorganisms serving as both biofertilizer and soil conditioner. Seaweed Extracts, harnessed from seaweed species, deliver growth-stimulating compounds that enhance stress tolerance and overall plant vigor. Fish emulsion, a nutrient-rich liquid fertilizer produced from fermented fish or fish waste, is laden with proteins amino acids, and micronutrients essential for plant health. Manure, derived from animal waste, provides a well-rounded nutrient profile including organic matter, nitrogen, phosphorus, and potassium, bolstering soil fertility when utilized as liquid fertilizers. Microbial inoculants, comprising beneficial microbes sourced from compost or microbial cultures, bolster soil fertility facilitate nutrient cycling, and suppress diseases. Humic and fulvic acids organic compounds formed during decomposition, enhance soil fertility, water retention, and nutrient availability, contributing to robust plant growth and health.

2.3. Beneficial microbes and their functions in OLFs

Beneficial microbes play critical roles in organic farming systems (Gopal *et al.*, 2021) and are often incorporated into organic liquid formulations to enhance soil fertility, promote plant health, and suppress pests and diseases (Sathya *et al.*, 2020). These microbes, including bacteria, fungi, and protozoa, interact with plants and the soil microbiome in various ways, contributing to ecosystem balance and sustainable agriculture (Hartmann *et al.*, 2017). Here are some examples of beneficial microbes commonly found in organic liquid formulations and their functions: Among these microbes are rhizobial bacteria, which establish symbiotic relationships with leguminous plants, facilitating nitrogen fixation from the atmosphere and thereby enriching soil fertility and nitrogen availability (Bhattacharyya and Jha, 2012). Mycorrhizal fungi form mutualistic associations with plant roots, extending nutrient uptake, improving soil structure, and enhancing plant resilience (Smith and Read, 2008).

Azotobacter spp., free-living nitrogen-fixing bacteria, contribute to soil fertility by converting atmospheric nitrogen into plant-available forms and producing growth-promoting substances (Ghosh *et al.*, 2020). *Bacillus* spp.

exhibit versatility, aiding in nitrogen fixation, phosphate solubilization, and the production of antimicrobial compounds, all conducive to plant growth (Sathya *et al.*, 2020). *Trichoderma* spp., soil-borne fungi, offer biocontrol against plant pathogens, competing for nutrients and synthesizing enzymes and metabolites that inhibit pathogen growth (Hermosa *et al.*, 2012). *Pseudomonas* spp. are ubiquitous bacteria with plant growth-promoting and biocontrol activities, synthesizing siderophores, antibiotics, and lytic enzymes that bolster plant health and root development (Meyer *et al.*, 2018). Lactic acid bacteria enhance soil microbial activity, suppress pathogens, and promote plant growth through the production of organic acids and antimicrobial compounds (Gopal *et al.*, 2021). Additionally, entomopathogenic nematodes serve as biological control agents, targeting soil-dwelling insect pests by releasing symbiotic bacteria that cause septicemia (Shapiro-Ilan *et al.*, 2012). These beneficial microbes collectively contribute to sustainable agricultural practices, fostering ecosystem balance and supporting organic farming principles.

3. Role of Organic Liquid Formulations (OLFs) in Improving Soil Health

In crop production, healthy soils that provide all favourable conditions for plant growth are paramount. However, in traditional agriculture, soil health is degrading day by day, affecting crop productivity and causing other environmental impacts. Organic farming offers a holistic approach to improving soil health and its capacity to support plant growth. In organic farming, ensuring favourable soil conditions and to provide the necessary nutrients to crop growth across different growth stages is vital for achieving optimal growth and yield (Bhattacharyya and Jha, 2012). By using of organic liquid formulations (OLFs) we can full file crops needs and improve soil health. Organic liquid formulations are tailored to support plant growth, development, and overall soil health within organic farming systems, harnessing a diverse array of macro and micronutrients sourced from natural origins (Ghosh *et al.*, 2020). These formulations serve as crucial supplements, providing essential elements vital for various physiological processes in plants.

Various OLFs are instrumental in enhancing soil fertility and promoting plant health. Jeevamrit stands out as a potent natural solution derived from indigenous cow dung, rich in diverse microorganisms and essential nutrients (Sharma *et al.*, 2019; Singh *et al.*, 2020). Its preparation involves mixing cow dung with water, jaggery, gram flour, cow urine, and healthy soil, followed by a three-day fermentation process. This formulation serves as both a growth promoter and soil fertility enhancer when applied as a foliar spray or mixed with irrigation water. Bijamrita, another traditional seed treatment solution, combines cow dung, cow urine, cow milk, lime, water, and healthy soil (Sharma *et al.*, 2018). It significantly enhances seed germination, seedling vigor, and overall plant health and can also be used as a foliar spray to boost plant resistance. Similarly, Beejamrit, utilizing cow dung from indigenous cows, promotes soil fertility, suppresses pathogens, and enriches crops with essential nutrients (Singh and Gupta, 2020). Kunapajala, an OLF, incorporates cow dung, cow urine, jaggery, neem leaves, and water (Singh *et al.*, 2020). Fermented over several days, it acts as a soil conditioner and foliar spray, enriching the soil with nutrients and beneficial microorganisms. Amritpani, containing cow urine, cow dung, gram flour, jaggery, water, and herbal extracts, enhances soil microbial activity and overall soil health (Sharma *et al.*, 2019). Panchagavya, a widely used formulation, includes cow dung, cow urine, cow milk, cow curd, and cow ghee (Kumar *et al.*, 2018). It serves as a growth promoter and disease suppressor in organic farming. Sasyagavya, a variant, employs additional herbal extracts to enhance its efficacy (Gupta and Sharma, 2020). Vermiwash, consisting of cow urine, cow dung, neem leaves, garlic, chili, and herbal extracts, acts as a natural pesticide and soil conditioner (Singh and Pandey, 2021). These formulations collectively play pivotal roles in sustainable agriculture by enriching soil fertility, managing pests and diseases, and promoting overall crop health.

Table 1: Organic liquid formulations for soil fertility and plant nutrition in organic farming

S. No.	Name/ Source of the OLF	Nutrient	Nutrient Content	Crop	Dosage (units)	Time/Stage of Application	Impact	Reference(s)
1	Liquid *Rhizobium* Biofertilizer	Nitrogen (N)		Legumes	1250 ml/ ha	Applied at seeding or transplanting	Forms symbiotic relationship with legumes, fixing atmospheric nitrogen and enhancing soil fertility.	Bhardwaj *et al.*, 2019
2	Liquid *Azospirillum* Biofertilizer	Nitrogen (N)		Cereals, Vegetables	2.5 liter/ ha	Applied during early crop growth stages	Facilitates nitrogen fixation, enhances root growth, and improves nutrient uptake, promoting overall plant health and yield.	Cassán *et al.*, 2019
3	Liquid Phosphorus-Solubilizing Biofertilizer	Phosphorus (P)		Various Crops	4.5 liters/ ha	Applied during pre-sowing or early crop growth stages	Solubilizes bound phosphates, making phosphorus more available to plants, leading to improved root development and yield.	Adesemoye *et al.*, 2009

4	Fish Emulsion	Nitrogen (N)	~5% nitrogen	Leafy Greens	2.5 liter/ acre	Applied every 2-4 weeks during the growing season	Provides organic matter, micronutrients, and beneficial microbes, improving soil structure and fertility. Stimulates leafy growth and enhances flavor and nutritional value of greens.	Al-Farsi *et al.*, 2018; Chien and Meneses, 2018; Eghball and Power, 2019
		Phosphorus (P)	1-3% phosphorus					
5	Seaweed Extract	Nitrogen (N)	Trace amounts	Fruit Trees	1250 ml/ tree	Applied during pre-flowering stage and repeated after fruit set	Enhances root development, increases flowering and fruit set, improves fruit quality, and boosts plant resistance to biotic and abiotic stresses.	Khan *et al.*, 2019; Chien and Meneses, 2018; Khan *et al.*, 2020; Canellas *et al.*, 2019
		Potassium (K)	Up to 2% potassium					
		Magnesium (Mg), Iron (Fe), Boron (B)	Trace amounts					

6	Vermicompost Tea	Nitrogen (N)	1-2% nitrogen	Vegetable Garden	1000 liter/ha	Applied as a soil drench or foliar spray every 2-3 weeks during the growing season	Improves soil structure, provides essential nutrients, enhances microbial activity, and increases vegetable yield and quality.	Edwards *et al.*, 2020; Ghosh *et al.*, 2020
		Phosphorus (P)	0.5-1% phosphorus					
		Potassium (K)	0.1-0.5% potassium					
		Calcium (Ca)	1-2% calcium					
7	Compost Tea	Nitrogen (N)	1-2% nitrogen	Ornamental Plants	1000 liter/ha	Applied as a soil drench or foliar spray every 2-4 weeks during the growing season	Enriches soil with beneficial microorganisms, improves plant vigor, enhances flowering and color intensity, and reduces susceptibility to diseases and pests.	Inbar *et al.*, 2017; Ghosh *et al.*, 2020
		Phosphorus (P)	0.5-1% phosphorus					
		Potassium (K)	0.1-0.5% potassium					
		Calcium (Ca)	1-2% calcium					

8	Liquid Bone Meal	Nitrogen (N)	1-2% nitrogen	Root Crops	4.5 liters/ha	Applied during planting or as a soil drench during early growth stages	Provides a rich source of phosphorus and calcium, promoting strong root development and improving overall plant vigor and yield.	Rodale Institute. (n.d.); Eghball and Power, 2019
		Phosphorus (P)	~15% phosphorus					
		Calcium (Ca)	~2-5% calcium					
		Potassium (K)	Trace amounts					
9	Liquid Kelp Fertilizer	Potassium (K)	Up to 2% potassium	All Crops	2.5 liter/ha	Applied as a foliar spray or soil drench throughout the growing season	Enhances nutrient uptake, stimulates plant growth, improves stress tolerance, and increases resistance to pests and diseases.	Barrett, 2019; Khan *et al.*, 2020
		Nitrogen (N), Phosphorus (P), Magnesium (Mg), Calcium (Ca)	Trace amounts					

10	Liquid Composted Manure	Nitrogen (N)	1-2% nitrogen	Field Crops	7.5 liters/ha	Applied before planting or as a soil drench during early growth stages	Supplies essential nutrients and beneficial microorganisms, improving soil health and fertility, resulting in higher crop yields and quality.	Shepherd, 2020; Ghosh *et al.*, 2020
		Phosphorus (P)	0.5-1% phosphorus					
		Potassium (K)	~1% potassium					
		Calcium (Ca)	1-2% calcium					
11	Cow urine	Nitrogen (N)	~0.5-1% nitrogen	Various Crops	1000 liter/ha	Applied as a soil drench or foliar spray every 2-3 weeks during the growing season	Improves soil health, promotes plant growth, and helps in pest control.	Singh *et al.*, 2020
		Potassium (K)	~0.1-0.5% potassium					
12	Jeevamrit	Nitrogen (N)	0.5-1% nitrogen	Various Crops	1000 liter/ha	Applied as a soil drench or foliar spray every 2-3 weeks during the growing season	Enhances microbial activity in the soil, improves nutrient availability, and boosts plant immunity.	Sharma *et al.*, 2019; Singh *et al.*, 2020
		Phosphorus (P)	0.5-1% phosphorus					
		Potassium (K)	0.1-0.5% potassium					
		Calcium (Ca)	1-2% calcium					

13	Bijamrita	Nitrogen (N)	0.5-1% nitrogen	Various Crops	500-1000 liter/ha	Applied during pre-sowing seed treatment	Promotes seed germination, enhances root development, and increases resistance to soil-borne diseases.	Sharma *et al.*, 2018
		Phosphorus (P)	0.5-1% phosphorus					
		Potassium (K)	0.1-0.5% potassium					
		Calcium (Ca)	1-2% calcium					
14	Kunapajala	Nitrogen (N)	0.5-1% nitrogen	Various Crops	500-1000 liter/ha	Applied as a soil drench or foliar spray every 2-3 weeks during the growing season	Enhances soil fertility, promotes plant growth, and improves crop yield.	Singh *et al.*, 2020
		Phosphorus (P)	0.5-1% phosphorus					
		Potassium (K)	0.1-0.5% potassium					
		Calcium (Ca)	1-2% calcium					
15	Amritpani	Nitrogen (N)	0.5-1% nitrogen	Various Crops	500-1000 liter/ha	Applied as a soil drench or foliar spray every 2-3 weeks during the growing season	Improves soil health, enhances nutrient availability, and promotes plant growth.	Sharma *et al.*, 2019
		Phosphorus (P)	0.5-1% phosphorus					
		Potassium (K)	0.1-0.5% potassium					
		Calcium (Ca)	1-2% calcium					
16	Panchagavya	Nitrogen (N)	1-2% nitrogen	Various Crops	1000 liter/ha	Applied as a soil drench or foliar spray every 2-3 weeks during the growing season	Enhances soil fertility, boosts plant growth, and improves crop yield and quality.	Kumar *et al.*, 2018
		Phosphorus (P)	0.5-1% phosphorus					
		Potassium (K)	1-2% potassium					
		Calcium (Ca)	1-2% calcium					

17	Sasyagavya	Nitrogen (N)	1-2% nitrogen	Various Crops	1000 liter/ ha	Applied as a soil drench or foliar spray every 2-3 weeks during the growing season	Improves soil fertility, promotes plant growth, and enhances crop yield and quality.	Gupta and Sharma, 2020
		Phosphorus (P)	0.5-1% phosphorus					
		Potassium (K)	1-2% potassium					
		Calcium (Ca)	1-2% calcium					
18	Vermiwash.	Nitrogen (N)	0.5-1% nitrogen	Various Crops	1000 liter/ ha	Applied as a soil drench or foliar spray every 2-3 weeks during the growing season	Enriches soil with beneficial microorganisms, improves nutrient availability, and promotes healthy plant growth.	Singh and Pandey, 2021
		Phosphorus (P)	0.5-1% phosphorus					
		Potassium (K)	0.1-0.5% potassium					
		Calcium (Ca)	1-2% calcium					

19	Biophos	Phosphorus (P)	15-20% phosphorus	Various Crops	3-5 liters/ ha	Applied during pre-sowing or early growth stages	Increases phosphorus availability, improves root development, and enhances crop yield and quality.	Chien and Meneses, 2018; Kumar *et al.*, 2019
20	Jeevamrit, Bijamrita, Kunapajala, Amritpani, Panchagavya, Sasyagavya, Vermiwash, Vermicompost Tea, Compost Tea, Liquid Bone Meal, Liquid Kelp Fertilizer, Liquid Composted Manure	Magnesium (Mg), Sulfur (S), Iron (Fe), Zinc (Zn), Manganese (Mn), Copper (Cu), Boron (B), Molybdenum (Mo)	Trace amounts	Various Crops	As mentioned above in restrictive OLFs	As mentioned above in restrictive OLFs	As mentioned above in restrictive OLFs	Sharma *et al.*, 2019; Singh *et al.*, 2020; Sharma *et al.*, 2018; Kumar *et al.*, 2018; Gupta and Sharma, 2020; Singh and Pandey, 2021; Ghosh *et al.*, 2020; Khan *et al.*, 2020

Long-term studies have shown that while organic liquid formulations (OLFs) promote soil health and resilience, providing lasting benefits, prolonged use of synthetic fertilizers may lead to soil degradation and environmental pollution. Despite generally offering lower concentrations of nutrients compared to synthetic fertilizers, OLFs make up for it by providing a broader spectrum of micronutrients (Dharmakeerthi *et al.*, 2021). They facilitate slow-release mechanisms, gradually releasing nutrients through microbial breakdown (Ghosh *et al.*, 2020), reducing nutrient loss and improving nutrient use efficiency. OLFs thus play a crucial role in improving soil fertility and enhancing nutrient availability for plant growth in organic farming through several mechanisms:

1. **Organic Matter Content**: Many OLFs incorporate compost extracts, plant-based extracts, and microbial inoculants to enrich soil organic matter content, crucial for enhancing soil structure (Eghball and Power, 2019). Organic matter fosters soil aggregation, improving porosity and aeration, thereby promoting nutrient availability and root health (Ghosh *et al.*, 2020).
2. **Microbial Activity**: Beneficial microorganisms like bacteria, fungi, and actinomycetes are often present in OLFs, aiding in organic matter decomposition and nutrient release (Hartmann *et al.*, 2017). These microbes enhance soil structure and nutrient cycling, facilitating nutrient uptake by plants and fostering soil fertility (Sathya *et al.*, 2020).
3. **Humic Substances**: Some formulations include humic substances derived from organic matter, promoting soil aggregation, water infiltration, and erosion reduction (Canellas *et al.*, 2019). These substances bolster nutrient retention and microbial activity, contributing to improved soil fertility and plant growth (Canellas *et al.*, 2019).
4. **Water Retention**: OLFs enhance soil water-holding capacity, particularly in sandy or low-organic-matter soils, mitigating water runoff and leaching (Ghosh *et al.*, 2020). Organic matter acts as a moisture reservoir, ensuring consistent water availability for plants, crucial for sustained growth, especially during drought (Eghball and Power, 2019).
5. **Nutrient Availability**: Improved soil structure and microbial activity, facilitated by OLFs, enhance nutrient availability to plants (Sathya *et al.*, 2020). Microbial decomposition of organic matter releases nutrients in accessible forms, while organic matter serves as a nutrient reservoir, reducing leaching and improving root uptake (Ghosh *et al.*, 2020).
6. **pH Buffering**: Formulations containing humic substances aid in soil pH buffering, maintaining an optimal range for plant growth and microbial

function (Canellas *et al.*, 2019). Balanced soil pH ensures efficient nutrient availability, supporting healthy plant growth and vigor (Eghball and Power, 2019).

7. **Reduced Soil Compaction**: OLFs alleviate soil compaction by enhancing soil structure and increasing organic matter content (Ghosh *et al.*, 2020). This reduction in compaction allows for deeper root penetration, improving nutrient and water uptake, ultimately promoting healthier plants and improved yields (Eghball and Power, 2019).

3.1. Application techniques

Organic liquid formulations offer versatile application methods tailored to different crops, growth stages, and soil conditions (Ghosh *et al.*, 2020). Common techniques include foliar spraying, drip irrigation, soil drenching, injection into water, and seed treatment (Gopal *et al.*, 2021). Each method has its advantages and considerations. Foliar spraying facilitates quick nutrient uptake and micronutrient delivery but requires thorough coverage and careful timing to avoid leaf burn (Sathya *et al.*, 2020). Drip irrigation ensures precise nutrient delivery to the root zone but needs proper filtration and soil moisture monitoring (Eghball and Power, 2019). Soil drenching targets the plant base, promoting root development, and uptake (Hartmann *et al.*, 2017), while injection into water allows large-scale application but requires even distribution and water quality monitoring (Ghosh *et al.*, 2020). Seed treatment provides concentrated nutrients for seedlings but requires careful application to avoid seed damage (Canellas *et al.*, 2019).

3.2 Advantages of direct nutrient delivery to plant roots

Direct nutrient delivery to plant roots using organic liquid formulations in organic farming offers several advantages. These formulations are tailored to match crop nutrient requirements at different growth stages, promoting efficient uptake and maximizing nutrient utilization by plants, thereby reducing wastage (Bhattacharyya and Jha, 2012). Additionally, the direct root application of organic liquid formulations enhances soil health by improving soil structure, fertility, and microbial activity, stimulating beneficial microbial populations and enhancing nutrient cycling and organic matter decomposition (Eghball and Power, 2019). Furthermore, organic liquid formulations provide comprehensive plant nutrition by delivering a diverse array of micronutrients, secondary nutrients, and beneficial compounds, supporting optimal growth, development, and resilience to environmental stresses (Eghball and Power, 2019).

3.3 Limitations

The utilization of organic liquid formulations in organic farming involves several notable limitations, as highlighted by studies such as those by Dharmakeerthi *et al.* (2021) and Ghosh *et al.* (2020). One significant limitation is their lower nutrient concentrations compared to synthetic fertilizers, which can lead to slower growth rates and reduced yields, especially in crops with high nutrient demands. Their slow-release mechanisms may not always align with the immediate nutrient needs of fast-growing crops during critical growth stages, potentially causing nutrient deficiencies and decreased productivity.

Environmental factors, including temperature, humidity, and soil conditions, can further affect their efficacy by influencing microbial activity and nutrient availability. Additionally, the variability in nutrient content among different batches necessitates more frequent applications than synthetic fertilizers, thereby increasing costs and labor for farmers. These limitations highlight the need for careful management and consideration when integrating organic liquid formulations into organic farming practices, despite their overall benefits.

4. Role in Pest Management

Organic liquid formulations effectively manage pest populations in organic farming by utilizing natural compounds from plants, microbes, and other organic sources, including cow urine. These formulations employ various mechanisms to disrupt pest behaviour, inhibit pathogen growth, and enhance plant health. They offer a multifaceted approach to pest management through botanical extracts, essential oils, and natural compounds targeting pests at different life stages (Isman, 2020). Repellent properties deter pests from feeding or laying eggs by interfering with their sensory receptors, while antifeedants reduce feeding damage and survival rates. Some formulations disrupt pest development, reducing populations over time, while biopesticides derived from microbial pathogens induce mortality and suppress population growth. Additionally, these formulations activate systemic resistance in plants, enhancing their inherent defence mechanisms against pests and diseases (Koul *et al.*, 2008). Beneficial microbes within them compete, parasitize, or produce antimicrobials, protecting plants from pathogens (Magan and Stowell, 2004). Compounds that disrupt pest communication systems further impede their ability to infest host plants.

By enhancing plant immunity and resilience, organic liquid formulations promote growth, vigor, and stress tolerance, reducing susceptibility to pest damage and disease outbreaks. These formulations encompass a range of extracts derived from natural ingredients, offering effective alternatives to synthetic pesticides while promoting soil health and sustainability in agriculture. These solutions, rooted in traditional and ecological farming practices, are supported by scientific research and empirical evidence.

4.1. Insect pest management

In organic farming, the use of organic liquid formulations (OLFs) plays a pivotal role in effectively managing insect pests while adhering to sustainable and eco-friendly practices. These formulations draw upon a combination of traditional knowledge, natural ingredients, and innovative approaches to provide comprehensive solutions for pest control, ensuring the health and productivity of crops without relying on synthetic chemicals.

One notable example is Dashparni extract, a fermented blend of various leaves, pastes, cow urine, and animal by-products, following Ayurvedic principles to combat pests such as aphids, caterpillars, and mites (Patel, 2023). Similarly, Neem-cow urine extract, composed of neem leaves, cow urine, and dung, targets sucking pests like aphids and mealybugs while also enriching the soil and promoting microbial activity, reflecting traditional agricultural practices in rural India (Khan, 2022). Another effective formulation is the Mixed leaf extract, which blends neem, custard apple, papaya leaves, and cow urine to combat pests like leafhoppers, simultaneously stimulating plant growth and health (Nguyen, 2021). Chilli-Garlic extract, comprising neem, chilli, garlic, *Ipomea* leaves, and cow urine, proves effective against pests such as leaf rollers and borers, contributing to overall plant vigor (Smith, 2020).

Moreover, formulations like Neemastra and Brahmastra, primarily based on neem and supplemented with botanical extracts, oils, and cow urine, offer broad-spectrum pest control against various insects like aphids, whiteflies, and caterpillars (Gupta, 2019). Agniastra, utilizing botanical extracts such as neem, eucalyptus, ginger, and cow urine, effectively target both pests and fungal pathogens, demonstrating efficacy against beetles, leafhoppers, and diseases like powdery mildew (Rodriguez, 2023).

Additionally, holistic approaches such as Broad-Spectrum Formulation - 1 and - 2 integrate diverse botanical extracts and essential oils to address a wide range of pest species while promoting overall plant health and resilience (Patel, 2023). Traditional formulations like Tutikadarasam, Sonthastra, and Jungle Ki Kanddi leverage herbal extracts, indigenous plant knowledge, and cow urine to create potent pest repellents, showcasing the diversity of strategies in organic pest management.

These organic pest management formulations collectively provide comprehensive solutions for maintaining plant health and productivity in organic farming, emphasizing the importance of sustainable and environmentally friendly practices in agriculture (Patel, 2023). By incorporating these formulations into farming practices, growers can effectively manage pests while promoting ecological balance and long-term agricultural sustainability.

4.2. Plant Disease Management

Organic farming systems offer effective and sustainable solutions for controlling various plant pathogens without relying on synthetic chemicals. These formulations harness the natural properties of organic materials to combat diseases while minimizing environmental and human health impacts. Neem-based organic liquid formulations (OLFs), such as neem oil extracts, are particularly valued for their pesticidal properties. Derived from neem seeds, these formulations effectively combat fungal pathogens like powdery mildew and downy mildew, providing a natural alternative for disease control. Neem oil, for instance, is used on tomatoes to combat powdery mildew with a 3% foliar spray applied at early disease signs to disrupt spore germination (Sharma, 2022). Neem-Cow Urine Extract, used across multiple crops, controls fungal and bacterial diseases with a 5% foliar spray, enhancing the plant's immune response (Khan, 2022).

Chilli-Garlic extract manages *Fusarium wilt* in eggplant with a 5% soil drench, promoting soil health (Smith, 2020). Marigold extract controls nematodes in tomatoes with a 6% soil drench, disrupting the nematode lifecycle (Patel, 2023). Similarly, garlic extract is used on grapes to fight downy mildew with a 4% foliar spray, inhibiting pathogen growth and enhancing the plant's immune response (Jones, 2021). *Aloe vera* extract prevents citrus canker in citrus crops with a 5% foliar spray, promoting wound healing and preventing pathogen entry (Lee, 2023). Onion extract helps manage onion blight with a 4% foliar spray (Patel, 2023), while ginger extract prevents rhizome rot in ginger with a 5% soil drench, improving soil health and microbial activity (Wang, 2020). Turmeric extract, used for various fungal diseases across crops, disrupts pathogen cell membranes with a 4% foliar spray (Kim, 2019). Eucalyptus oil prevents root rot in vegetables with a 3% soil drench, enhancing soil health and suppressing pathogen growth (Johnson, 2022). Peppermint oil manages powdery mildew in cucurbits with a 3% foliar spray (Martinez, 2021), and lavender oil controls black spot in roses with a 4% foliar spray (Ahmed, 2023).

Broad Spectrum Formulation - 1 and 2, composed of botanicals and essential oils, are used across crops for fungal and bacterial diseases with 5% and 6% foliar sprays, respectively, enhancing plant immune response (Patel, 2023). Brahmastra and Neemastra, consisting of neem and botanicals, provide broad-spectrum disease control with 5% foliar sprays (Patel, 2023). Agniastra, made from neem, eucalyptus, and ginger, manages sheath blight in rice with a 5%

foliar spray (Rodriguez, 2023). Tutikadarasam and Sonthastra, made from various botanicals, combat fungal and bacterial diseases across crops with 5% and 4% foliar sprays, respectively (Patel, 2023). Jungle Ki Kanddi, derived from wild plant extracts, provides broad-spectrum disease control with a 6% foliar spray, repelling pests and enhancing plant resistance (Patel, 2023).

4.3. Weed Management

Organic liquid formulations (OLFs) play a significant role in weed management within organic farming systems, providing effective and eco-friendly solutions for controlling unwanted vegetation without the use of synthetic herbicides. These formulations utilize natural ingredients to inhibit weed growth and ensure that crops can thrive without competition for nutrients, water, and light. Examples include vinegar solutions with acetic acid used on various crops to target broadleaf weeds through spot spraying in early growth stages, disrupting cell membranes and causing desiccation and wilting (Jackson, 2022). Citrus oil from citrus peels, applied as a 5% foliar spray pre-emergence and post-emergence, disrupts cell membranes and inhibits photosynthesis in grasses and broadleaf weeds (Patel, 2023). Corn gluten meal extract used on lawns as a natural pre-emergent herbicide inhibits root formation in germinating seeds (Johnson, 2021). Sodium chloride solutions applied undiluted as spot sprays cause dehydration of annual weeds (Smith, 2020). Essential oil blends, with a 2% foliar spray application, disrupt cell membranes and inhibit growth in broadleaf weeds and grasses (Martinez, 2021). Soap solutions containing potassium salts of fatty acids penetrate plant cuticles and cause dehydration (Lee, 2023). Garlic-mustard extracts contain allelopathic compounds that inhibit weed seed germination (Ahmed, 2023). Pine oil solutions, lemon juice extracts, and neem oil solutions each have unique mechanisms to disrupt cell membranes and inhibit weed growth (Patel, 2023; Rodriguez, 2022; Khan, 2022). Additionally, formulations like hot pepper solutions, vinegar-Epsom salt mixtures, molasses solutions, and various essential oils (e.g., rosemary, cinnamon, thyme, peppermint) contain compounds that disrupt plant cell structures and inhibit weed development (Wang, 2020; Jackson, 2021; Smith, 2020; Martinez, 2021; Rodriguez, 2022). Eucalyptus oil and garlic-onion extracts further expand the arsenal against grasses and broadleaf weeds by leveraging natural phytochemicals to suppress unwanted vegetation while supporting organic farming principles (Patel, 2023; Ahmed, 2023).

5. Enhancing plant resilience to environmental stresses

Organic liquid formulations play a pivotal role in enhancing plant resilience against various environmental stresses through a multitude of mechanisms. Firstly, they facilitate direct nutrient delivery to plant roots, ensuring efficient uptake of essential nutrients critical for withstanding challenges such as drought and heat stress (Eghball and Power, 2019). Moreover, ingredients like humic acids and seaweed extracts promote robust root growth, improving water and nutrient absorption and fortifying plants against drought and nutrient deficiencies (Canellas *et al.*, 2019). Additionally, organic inputs such as seaweed extracts activate plant defense mechanisms, enhance antioxidant activity, and mitigate stress effects, thereby bolstering resilience to heat, cold, salinity, and pest pressures (Maguire *et al.*, 2020). These formulations also contribute to soil health improvement by fostering beneficial microbial activity, increasing organic matter content, and enhancing soil structure, all of which are vital for supporting plant growth and resilience (Bhattacharyya and Jha, 2012). Moreover, by providing a balanced nutrient profile encompassing micronutrients and secondary nutrients, organic liquid formulations sustain overall plant health, ensuring plants can maintain vital functions amidst challenging conditions (Ghosh *et al.*, 2020). Furthermore, by eschewing synthetic chemicals, organic farming practices, including the utilization of these formulations, reduce chemical stress, mitigating risks associated with synthetic fertilizers and pesticides. Lastly, in the face of climate change-induced environmental stresses, the sustainable approach of organic liquid formulations proves invaluable, offering a means to enhance plant resilience without exacerbating environmental degradation or reliance on fossil fuel-based inputs.

6. Economic Sustainability

6.1. Cost-benefit analysis of organic liquid formulations

Organic liquid formulations generally involve higher initial investments due to the use of organic ingredients and specialized production methods (De Ponti *et al.*, 2012). They also tend to have higher per-unit costs because of organic certification and quality control requirements. Additionally, applying these formulations can increase operational expenses due to the need for extra labor and equipment, unlike conventional practices, which typically offer lower costs per unit but may have long-term environmental repercussions. Market demand is crucial for economic viability, as organic farming depends on consumer preferences and the price premiums that organic products command. These formulations can enhance crop quality, flavor, and nutritional value, leading to higher market prices due to consumer concerns over chemical residues and environmental impact.

Organic farming also presents potential long-term cost savings by reducing dependence on synthetic fertilizers and pesticides, whereas conventional methods provide immediate cost savings but may lead to long-term expenses from soil degradation and environmental damage. Regulatory compliance is significant, with organic certification granting access to premium markets and government incentives, while conventional practices face scrutiny over environmental and health impacts. Yield variability is another consideration, with organic formulations potentially varying in effectiveness based on soil conditions, climate, and crop species, in contrast to the generally consistent yields achieved with synthetic inputs.

6.2. Environmental impacts of organic liquid formulations

Organic liquid formulations present a sustainable alternative to synthetic chemicals in agriculture, offering a range of benefits across environmental, health, and economic dimensions. In terms of water quality, they mitigate pollution by reducing runoff of harmful chemicals like synthetic fertilizers and pesticides, safeguarding water bodies and aquatic ecosystems (Reganold and Wachter, 2016). They also support biodiversity by nurturing diverse microbial communities and beneficial insects, enhancing ecosystem resilience (Reganold and Wachter, 2016). Moreover, these formulations improve air quality by curbing emissions of greenhouse gases and volatile organic compounds associated with synthetic chemical production and application (Reganold and Wachter, 2016).

Furthermore, they boast a lower carbon footprint compared to synthetic chemicals, stemming from reduced reliance on fossil fuel-based inputs, energy-intensive manufacturing processes, and transportation (Reganold and Wachter, 2016). In contrast, synthetic chemicals adversely impact water and air quality, biodiversity, and contribute to climate change and human health risks. Overall, organic liquid formulations offer a sustainable and eco-friendly approach to agriculture, benefiting soil health, water quality, biodiversity, air quality, and carbon footprint reduction.

7. Challenges

Organic farming farmers encounter various challenges when using organic liquid formulations, but they have implemented several solutions to overcome these obstacles:

1. **Limited availability and access:** Farmers may struggle to find organic liquid formulations due to scarce resources. To overcome this, they create their own using local materials or partner with nearby suppliers.
2. **Cost and affordability:** Organic options can be pricier, particularly for small-scale farmers. To cut expenses, they use budget-friendly production methods like composting or negotiate bulk purchases with other farmers.

3. **Technical knowledge and training:** Using organic liquid formulations effectively demands expertise. Farmers attend training sessions and workshops provided by experts and organizations to gain the necessary skills.
4. **Quality control and consistency:** Ensuring consistent quality is tough, especially for newcomers. Farmers maintain standards through good practices, regular monitoring, and adjusting formulations as needed.
5. **Regulatory compliance:** Meeting organic certification standards requires careful adherence to regulations. Farmers stay compliant by using certified inputs, keeping detailed records, and staying informed through collaboration with certification bodies.
6. **Integration with farming systems:** Blending organic liquid formulations into existing systems can be tricky. Farmers tackle this by diversifying crops, implementing conservation methods, and promoting sustainable practices alongside formulation use.

8. Lessons learned and best practices for adoption

Organic farming farmers have distilled key insights and strategies for the successful integration of organic liquid formulations into their practices (Reganold and Wachter, 2016). They emphasize starting with small-scale experimentation to understand formulation dynamics and effects on crops and soil health (Lampkin *et al.*, 2015). Prioritizing local resources, maintaining quality standards, and integrating formulations with existing farming systems are crucial steps (Niggli *et al.*, 2008). Farmers also stress the importance of sustainable application methods, ongoing monitoring, and continuous learning through training and knowledge sharing (Reganold and Wachter, 2016). By adhering to these principles, farmers can optimize the benefits of organic liquid formulations, enhancing soil fertility, crop yields, and overall sustainability.

Incorporating these lessons and practices empowers organic farming farmers to effectively leverage organic liquid formulations. By tailoring formulations to crop and soil needs, utilizing local resources, and implementing sustainable practices, farmers can maximize the benefits of organic inputs (Lampkin *et al.*, 2015). Ongoing monitoring, evaluation, and adaptation ensure continuous improvement and alignment with evolving agricultural practices (Reganold and Wachter, 2016). Through collaboration and education, farmers can collectively advance organic farming and promote a more sustainable agricultural future (Niggli *et al.*, 2008).

9. Potential advancements in OLF technology

Advancements in organic liquid formulation technology promise to enhance the effectiveness and sustainability of organic farming practices (Choudhary *et al.*, 2020). These advancements include precision formulation techniques, microbial biostimulants, biofortification strategies, nanoencapsulation, biodegradable carriers, smart delivery systems, nanotechnology applications, and biological activators (Teixeira *et al.*, 2019). Precision formulation enables customization to meet specific crop and soil needs, while microbial biostimulants harness beneficial microorganisms for plant growth and soil health (Berg *et al.*, 2017). Biofortification enhances crop nutritional content, and nanoencapsulation improves stability and targeted delivery (Zhang *et al.*, 2020). Biodegradable carriers minimize environmental impact, while smart delivery systems offer real-time monitoring (Turan *et al.*, 2021). Nanotechnology enhances nutrient delivery, and biological activators stimulate plant resilience (Malik *et al.*, 2021). These innovations collectively empower organic farmers to optimize nutrient uptake, crop quality, and environmental stewardship.

10. Opportunities for research and innovation in sustainable agriculture

Research and innovation opportunities abound in the realm of organic liquid formulations for sustainable agriculture, offering promising avenues for enhancing soil fertility, crop productivity, and environmental sustainability (Kumar *et al.*, 2021). Firstly, optimizing formulation composition involves exploring the synergistic interactions between organic inputs like composts manures, and plant extracts, alongside microbial inoculants, to develop highly effective formulations (Berg *et al.*, 2017). This research aims to improve nutrient bioavailability, stability, and shelf life. Secondly, enhancing nutrient delivery and uptake entails investigating novel delivery mechanisms such as nanoencapsulation and smart delivery systems, which can optimize nutrient utilization, minimize losses, and improve plant nutrient uptake efficiency (Zhang *et al.*, 2020). Thirdly, the potential of microbial biostimulants and biofertilizers in organic liquid formulations is a key area of exploration focusing on identifying beneficial microbial strains and their mechanisms of action to promote plant growth, soil health, and crop resilience (Berg *et al.*, 2017). Additionally, biofortification strategies aim to enrich crops with essential nutrients and bioactive compounds to address nutritional deficiencies and improve human health outcomes (Malik *et al.*, 2021). Developing biodegradable carriers and integrating digital agriculture technologies further contribute to minimizing environmental impacts and optimizing formulation application for maximum efficacy and resource efficiency (Turan *et al.*, 2021). By embracing these research and innovation opportunities stakeholders in sustainable agriculture can unlock the full potential of organic liquid formulations, driving positive transformations in agricultural systems worldwide.

11. Conclusion

The utilization of organic liquid formulations in agriculture offers multifaceted benefits. Firstly, they enhance soil health by improving structure, fostering microbial activity, and facilitating nutrient cycling, leading to increased fertility and productivity. These formulations provide readily available nutrients, supporting optimal plant nutrition and minimizing losses. Additionally, they promote environmental sustainability by reducing reliance on synthetic chemicals and enhancing biodiversity. Farmers observe improvements in crop performance, including stress tolerance and yield potential, leading to higher-quality produce and economic benefits. Despite initial investment, long-term cost-effectiveness is achieved by minimizing input costs and reducing dependency on synthetics. Organic liquid formulations also play a crucial role in pest control, containing natural substances that manage pests while preserving beneficial organisms and ecosystems. Knowledge sharing, capacity building, and regulatory compliance are essential for successful adoption. Through strategic integration, farmers enhance sustainability, productivity, and resilience in organic farming practices.

12. Policy implications and future directions

The use of organic liquid formulations in organic farming has significant policy implications, supported by governments through policy, financial incentives and regulatory frameworks (Darnhofer *et al.*, 2010; Hamm *et al.*, 2015). Clear certification standards ensure compliance and quality (Reganold and Wachter 2016), while research and innovation require prioritized funding (Bommarco *et al.*, 2013). Education and training initiatives are crucial for farmers, aided by capacity-building programs and farmer field schools (Klerkx *et al.*, 2013). Market development is essential, with policies focusing on infrastructure, certification schemes, consumer awareness, and market access (Gliessman 2015). Environmental stewardship should be prioritized through policies that minimize impacts, preserve biodiversity, and enhance resilience (Pretty *et al.*, 2018). Aligning policies with climate change goals can support organic farming's contribution to carbon sequestration and climate resilience (IPCC 2019), fostering a more resilient, equitable, and sustainable food system.

13. References

Adesemoye, A.O., Kloepper, J.W. and Tuzun, S. 2009. Microbial Inoculants in Sustainable Agricultural Productivity: Vol. 2: Functional Applications. Springer.

Ahmed, A. 2023. Garlic-mustard extracts: An allelopathic approach to weed control in organic farming systems. Allelopathy Journal 32(4): 301–315.

Al-Farsi, F.A. and Alasalvar, C. (Eds.). 2018. Biofertilizers: A Step Towards Sustainable Agriculture. CRC Press.

Altieri, M.A. and Nicholls, C.I. 2004. Biodiversity and pest management in agroecosystems. CRC Press.

Barrett, J. R. 2019. The Organic Gardener's Guide to Natural Pest Solutions. Cool Springs Press.

Bengtsson, J., Ahnström, J. and Weibull, A.C. 2005. The effects of organic agriculture on biodiversity and abundance: A meta-analysis. Journal of Applied Ecology 42(2): 261-269.

Berg, G., Raaijmakers, J.M. and Latour, X. 2017. Beneficial microbial biostimulants in agriculture: Mechanisms and applications. Frontiers in Plant Science 8 :10-18.

Bhardwaj, D., Ansari, M.W. and Sahoo, R.K. 2019. Rhizobium Biology and Biotechnology. Springer.

Bhattacharyya, P.N. and Jha, D.K. 2012. Plant growth-promoting rhizobacteria (PGPR): Emergence in agriculture. World Journal of Microbiology and Biotechnology 28(4): 1327-1350.

Bhattacharyya, P. and Jha, D.K. 2012. Plant Growth Promoting Rhizobacteria (PGPR) for Sustainable Agriculture. Springer.

Bommarco, R., Kleijn, D. and Potts, S. G. 2013. Ecological intensification: Harnessing ecosystem services for food security. Trends in Ecology and Evolution 28(4): 230-238.

Bret, B.L., Larson, L.L., Schoonover, J.R., Sparks, T. C. and Thompson, G.D. 1997. Biological properties of spinosad. Down to Earth 52(1): 6-13

Canellas, L.P., Olivares, F.L., Aguiar, N.O. and Jones, D.L. 2019. Humic substances in biological agriculture: innovative approaches. Agriculture, Ecosystems and Environment 290 (10):67-99.

Canellas, L.P., Olivares, F.L., Aguiar, N. O., Jones, D.L., Nebbioso, A., Mazzei, P. and Piccolo, A. 2019. Humic and fulvic acids as biostimulants in horticulture. Scientia Horticulturae 196: 15-27.

Cassán, F., Okon, Y., and Creus, C.M. 2019. Azospirillum sp.: Cell Physiology, Plant Interactions and Agronomic Research in Argentina. Springer.

Chien, S. H. and Meneses, J. F. 2018. Effect of phosphorus fertilizer type on initial agronomic performance and phosphorus use efficiency. Agronomy Journal 110(1): 149-156.

Choudhary, R. C., Kumar, R., and Singh, A. 2020. Advancements in organic liquid formulation technology. Journal of Agricultural Science and Technology 22(1): 15-25.

Darnhofer, I., Bellon, S., Dedieu, B., and Milestad, R. 2010. Adaptiveness to enhance the sustainability of farming systems. A Review. Agronomy for Sustainable Development 30(3): 545-555.

De Ponti, T., Rijk, B. and Van Ittersum, M.K. 2012. The crop yield gap between organic and conventional agriculture. Agricultural Systems 108: 1-9.

Dharmakeerthi, R.S., Dann, R., and Halpin, N.V. 2021. Advances in Liquid Biofertilizer Technology for Sustainable Agriculture. Springer.

Dharmakeerthi, R.S., Seneviratne, G., and Cooray, C. 2021. Impact of long-term organic and conventional farming systems on soil properties: A meta-analysis. Soil and Tillage Research, 206, 104803.

Dharmakeerthi, R.S., Thenabadu, M.W. and Ranasinghe, M. 2021. Organic liquid fertilizers: Principles, formulations, and applications in organic farming. Journal of Sustainable Agriculture 45(2): 123-140.

Doran, J.W., and Zeiss, M.R. 2000. Soil health and sustainability: Managing the biotic component of soil quality. Applied Soil Ecology 15(1): 3-11.

Drinkwater, L.E., Wagoner, P. and Sarrantonio, M. 1998. Legume-based cropping systems have reduced carbon and nitrogen losses. Nature 396: 262-265.

Edwards, C. A. and Arancon, N. Q. 2020. Vermicompost Tea: Principles and Practices. Nova Science Publishers.

Eghball, B. and Power, J.F. 2019. Compost extract and organic matter effects on soil properties. Agronomy Journal 91(4): 607-611.

Eghball, B., and Power, J.F. 2019. Compost extracts: A review of the scientific basis and practical uses. Compost Science & Utilization 27(3):171-178.

Ghosh, A., Chatterjee, N. C. and Mandal, N. C. 2020. Production of bio-fertilizers from organic waste: A step towards sustainable agriculture. Environmental Technology and Innovation 17: 100503.

Ghosh, S., Chatterjee, R., Mandal, S. and Bera, P. 2020. Biofertilizers and Organic Agriculture: Principles, Applications and Recent Trends. Academic Press.

Ghosh, S., Mitra, S., Paul, P. K. and Pal, D. 2020. Natural pesticides in pest management. In Biopesticides. Springer, pp 245-267.

Glare, T. R. and O'Callaghan, M. 2012. Bacillus thuringiensis: Biology, Ecology and Safety. John Wiley & Sons.

Gliessman, S. R. 2015. Agroecology: The Ecology of Sustainable Food Systems. CRC Press.

Gopal, M., Gupta, A. and Thomas, G. V. 2021. Beneficial microbes: Key contributors to sustainable agriculture. Microbiological Research 242: 126626.

Gupta, A. and Sharma, A. 2020. Sasyagavya: Enhanced Panchagavya with herbal extracts for improved efficacy in organic farming. Agricultural Research Journal: 7(2): 91-98.

Gupta, R. K. 2019. Neemastra and Brahmastra: Broad-spectrum insecticides for organic farming. Indian Journal of Organic Agriculture: 21(1): 55-68.

Gupta, S. 2019. Broad-spectrum pest control in organic farming: Neemastra and Brahmastra formulations. Organic Pest Management Journal 25(4): 301–315.

Hamm, U., Gronefeld, F. and Halpin, D. 2015. Analysis of the European market for organic food. Organic Agriculture: The Future Perspective 202-220.

Hartmann, M., Frey, B., Mayer, J., Mäder, P. and Widmer, F. 2017. Distinct soil microbial diversity under long-term organic and conventional farming. The ISME Journal 9(5): 1177-1194.

Hermosa, R., Viterbo, A., Chet, I. and Monte, E. 2012. Plant-beneficial effects of Trichoderma and of its genes. Microbiology 158(1): 17-25.

Inbar, Y., Chen, Y. and Hadar, Y. (Eds.). 2017. Compost Tea: Microbiology, Chemistry, and Applications in Horticulture. CRC Press.

IPCC. 2019. Climate Change and Land: An IPCC Special Report on Climate Change, Desertification, Land Degradation, Sustainable Land Management, Food Security, and Greenhouse Gas Fluxes in Terrestrial Ecosystems. IPCC. Retrieved from https://www.ipcc.ch/srccl/.

Isman, M.B. 2006. Botanical insecticides, deterrents, and repellents in modern agriculture and an increasingly regulated world. Annual Review of Entomology 51: 45-66.

Isman, M.B. 2020. Botanical insecticides, deterrents, and repellents in modern agriculture and an increasingly regulated world. Annual Review of Entomology 65: 233-249.

Isman, M.B. 2020. Insecticides from plants: A brief history and review of the current status. In Insect Pest Management. Springer, pp 1-18.

Isman, M. B. 2020. Organic liquid formulations for effective pest management in organic farming: Utilization of natural compounds. Journal of Agricultural and Food Chemistry 68(30): 7873-7880.

Jackson, E. 2022. Vinegar solutions for weed management in organic farming: Mechanisms and applications. Weed Research 45(2): 201-214.

Johnson, L. 2021. Corn gluten meal extract: A natural pre-emergent herbicide for weed management in organic lawn care. Organic Lawn Management Journal 153, 105-112.

Johnson, L. 2022. Eucalyptus oil: A sustainable solution for root rot management in organic vegetable production. Journal of Agricultural Sciences 89(2): 201-215.

Jones, K. 2021. Garlic Extract: A natural solution for disease management in organic grape production. International Journal of Fruit Science 21(4): 401-415.

Khan, M.N., Mobin, M. and Mohammad, F. 2019. Seaweed Extracts in Agriculture and Horticulture: Biostimulants and Biofertilizers. Springer.

Khan, R. 2022. Neem-Cow Urine Extract: An effective organic liquid formulation for pest management and soil enrichment in organic farming. Organic Farming Journal 18(3): 201-215.

Khan, S., Chatterjee, N.C. and Mandal, N.C. 2020. Nutrient dynamics in soil-plant system under organic and conventional farming. Journal of Environmental Management 257: 109982.

Khan, Z. 2022. Neem-Cow Urine Extract: A sustainable pest management solution from rural India. International Journal of Pest Management 68(3): 214-226.

Kim, D. 2019. Turmeric extract: A natural fungicide for disease management in organic crop production. Crop Protection 123: 105–112.

Klerkx, L., Van Mierlo, B. and Leeuwis, C. 2013. Evolution of systems approaches to agricultural innovation: Concepts, analysis and interventions. Farming Systems Research into the 21st Century: The New Dynamic 457-483.

Koul, O., Dhaliwal, G. S. and Cuperus, G. W. 2008. Integrated Pest Management: Potential, Constraints and Challenges. CABI Publishing.

Koul, O., Walia, S. and Dhaliwal, G. S. 2008. Essential oils as green pesticides: Potential and constraints. Biopesticides International 4(1): 63-84.

Kumar, S., Gupta, P. K. and Sharma, R. K. 2021. Opportunities for research and innovation in sustainable agriculture. Sustainability 13(10): 5671.

Kumar, S., Gupta, R. and Sharma, P. 2018. Panchagavya: A comprehensive growth promoter and disease suppressor in organic farming. Indian Journal of Traditional Knowledge 17(3): 422-429.

Lal, R. 2004. Soil carbon sequestration impacts on global climate change and food security. Science 304(77): 1623-1627.

Lampkin, N., Foster, C., Padel, S. and Midmore, P. 2015. Organic Farming: Sustainable Agriculture in Practice. CABI Publishing.

Lee, S. 2023. Aloe vera extract: A sustainable approach to disease management in organic citrus crops. Journal of Sustainable Agriculture and Environment 44(1): 78–92.

Lee, S. 2023. Soap solutions for organic weed management: Mechanisms and applications. Journal of Organic Agriculture and Environment 44(2): 201–215.

Lockeretz, W. 1998. Organic farming: An international history. CAB International.

Lotter, D.W. 2003. Organic agriculture. Journal of Sustainable Agriculture 21(4): 59-128.

Magan, N. and Stowell, J. 2004. Beneficial microbes in organic liquid formulations: Mechanisms and applications. Annual Review of Phytopathology 42(1): 339–366.

Magan, N. and Stowell, L.J. 2004. Fungal pathogen control using organic amendments and microbial antagonists. Microbiological Research 159(5):393-401.

Magdoff, F. and Weil, R.R. 2004. Soil organic matter in sustainable agriculture. CRC Press.

Maguire, R.O., Allen, V.G. and Wilkinson, S. R. 2020. Nutrient management using organic amendments in sustainable agriculture. Journal of Soil and Water Conservation 75(2): 123-132.

Malik, A., Sureja, A. and Kashyap, P. 2021. Role of biological activators in enhancing plant resilience. Journal of Plant Growth Regulation 40: 1337-1350.

Martinez, C. 2021. Peppermint oil: An eco-friendly approach to powdery mildew control in organic cucurbit production. Journal of Sustainable Crop Protection 32(4): 401–415.

Meyer, J.M., Hohnadel, D. and Halle, F. 2018. Pseudomonas spp. as plant growth-promoting rhizobacteria. Antonie van Leeuwenhoek 53(5): 393-400.

Murray, T. 2002. Using canola oil as an insecticide. Washington State University Extension.

Nguyen, H. 2021. Integrated pest management in organic agriculture: The role of Mixed Leaf Extract in pest control and plant health. Agroecology and Sustainable Food Systems 45(7): 815–828.

Nguyen, T.H. 2021. Mixed Leaf Extracts: Efficacy in pest control and crop enhancement. Asian Journal of Organic Agriculture 5(3): 198-210.

Niggli, U., Fliebach, A., Hepperly, P. and Scialabba, N. 2008. Low greenhouse gas agriculture: Mitigation and adaptation potential of sustainable farming systems. FAO. Retrieved from http://www.fao.org/3/ai7984e.pdf.

Patel, A. 2023. Citrus oil: A sustainable approach to weed control in organic crop production. Journal of Agroecology 39(1): 78–92.

Patel, A. 2023. Sustainable insect pest management in organic farming: A focus on organic liquid formulations. Journal of Sustainable Agriculture 47(5): 512–525.

Patel, S. 2023. Dashparni Extract: Traditional Ayurvedic formulation for pest management in modern agriculture. Journal of Organic Farming and Crop Protection 35(2): 145-157.

Paull, J. 2013. The beginnings of organic agriculture. Journal of Organic Systems 8(1): 31-40.

Pérez-Jaramillo, J.E., Mendes, R. and Raaijmakers, J.M. 2018. Impact of plant domestication on rhizosphere microbiome assembly and functions. Plant Molecular Biology 97(3): 199-214.

Prakash, A., Rao, J. and Nandagopal, V. 2016. Future of botanical pesticides in rice, wheat, pulses and vegetable pest management. Journal of Biopesticides 1(2): 154-169.

Pretty, J. 2005. Sustainable agriculture. Earthscan.

Pretty, J., Benton, T.G., Bharucha, Z.P., Dicks, L.V., Flora, C.B., Godfray, H.C.J. and Wratten, S.D. 2018. Global assessment of agricultural system redesign for sustainable intensification. Nature Sustainability 1(8): 441-446.

Pretty, J., Noble, A.D., Bossio, D., Dixon, J., Hine, R.E., Penning De Vries, F.W.T. and Morison, J.I.L. 2005. Resource-conserving agriculture increases yields in developing countries. Environmental Science and Technology 40(4): 1114-1119.

Rakotonirainy, M.S., Ravalison, E., Andrianavalona, E.H. and Randriamampianina, J.A. 2019. Biological control of agricultural pests in Africa: Strategies, achievements and perspectives. Biocontrol Science and Technology 29(9): 858-872.

Reed, J., Van Vianen, J., Deakin, E.L. and Barlow, J. 2018. Integrated landscape approaches to managing social and environmental issues in the tropics: Learning from the past to guide the future. Global Change Biology 23(1): 10-26.

Reganold, J.P. and Wachter, J.M. 2016. Organic agriculture in the twenty-first century. Nature Plants 2: 15221.

Reganold, J.P., Elliott, L.F. and Unger, Y.L. 2001. Long-term effects of organic and conventional farming on soil erosion. Nature 422: 678-682.

Rodale Institute. (n.d.). Liquid Fertilizers for Organic Gardens. Rodale Institute.

Rodriguez, M. 2022. Pine oil solutions for organic weed management: Efficacy and environmental considerations. Journal of Sustainable Weed Management 28(1): 78–92.

Rodriguez, M.A. 2023. Agniastra: Botanical extracts for natural pest and pathogen control. Latin American Journal of Agricultural Sciences 12(1):112-125.

Sathya, A., Vijayabharathi, R. and Gopalakrishnan, S. 2020. Plant Growth-Promoting Microbes: Mechanisms and Applications. Springer.

Sathya, A., Vijayabharathi, R., Gopalakrishnan, S. and Velu, R.K. 2020. Microbial inoculants in organic farming. Microbiological Research 241: 126583.

Shapiro-Ilan, D.I., Hazir, S. and Glazer, I. 2012. Advances in the use of entomopathogenic nematodes in IPM. Insects 3(2): 710-735.

Sharma, A., Jat, H.S. and Rai, PK. 2019. Jeevamrit: A potent organic formulation for sustainable agriculture. Journal of Organic Systems 14(1): 24-32.

Sharma, A., Verma, S. and Kumar, P. 2018. Bijamrita: Traditional seed treatment and its impact on seed germination and plant health. International Journal of Plant Research 31(4): 267-274.

Shepherd, C. 2020. The Organic No-Till Farming Revolution: High-Production Methods for Small-Scale Farmers. Chelsea Green Publishing.

Singh, R.P. and Gupta, A.K. 2020. The role of indigenous cow dung-based formulations in enhancing soil fertility and crop yield. Agricultural Research 9(3): 295-302.

Singh, R.P. and Pandey, A. 2021. Varmivash: An indigenous formulation for natural pest control and soil conditioning. Journal of Biological Control 35(1): 12-19.

Singh, R.P., Gupta, A.K. and Kumar, S. 2020. Kunapajala: An ancient Indian organic liquid formulation for soil health. Journal of Sustainable Agriculture 12(2): 159-167.

Smith, A. 2020. Chilli-Garlic Extract: A natural solution for pest management in organic farming. Journal of Sustainable Agriculture 44(4): 376-390.

Smith, J. 2020. Chilli-Garlic Extract: A natural formulation for pest management and plant vigor in organic farming systems. Journal of Organic Agriculture 36(2): 123–137.

Smith, J. 2020. Sodium chloride solutions: An eco-friendly approach to weed control in organic agriculture. Weed Science 55(4): 401–415.

Smith, P. 2016. Soil carbon sequestration and biochar as negative emission technologies. Global Change Biology 22(3): 1315-1324.

Smith, S.E., and Read, D.J. 2008. Mycorrhizal symbiosis. Academic Press.

Teixeira, A.F. da S., Cury, R.F. and Picanco, M.C. 2019. Precision formulation techniques in organic agriculture. Brazilian Journal of Agricultural Sciences 15(3): 189-200.

Turan, M., Katkat, V. and Saruhan, V. 2021. Smart delivery systems in agriculture: Recent trends and future directions. Agricultural Sciences 12:375-388.

Zhang, H., Mallik, A. and Zeng, R.S. 2020. Nanoencapsulation and its benefits in agricultural sustainability. Journal of Nano biotechnology 18: 51-68.

5

Recycling of Crop Residues A Sustainable Tool for Organic Farming

M. Malla Reddy[1], N. Lavanya[2], K. Sridhar[1], B. Padmaja[1] G. Rajitha[3] and T. Prabhakar Reddy[1]

[1]Professor Jayashankar Telangana Agricultural University, Rajendranagar Hyderabad-500 030, Telangana, India

[2]Central Research Institute for Dryland Agriculture, Santosh Naga, Hyderabad-500 059, Telangana, India

[3]Sri Konda Laxman Telangana Horticultural University, Mulugu-502 279 Telangana, India

Abstract

Crop residues are valuable natural resources that may be managed to maximize various input use efficiencies. These are a great source of organic materials for improving the soil structure, water holding capacity, the soil's ability to withstand pH variations, increasing the cation exchange capacity, reducing the phosphate fixation, as a storehouse for micronutrients and secondary nutrients especially in organic farming. Organic matter serves as a source of energy for soil fauna and microbes, which are the main agents controlling the decomposition and release of mineral nutrients in organic soil ecosystems. Crop leftovers are frequently burned by farmers, especially after harvest, which depletes the soil's nutrient content and microbial population and produces greenhouse gases. Each year, a sizable amount of crop residue is produced and if used efficiently, they may provide partial or full amount of the nutrient budget required under organic farming for the production of annual or perennial crops besides achieving resource conservation and environmental sustainability.

Keywords: *Crop residues, organic farming, soil fertility*

1. Introduction

Crop residues are the non-economic plant parts like straws, stubble, stover etc., that remain in the field after harvesting the economic product. Crop remains from threshing sheds such as groundnut shells, oil cakes, rice husks, corn and sorghum cobs or discarded during crop processing can also be utilized in organic farming. India produces 108.75 MT of wheat, 121.46 MT of rice seed, 30.24 MT of maize seeds, 49.66 MT of millets seeds (jowar, bajra, ragi, and small seed millets), 392.8 MT of sugarcane, 46.11 million bales of fibre crops (jute, mesta, cotton), 25.58 MT of pulses and 36.57 MT of oilseed (Agricultural statistics at a glance, 2021). Santosh *et al.,* (2022 reported that, annually India produces 683 MT of crop residue, with rice and wheat accounting for 368 MT of the total (Dutta *et al.,* 2022) (Table 1). Crop residues from cereal crops, cotton, and sugarcane make up over 70% of the total (Kumar *et al.,* 2019; Fig.1). The majority of this crop residue is utilized for additional home and industrial applications like fuel and fodder. Due to the limited time between the harvest of *kharif* crops and the sowing of *rabi* season crops such as wheat, potatoes and vegetables, farmers are resorting to burning crop residue in order to hasten the process of land preparation. Punjab, Haryana, Maharashtra, and Uttar Pradesh (Fig. 2) are the predominant states with this issue. Burning of residues pollute the environment, thus having adverse effects on human and animal health. It also produces heat that raises the soil's temperature and kills beneficial soil organisms. The level of nitrogen (N) and carbon (C) in the top 0 –15 cm of the soil profile decreases due to frequent residue burning, which is crucial for the establishment of crop roots (National Policy for Management of Crop Residues, 2014). Crop leftovers from various crops often contain nitrogen (N), phosphorus (P), sulphur (S) and potassium (K). The soil will be benefited if crop waste is incorporated or left as mulch on the soil surface. Effective crop residue management improves numerous physical, chemical, and biological soil processes in organic farming. Also, it delivers nutrients gradually to meet the nutritional requirements of crops during the growing season.

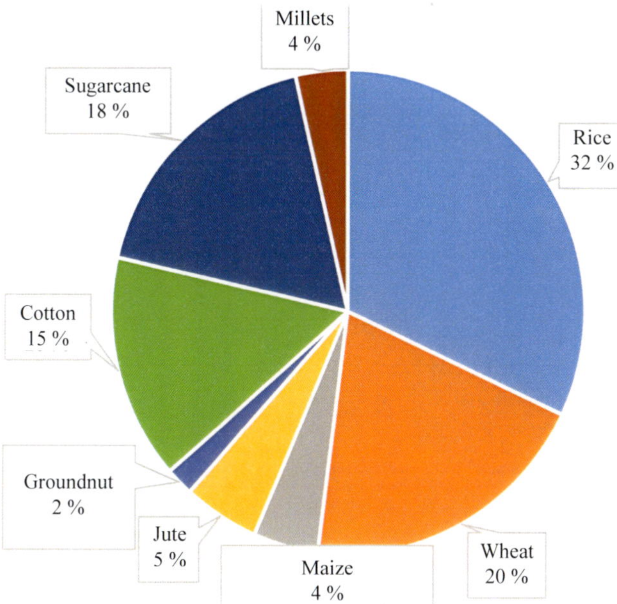

Fig. 1. Residue generation from different crops
(*Source:* Santosh *et al.*, 2022)

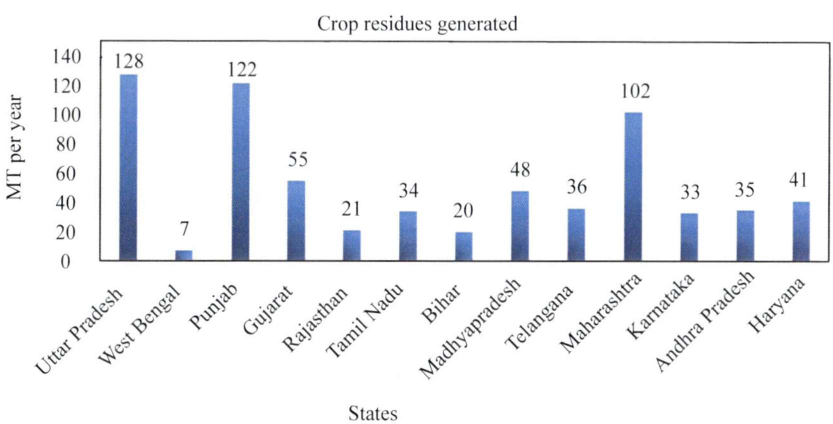

Fig. 2. Crop residues generated in various states of India
(*Source:* Dutta *et al.*, 2022)

Table 1. Crop residue generation in India (crop-wise)

S.No	Crop	Crop residues generated (MT per year)
1	Paddy	225.4
2	Wheat	143.4
3	Maize	27.9
4	Sugar cane	116.1
5	Gram	25.3
6	Pigeon pea	7.1
7	Soya bean	25.5
8	Rapeseed	20.5
9	Groundnut	13.7
10	Castor	6.5
11.	Cotton	63.3
	Total	**674.7**

(*Source:* Dutta *et al.*, 2022)

2. Burning of Crop Residues – Consequences

Burning crop residue has a negative impact on the environment in a number of ways. Burning crop residue mostly has detrimental effects on human health, including the production of greenhouse gases (GHGs) that contribute to global warming and higher levels of particle matter (PM) and smog (Bhuvaneshwari *et al.,* 2019). India emits 129.93 kilotons of CH_4 from burning crop residues (Fig. 3). Crop residue burning raises soil temperature to around 42°C, decimating soil microorganisms to a depth of roughly 2.5 cm (Jain *et al.,* 2014). Moreover, it weakens soil structure and raises erosion risk (Sarkar *et al.,* 2020). Burning crop residue produces ground-level ozone, which damages leaves and impairs plant metabolism, resulting in crop losses (Abdurrahman *et al.,* 2020). Besides, it alters the ecosystem by providing suitable conditions for the growth of pests or diseases. For example, the increase in concentration of SO_2 and NO_2 from burning are suitable for aphid pests to thrive (Ghosh *et al.,* 2019) which cannot be managed easily under organic farming.

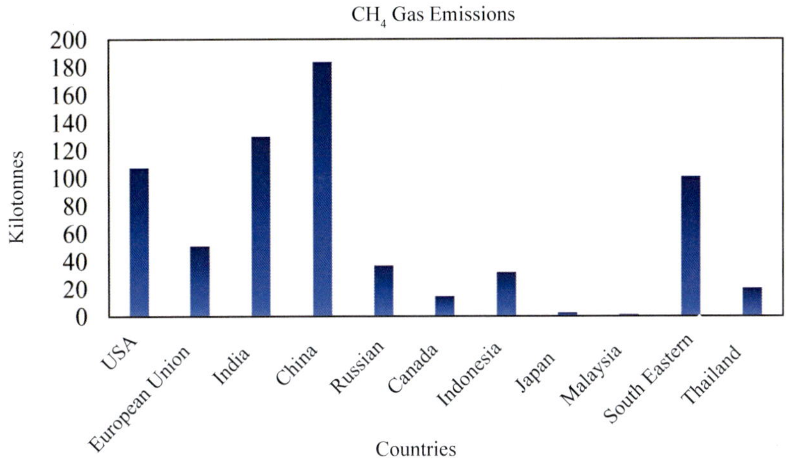

Fig. 3. CH_4 emissions (in kilotons) from burning crop residues among different countries (*Source:* Agricultural Statistics at a Glance, 2021)

3. Nutrient Content in Crop Residues

As a carbon-rich biomass, crop residues contain carbon (40%-45%), nitrogen (0.6%-1%), phosphorus (0.45%-2%), potassium (14%-23%) and microelements, which are necessary for crop growth (Table 1) (Grzyb *et al.,* 2020). They help in alleviating imbalances of nutrients in farm soil and making up for drawbacks of inorganic fertilizers. The features of crop residues (C: N ratio and chemical composition), the climate (temperature and wetness), the soil conditions (pH and water content), and the application method (direct and indirect) are all related to the release rate and content of nutrients (Wạng *et al*., 2020). Usually, it is assumed that C: N ratio greater than 25:1 leads to the rapid immobilization of inorganic nitrogen while lower C: N ratio results in mineralization. Warm temperature and appropriate soil moisture can potentially improve the decomposition of crop residues and nitrogen release (Singh *et al.,* 2020).

Table 2. Composition of different crop residues (dry basis)

Nutrient (%)	Rice straw	Wheat straw	Corn stover	Cotton stalk
C (%)	40.74 ± 1.76	42.1 ± 1.19	43.86 ± 1.31	45.83 ± 1.71
N (%)	0.79 ± 0.27	0.60 ± 0.15	1.00 ± 0.25	1.12 ± 0.23
P (%)	1.97 ± 2.17	0.45 ± 0.22	0.95 ± 0.49	1.40 ± 0.58
K (%)	17.41 ± 6.38	23.45 ± 7.68	17.73 ± 7.00	12.64 ± 3.98
S (%)	0.36 ± 0.12	0.37 ± 0.11	0.39 ± 0.22	0.42 ± 0.24

(*Source:* Grzyb *et al.*, 2020)

4. Crop residues- Role in organic farming

4.1. Soil physical properties

It has been demonstrated that adding leguminous crop residues improves the physical characteristics of the soil, including its ability to retain water and its permeability. The availability of nutrients in the root zone of the crops is also increased by using leftover leguminous crops, which promote crop growth and yield (Smitha *et al.,* 2019). Heavy machinery and farm implements including planters, zero-tillage implements, reapers, and combine harvesters are often used in today's input-intensive agriculture. Carlesso *et al.* (2019) stated that indiscriminate use of this heavy machinery compacts the soil and degrades its natural capacities for storing water, airflow, and infiltration. Lou *et al.,*2011 showed that straw coverage decreased the temperature of the topsoil (0–5 cm depth) and increased the soil moisture from 12.3% to 16.6% in 2008 and 2009, respectively, compared to the straw removal treatment in a 12-year maize field experiment. Zhao *et al.,* 2019 also demonstrated that the soil water content markedly increased by 19% after two-year straw incorporation compared with no straw treatment. Applying agricultural residues made from ryegrass, straw, and mixed litter will greatly increase the soil's porosity and water-holding capacity, which would ultimately increase the soil's productivity. Another study conducted by Gao *et al.,*2020 showed that the total soil porosity increased by 23.0% at the physiological maturity stage under all summer maize stalks return treatment. Zheng *et al.,*2019 also found that the total porosity of the soil increased by 21.7% with application of straw (4500 kg ha^{-1} a^{-1}) in a five-year fixed-site field experiment conducted in a paddy field in Northeast China. It has been discovered that using crop wastes in conjunction with conservation tillage, improves the soil's aggregate and carbon storage in rice-based farming systems (Wang *et al.,* 2019). Santosh *et al.,* 2022 stated that, the bulk density was often lower when there was least amount of soil disturbance and agricultural waste was maintained on the soil surface or incorporated into the soil. Zhao *et al.,*2018 observed that the soil bulk density in the 0–20 cm layer decreased by 5.7% after 7 years of input of maize and wheat straw. In addition, straw return treatments decreased the soil bulk density in both the soil layers (0–20 cm, 20– 40 cm). Mulching the crop leftovers and stubbles on the soil surface reduces the evaporative loss of water by shielding the soil surface from direct sunlight and acting as a barrier to air passage over the surface in contrast to tilled soils (Abid and Lal, 2008). This is more relevant in organic farming.

4.2. Soil chemical properties

Crop residue burning increases the soil's pH and electrical conductivity (EC). Rajput *et al.* (2017) reported that between the pre-burning and post-burning periods, the soil pH and EC values raised from 7.94, 699 S cm^{-1} to 8.46, 800 S cm^{-1}, respectively. Cao *et al.* 2021 showed that straw coverage with no-till and rotary tillage significantly reduced the soil pH from 7.7 to 7.4 and 7.2. Crop residues can be effectively managed in a way that enhances the soil chemical characteristics like cation exchange capacity (CEC), electrical conductivity and pH as well, as the transformation of various primary and secondary plant nutrients. The application of crop residue is positively correlated with the soil carbon pool (both total and labile pool). An irrigated maize production system's total and labile carbon pool is said to be greatly increased by applying wheat residue as a mulch at the rate of 0.1 t ha^{-1} along with 75 kg N ha^{-1} (Chatterjee *et al.,* 2018). In a rice-maize cropping system with residue management, conservation tillage can be used to increase system production, K use efficiency, and apparent K balance. Moreover, residue integration can cut the need for external K supplies up to 15% by recycling the soil's available K (Singh *et al.,* 2018). Crop residues can be used as an alternative to chemical amendments for restoration of optimal soil chemical properties in organic farming.

4.3. Soil microbial activity

Continuous burning of crop residues raises soil temperature, which adversely lowers soil microbial population (Chauhan *et al.,* 2012). Burning also decreases the dehydrogenase enzyme, along with the activity of amylase, cellulose, invertase, and alkaline phosphatase, which are crucial for preserving the fertility and health of the soil (Stawinski *et al.,* 2015). As a result, residue retention is a crucial aspect in fostering improved soil health through increased soil microbial activity, which helps to sustain soil temperature in difficult circumstances. According to Samui *et al.* (2020) and Mondal *et al.* (2020), crop residue mulching boosted microbial activity in the top layer of soil. These results might be explained by alterations in the plant-soil microclimate, greater availability of nutrients, water and regulation of soil temperature. According to Chatterjee *et al.* (2018), application of wheat crop residue considerably increased soil microbial biomass carbon (MBC). Leguminous crop residues like cluster bean should be incorporated before planting the following crop, as this has been shown to improve soil microbial biomass and dehydrogenase activity (DHA) compared to the control (with the application of crop residue) (Smitha *et al.,* 2019). Jat *et al.,* 2020 reported that conservation agriculture practices resulted in 43 % and 68 % higher β-glucosidase activity at surface and sub-surface soil depths, respectively, over conventional methods. Increased soil respiration (plant root respiration, soil microbial respiration and soil organic matter decomposition) is also facilitated by increased soil enzymatic activity and soil microbial biomass carbon (Yang *et al.*, 2017). Sharma *et al.* 2020 observed that the dehydrogenase (DHA), fluorescein diacetate (FDA) and alkaline phosphatise (Alk-P) activity increased by 49.4, 57.8, and 99.9%, ~45.1, 47.6, and 34.1% and ~65.2, 71.9, and 27.5%, respectively in rice straw incorporation @ 10 Mg ha-1 with Nitrogen fertiliser @ 90, 120 and 150 kg ha^{-1} treatments as compared to control plots (rice straw incorporation @ 0 Mg ha-1 with Nitrogen fertiliser @ 0 kg ha-1) under rice-wheat cropping system. Escalante *et al.* (2020) observed that retention of residue under conservation agriculture has also reduced the population of soil nematodes in a wheat-soybean cropping system. Increased microbial activity under organic farming is highly essential for enhanced availability of nutrients, increasing the efficacy of bio agents for the control of pests and diseases. Various benefits of crop residues are detailed below and also depicted in Fig. 4.

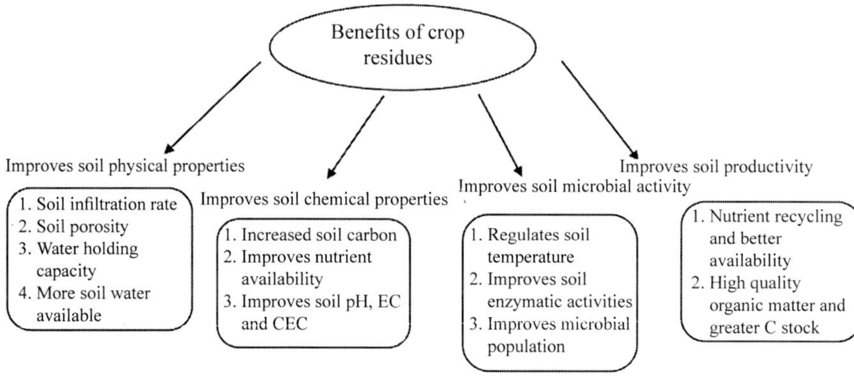

Fig. 4. Schematic representation of various benefits of crop residues

4.4. Soil organic carbon

The incorporation of agricultural residues into the soil significantly changes the labile pool of soil organic carbon. According to residue management techniques, increases in heavy-fraction carbon (>1.6 g cm^3) are what cause changes in total organic carbon in long-term practices (Wang *et al.*, 2005). Intensive farming techniques encourage the ongoing loss of organic matter, which has a negative impact on the global carbon balance and soil quality. When compared to cereals, the integration of residue from legumes ensures high biomass production and verifies a net increase in soil carbon (Tiemann *et al.*, 2015). Beyond 30 cm of soil depth, leguminous stubbles add 60% more soil organic carbon (SOC) and have carbon and nitrogen contents that are 49% and 133% greater than those of control plots (which do not incorporate residue). However, due to their low lignin content and restricted C: N ratio (12–13:1), green manuring of legumes has little effect on the soil carbon storage and this is one of the factors that contributes to the rapid decomposition of the legumes. As compared to control plots (no residue incorporation), sunhemp residue retention produced 0.92% more soil organic carbon (SOC) and 0.64% less soil inorganic carbon (SIC) (Wang *et al.,* 2005). The availability of N would have been enhanced by a low C: N ratio in the soil residue which is essential for rapid residue conversion from a C pool into a particulate C fraction. The ability to catch atmospheric N and then store it in the soil would have produced this effect. The changes in edaphic and meteorological conditions, as well as the residue management strategies, cause this change in soil organic C content to alter with different locales.

4.5 Soil nutrients status and availability

The nutrients from crop residues are initially unavailable to plants because they are organically bound and must be mineralized in order to become available (Singh and Bowden, 2006). Crop residues are decomposed and then mineralized after incorporation by a variety of soil microbes. As a result, the agricultural residues are transformed into simple monomers that are then physically, chemically, physiologically digested and finally converted into organic matter (Salas *et al.,* 2003). Moreover, some nutrients in crop residues exist as soluble inorganic compounds like K^+ and SO_4^{2} or are linked to mineralizable organic components like protein-bound S or phosphate ester (Singh and Bowden, 2006). Crop residue incorporation has the advantage of preserving nutrients in the soil as slowly accessible forms by temporarily immobilising the nutrients that are released into the soil via residues. As a result, plants cannot absorb all of the nutrients at once, but they may become available over the course of the crop's life or in the crop that follows. As a result, nutrient losses due to leaching or volatilization are reduced and the efficiency of nutrient utilization is increased. Pituello *et al.* (2016) and Piccoli *et al.* (2020) found that residue retention statistically increased the efficiency of nutrient utilization. Repeated residue breakdown considerably widens the nutrient distribution and prior studies found that using conservation tillage rather than conventional tillage results in larger organic and inorganic phosphorus accumulation in the soil surface (Du Preez *et al.,* 2001). By saturating P adsorption sites on soil colloids, it has been hypothesised that increased organic matter accumulation through conservation practices may increase the P content in available form. The availability of N also yielded a similar result (Salinas-Garcia *et al.,* 2001). Crop residues left on the soil surface show a considerable difference in soil nutrient losses when compared to clean tillage practices. They also regulate the soil's temperature, moisture content, and infiltration rate (Sarkar *et al.,* 2020). Retained residues in the soil surface lessen soil erosion, which lowers the loss of surface nutrients (Lal, 2005). Contrarily, wastes with a high rate of decomposition may result in increased N losses by leaching and denitrification, or occasionally through significant weed uptake of N. Because residues contain a lot of total K, removing them from fields increases the chance of K deficiency in soil (Sarkar *et al.,* 2020). The quality of the crop residue, particularly the amount of N, lignin and polyphenol determines the decomposition and release of nutrients and may have an impact on the dynamics of decomposition (Whitbread *et al.,* 2003). While crop residues from cereals with high C: N ratios may require additional nitrogenous fertilizer to reduce the demand for microbial N for decomposition, residues from legume crops have a high amount of N in the leaves and promote quick mineralization (Wang *et al.,* 2019).

4.6 Soil productivity

Crop residues have an impact on yield in relation to general aspects of fertility, such as the physical, chemical, and biological characteristics of the soil (Poeplau *et al.,* 2017). Compared to other crops, crop wastes have a greater impact on cereal productivity (Schjonning *et al.,* 2018). Increased residues are anticipated to boost the soil's carbon content, water and heat retention, microbial activity support and nutrient mobilization (Wei *et al.,* 2016). Piccoli *et al.* (2020) concluded that using crop leftovers as a source of organic matter resulted in 12% and 16% higher yields for maize and sugar beet respectively, than other sources. Nevertheless, Hijbeek *et al.* (2017) observed that the increase in maize output owing to residue retention was just 4%. Moreover, Mandal *et al.* (2004) found that straw left on the field increased wheat production by 12.3%. Long-term rice residue incorporation (13 years) resulted in 53.25% and 34.89% higher yield in wheat as compared to residue removal and residue burning, respectively. They also reported a similar trend in the case of maize and concluded that overall soil physical, chemical, and biological properties, particularly the increase in the soil organic matter was improved significantly. Major reasons behind this greater soil productivity were improved C stock and microbial activities.

4.7 Soil health and yield

Crop residues, whether they are incorporated into the soil or are left on the surface, can have a variety of advantageous effects on the physical, chemical, and biological characteristics of soil. Mulching the plant residues raises the minimum soil temperature in the winter and lowers it in the summer due to the shading effect since it reduces the upward heat flow from the soil. In addition to acting as a storehouse for plant nutrients, crop residues also lower nutrient leaching, increase cation exchange capacity (CEC), provide a favourable environment for biological N_2 fixation, increase microbial biomass and enhance the activities of enzymes like alkaline phosphatase and dehydrogenase.

Leaving sizable amounts of crop residues evenly distributed throughout the soil surface minimizes wind and water erosion, improves water infiltration, moisture retention and reduces surface sediment and water runoff. The crop residues play a major role in amelioration of soil acidity by releasing hydroxyls especially during the decomposition of residues with higher C: N ratio and soil alkalinity by applying residues of low C: N ratio crops like legumes and oilseeds as soil amendments are not allowed under organic farming.

Achieving a successful and efficient nutrient management aim in organic agriculture requires the establishment and maintenance of soil health. On or near the soil surface, conservation agriculture creates a layer of crop nutrients, particularly P. Reduced tillage and soil organic carbon accumulation help to maintain a stable soil structure, but an undisturbed soil structure creates macropores and preferred flow channels that can move nutrients, including P, downward into deeper soil layers. The processes including nutrient depletion and/or enrichment through production of biomass/crop wastes, demand for external inputs and environmental effect, the choice of cropping systems can have a significant impact on the soil quality. By reducing ammonium-nitrogen loss and organic-nitrogen loss with sediment, conservation tillage can lower overall nitrogen loss; however, it might not lower nitrogen leaching in the form of nitrate.

Yield response to residue management differs with the extent of management expertise, cropping patterns, native soil fertility and climate under organic farming. Higher yields with crop residue application result from enhanced soil characteristics, increased soil organic matter and earthworm activity and improved soil structure after a period of 4-7 years.

5. Crop Residues Management

Crop residues are usually managed by either disposal in landfills or by burning which increases emissions of pollutants into air. An efficient strategy to overcome the problem is through utilization of crop residues as a source of nutrients and organic matter in organic farming.

5.1. On farm management of crop residues

5.1.1 Surface mulching and retention

The advantages of leaving crop residue on the soil surface include less weed emergence, improved physical, chemical and biological properties of soils, plant nutrient recycling and nutrient availability in future crops under organic farming. Direct drilling in surface mulched residues is a technique whereby straw leftovers from a previous crop are left on the soil surface without incorporation. Formerly, after mixed harvesting, loose straw had to be removed, and the residues that had been removed had to be partially or entirely burned before rice residues could be seeded with ZT seed drills. The Indo-Gangetic plains (IGP) now allow rice-wheat farmers to drill wheat directly into the rice residue using a "Happy Seeder" method (Table 3). The accumulation of debris on the surface aids in preventing wind and water erosion, preserving the rich surface soil. In some circumstances, surface preservation of any or all residues could be the best choice. In comparison to burning, residues boost soil NO_3 concentration by 46%, N absorption by 29%, and yield by 37% by slowly decomposing on the surface and collecting organic carbon and total nitrogen in the top 5–15 cm of the soil (Mandal *et al.,* 2004). They also postpone erosion. Retention, on the other hand, offers a protected environment for both harmful and advantageous species as well as a carbon source for heterotrophic N-fixation, increased microbial activity, soil C and N, and reduced rice N fertilizer needs (Ladha *et al.,* 2022) which are of prime importance in organic farming.

Because field preparation must be done before crop residue may be incorporated into the soil, it is quicker and simpler to collect the residue and use it as mulch for the subsequent crop. For achieving this, there is a tonne of equipment accessible. A no-tillage drill increases wheat productivity, profitability, and resource use efficiency. Only when no-till is applied consistently and the soil surface is saturated with at least 30% prior crop residue will the full benefits of no-till be realised. Direct drilling in standing and loose residues will lead to increased use of new-generation planters like Happy seeder and spatial drill.

When compared to no mulch, rice straw mulch enhanced wheat grain output, reduced crop water use by 3–11% and improved water use efficiency by 25%. When compared to no mulch, root length densities were 40% greater as a result of mulch's ability to hold moisture in deeper soil layers. In no-till systems, managing rice residue has a number of advantages, such as preserving soil moisture, weed control, improved soil quality and a decrease in greenhouse gas emissions of almost 13 t ha^{-1} (Fabio *et al.,* 2014). It also controls canopy temperature at the grain-filling stage to lessen the effects of terminal heat on wheat. If mulched straw is effective at suppressing weeds as herbicide cannot be used in organic farming.

Table 3. Machinery used for crop residue management

S.No	Crop ecosystem	Machinery for crop residue	References
1	Paddy	Super SMS (Straw Management System)	Ramulu *et al.*, 2023
2	Paddy-wheat	Happy Seeder	
3	Paddy-wheat	Super seeder	
4	Paddy	Rice straw chopper-cum-incorporator	
5	Paddy-wheat	Zero tillage seed drill	

5.1.2 Crop residue incorporation

Residue incorporation enhances soil organic matter and the levels of N, P, and K. Ali *et al.* (2016) found increased uptake of nitrogen in residue incorporated treatments in rice. Rice residue can be managed in-situ with a rotavator. Compared to burning crop wastes in the field, *in-situ* assimilation of crop residues improves the quantity of P and K that is readily available (Hung *et al.*, 2019). According to Fataah and Addo (2016), cowpea residue incorporated plot showed the greatest maize yield which was comparable with the yields obtained from the fertilized plots. To adjust the high C: N ratio at the time of residue integration, more N fertilizer must be supplied because incorporation results in the temporary immobilization of nitrate N (Riggs, 2016).

5.1.3 Farm Mechanization

Farmers still burn crop despite being aware of its drawbacks because there is a shortage of equipment that is both economically viable and adequate for disposing of leftovers. However, adopting sophisticated equipment to prevent burning such as happy seeders for direct drilling, equipped combine harvesters and straw management systems, is a possibility. Manufacturers have lately improved crop residue management technology to enable finer chopping, wider and more uniform spreading and more seed soil contact for better crop emergence and production. The management of paddy residues has more promise for enhancing soil health, productivity, lowering pollution, and achieving sustainable agriculture thanks to farm equipment built on resource conservation technologies (RCTs).

The nation offers modern zero-till seed-cum-fertilizer drill/seed planters (happy seeder, spatial zero seed cum fertilizer drill) (Table 3) for direct planting of the following crop in loose and anchored straw loads up to 10 t ha^{-1}. These methods are useful for controlling weeds, managing agricultural waste and preserving soil moisture and nutrients as an effective alternative to herbicides in organic farming. The cheerful seeder technique is a revolution in paddy-wheat crop rotation in North West India.

A Straw Management System (SMS) is installed on the back of the combine harvester, right behind the straw walkers and below the chaffer sieves, to disperse rice straw uniformly after paddy harvesting by the combine harvester. The mechanism receives straw from the combine harvester's straw walkers on one side and expels it out of the housing on the other. The harvester's straw walker drops loose residues behind it, which are dispersed by the spinning discs. To equally spread the residues across the width of the combine harvester, the chopped material is blown off in a tangential direction and diverted by a deflector.

Happy Seeder-based systems are 10–20 percent more profitable than burning and this mechanization approach emerged as the most profitable and scalable residue management approach. In comparison to all other burning choices, this option has the greatest potential to reduce the environmental footprint of on-farm activities since it would eliminate air pollution and reduce greenhouse gas emissions per acre by more than 78 percent (Manpreet *et al.,* 2022).

5.2 Off-farm management of crop residues

5.2.1 Baling and removing the straw

Agricultural straw can be utilised in a variety of ways, including bedding for plants and livestock, fuel for vehicles, building materials and livestock feed. Commercially accessible straw baler machines are a promising technology for separating and collecting straw following combine harvesting and using the leftovers for off-farm activities.

5.2.2 Crop residues as livestock feed

The CR has historically been used as livestock feed in India. Conversely, crop leftovers are unattractive and have a poor level of digestibility; as a result, they cannot be used as the exclusive source of feed for cattle. Due to the high concentration of silica in rice leftovers, they are regarded as poor calf fodder (4-7%). It differs from other straws in that it has higher silica (12-16%) and less lignin (3-5%) than other crop straws (6.7 vs. 10-12 % in others) (Aquino *et al.,* 2020). There are several ways to improve the nutritional value of rice straws. Crop residues have undergone physical, chemical, and biological processes to weaken and dissolve lignin-cellulose linkages, improving their nutritional value. Although it requires more effort and expense, over 75% of wheat straw is used as animal fodder and is sliced into little pieces using a specialized cutting machine. If rice straw is to be fed to cattle, the rice crop should be harvested as close to the ground as is practical because stems contain less silica than leaves but are more digestible. To provide the animals with the nutrients they require, the wastes must be digested, enhanced with molasses, and supplemented with green forage as an organic feed to livestock in organic farming.

5.2.3 Composting

Agricultural wastes are used as bedding for animals before being dumped in dung pits to prepare compost. In the animal shed, each kilogram of straw absorbs about 2-3 kilograms of urine, enhancing it with nitrogen. Composting rice crop wastes from a hectare of land results in around 3 tonnes of manure as nutrient dense farmyard manure (FYM) which is the prime nutrient source in organic farming (Tarafdar and Yadav, 2017). To transform the compost from rice straw into value-added compost, phosphorus can be added utilizing an indigenous source of low-grade rock phosphate (Sidhu and Beri, 2005). The bio-physical composting processes, however, can be greatly enhanced by mechanical composting. The compost result can be applied to rice fields as a soil amendment or utilized as a growing medium for vegetables and other crops. It improves the soil's organic matter and nutritional content under organic farming specifically, its nitrogen and carbon concentrations.

5.2.4 Biochar production

Biochar is made up primarily of carbon, used as a soil additive to enhance soil fertility, carbon sequestration and water infiltration. It is made when organic materials or biomass thermally decompose between 500 and 700 °C in a tiny amount of oxygen. A brand-new, better carbonization technique is hydrothermal carbonization (HTC). By fully destroying the plant cell wall, HTC of lignocellulosic biomass enables fast conversion of biomass into a carbon-rich, lignin-like product (hydrochar). Compared to the basic substance, hydrochar has a much higher heating value. Rice straw can be used to make biochar, which has a lot of potential as a nutrient source in organic farming. Using biochar as a soil amendment has a lower carbon impact than using it as a fuel. Additionally, the risk of climate change brought on by Greenhouse gas (GHG) emissions in the atmosphere is reduced by carbon sequestration with biochar.

5.2.5 Biogas production

Agricultural residue biomass, produced through anaerobic digestion, gasification and pyrolysis processes is a reliable source of energy that immediately lowers the amount of CO_2 in the atmosphere. The remaining slurry can be used as manure and the procedure yields a high-quality gas with a methane level of between 55 and 60 percent. It guarantees a non-destructive method for harvesting manure for reusing in organic farming and obtaining high-quality fuel gas from agricultural waste.

5.3 Other ways of crop residue management

Crop residues can also be used in producing mushrooms, bioethanol, paper pulp and high-value industrial products for commercial use.

5.3.1 Microbial methods

The main controllers of soil carbon and nutrient cycle processes are soil-microbial communities. The preservation of carbon and the supply of crop nutrients in agroecosystems are subsequently influenced by variations in the composition of the microbial population in the soil (Schimel and Schaffer, 2012). A well-structured microbial community is necessary for efficient crop-residue decomposition because different microbial communities utilise distinct components within the complex composition of crop-residues differently (Fan *et al.*, 2014). In contrast to fungi and actinomycetes, which can break down resistant lignocellulose components and predominate in the later stage of decomposition, bacteria prefer to utilise labile residual C and dominate in the early stages of decomposition (Paterson *et al.*, 2008; Bai *et al.*, 2016). It is significant to highlight that different soil characteristics and residue kinds can affect how microbial groups control the breakdown process (Koranda *et al.*, 2014). It is therefore, necessary to use microbial consortia, which include powerful strains of fungus, bacteria and actinobacteria for the quick breakdown of crop wastes without any chemical pre-treatment (Kumar *et al.*, 2008). Moreover, groups of lignocellulolytic microorganisms have been shown to decompose crop leftovers more effectively than a single inoculum (Sahu *et al.*, 2020).

Depending on the type of residue, liquid culture (*Aspergillus flavus* Link, Aspergillus *terreus* Thom, *Penicillium pinophilum* Thom C., and *Alternaria alternate* (Fr.) keissl.) can be applied three times during crop growth: immediately following sowing, at the time of first irrigation and at second irrigation (Choudhary *et al.*, 2016). For boosting microbial breakdown, the culture can also be combined with a curd and jaggery solution and incubated for 12 hours before use.

A. niger isolate F44 and *Trichoderma viride* isolate F26 were created as a suitable lignocellulolytic fungal consortia in paddy for the quick decomposition of paddy straw's cellulose, hemicellulose, lignin and total carbon (Kausar *et al.*, 2010). The most environmentally friendly, economically feasible and advantageous option for crop residues is microbial decomposition which also has the added advantages of enhancing soil carbon content and enhancing soil health and productivity (Goswami *et al.*, 2020).

6. Constraints of Crop Residue Recycling in Organic Farming

- Farmers prefer clean cultivation; hence they go for burning of crop residues
- Short window between the harvest of *kharif rice* and the planting of *rabi* crops
- Because of the high C: N ratio, incorporation of residues results in short-term immobilization of N (Hofmann *et al.,* 2009)
- Too much residue slows down soil drying and warming in regions with cold, wet springs, delaying planting (Mallareddy *et al.,* 2021)
- Residue build-up in the furrow openers, traction issues with the seed drill's drive wheel, trouble in using fertilizer metering systems in the loose straw, and uneven sowing depth are all consequences of residue retention when sowing (Goswami *et al.,* 2020)
- Inadequate supply of custom hiring services for straw-management equipment (happy seeder, baler, chopper, Zero till seed drill)
- Relatively expensive as compared to burning
- Farmers' resistance to exert additional effort to compost the wastes
- Lack of technical expertise about *in-situ* straw management among farmers
- Lack of storage and logistics to handle the residues in a short period of time

7. Conclusion

Recycling nutrient rich crop residues in organic agriculture replenishes the soil health and productivity. It also improves the soil physical, chemical, microbial activity thus achieving resource conservation and environmental sustainability which are the key aspects in organic farming. Crop residues can be efficiently managed by using suitable farm machinery like happy seeders, zero till drill, balers, through options such as composting, biochar and bio gas production, use of microbial consortia for accelerated decomposition of crop residues etc. to assure the nation's food security through organic farming sustainable agriculture.

8. Way Forward

To fully understand and utilise the potential of these irreplaceable resources, all stakeholders involved in organic farming—farmers, supply chain and value chain service providers, researchers, extension agents, policymakers, civil servants and consumers must collaborate. Useful, reasonably priced farm machinery to facilitate the collection, volume reduction, transportation and application of residues to organic farms have to be developed. Windrowing should be designed for effective management of straw and a lower cutter bar should be created for harvesting straw at an appropriate height. For grain recovery, twin cutter bar type combine harvesters should be utilized to harvest the crop's top layer. The use of satellite-based remote sensing technology to monitor agricultural residue management on organic farms may be explored.

9. References

Abdurrahman, M.I., Chaki, S. and Saini, G. 2020. Stubble burning: Effects on health & environment, regulations and management practices. Environmental Advances 2: 100011.

Abid, M and Lal, R. 2008. Tillage and drainage impact on soil quality, aggregate stability, carbon and nitrogen pools. Soil and Tillage Research 100: 89–98.

Ali, N.M., Alam, P.M., Ahmad, E and Sah, A. 2016. Effect of different residue management practices of rice on growth and yield of wheat and soil health in rice-wheat system. International Journal of Bio-resource and Stress Management 7(4): 567-574.

Aquino, D., Barrio, A. D.., Trach, N. X., Hai, T., Khang, D. N., Toan, N. T. and Hung, N. V. 2020. Rice straw-based fodder for ruminants. In: Martin Gummert, Nguyen Van Hung, Pauline Chivenge, Boru Douthwaite (eds) Sustainable rice straw management. Springer nature pp111-129.

Bhuvaneshwari, B., Hiroshan, H. and Jay, N. 2019. Crop residue burning in India: policy challenges and potential solutions. International Journal of Environmental Research and Public Health 16: 832.

Cao, Q.J., Li, G., Yang, F.T. and Enping, Z. 2021. Eleven-year mulching and tillage practices alter the soil quality and bacterial community composition in Northeast China. Archives of Agronomy and Soil Science 1-16

Carlesso, L., Beadle, A., Cook, S.M., Evans, J., Hartwell, G., Ritz, K. and Murray, P.J. 2019. Soil compaction effects on litter decomposition in an arable field: implications for management of crop residues and headlands. Applied Soil Ecology 134: 31-37.

Chatterjee, S., Bandyopadhyay, K.K., Pradhan, S., Singh, R. and Datta, S.P. 2018. Effects of irrigation, crop residue mulch and nitrogen management in maize (Zea mays L.) on soil carbon pools in a sandy loam soil of Indo-Gangetic plain region. Catena 165: 207-216.

Chauhan, B.S., Mahajan, G., Sardana, V., Timsina, J. and Jat, M.L. 2012. Productivity and sustainability of the rice-wheat cropping system in the Indo-Gangetic Plains of the Indian subcontinent: Problems, opportunities and strategies. Advances in Agronomy 117: 315-369.

Choudhary, M., Sharma, P.C., Jat, H.S., Nehra, V., McDonald, A.J. and Garg, N. 2016. Crop-residue degradation by fungi isolated from conservation agriculture fields under rice-wheat system of North–West India. International Journal of Recycling of Organic Waste in Agriculture 5: 349-360

Agriculture Statistics at a Glance. 2021. Directorate of Economics and Statistics. Ministry of Agriculture. Government of India.

Dutta, A., Abhik, P., Kali, K.H., Chaitanya, P.N., Narendra, K. and Rakshith, A. 2022. A state of the art review in crop residue burning in India: Previous knowledge, present circumstances and future strategies. Environmental Challenges 8: 100581.

Du Preez, C.C., Steyn, J.T. and Kotze, E. 2001. Long-term effects of wheat residue management on some fertility indicators of a semi-arid Plinthosol. Soil and Tillage Research 63: 25–33.

Eriksen-Hamel, N.S., Speratti, A.B., Whalen, J.K., Legere, A. and Madramootoo, C.A. 2009. Earthworm populations and growth rates related to long-term crop residue and tillage management. Soil and Tillage Research 104: 311–316.

Escalante, L.E., Brye, K.R. and Faske, T.R. 2020. Nematode populations as affected by residue and water management in a long-term wheat-soybean double-crop system in Eastern Arkansas. Applied Soil Ecology 157: 103761.

Fabio, S., Angelica, G., Stefano, S., Giovanni, C. and Michele, P. 2014. Effects of straw mulch on growth and yield of durum wheat during transition to conservation agriculture in Mediterranean environment. Field Crops Research 167: 51-63.

Fan, F.L., Yin, C., Tang, Y.J., Li, Z.J., Song, A., Wakelin, S.A., Zou, J. and Liang, Y.C. 2014. Probing potential microbial coupling of carbon and nitrogen cycling during decomposition of maize residue by ^{13}C–DNA–SIP. Soil Biology and Biochemistry 70: 12–21.

Fataah, J. and Addo, J.S. 2016. Effect of cowpea residue nitrogen on maize growth and yield in the semi-arid deciduous region of Ghana. International Journal of Science and Technology 6(9): 21-23.

Gao, F., Hu, J., Ren, B.Z., Peng, Liu and Bin, Z. 2020. Improving soil properties and grains yield of winter wheat and summer corn under residue management strategies. Agronomy Journal 112 (5):4287-4302.

Garai, S., Mondal, M and Mukherjee, S. 2020. Smart practices and adaptive technologies for climate resilient agriculture. New Delhi Publishers, Kolkata, India, pp 327-358.

Ghosh, P., S. Sharma, I., Khanna, A., Datta, R., Suresh, S., Kundu, A., Goel, D. and Datt. 2019. Scoping study for South Asia air pollution. Energy Resource Institution 2:153-164.

Goswami, S.B., Mondal, R. and Mandi, S.K. 2020. Crop residue management options in rice-rice systems. Archives of Agronomy and Soil Science 66(9): 1218-1234.

Grzyb, A., Wolna-Maruwka, A. and Niewiadomska, A. 2020. Environmental factors affecting the mineralization of crop residues. Agronomy 10 (12):18.

Hijbeek, R., Van Ittersum, M.K., Ten Berge, H.F.M., Gort, G., Spiegel, H. and Whitmore, A.P. 2017. Do Organic Inputs Matter: A meta-analysis of additional yield effects for arable crops in Europe. Plant and Soil 411: 293–303.

Hofmann, A., Heim, A., Christensen, B.T., Miltner, A., Gehre, M. and Schmidt, M.W. 2009. Lignin dynamics in two ^{13}C-labelled arable soils during 18 years. European Journal of Soil Science 60(10): 250-257.

Hung, D.T., Hughes, H.J., Keck, M. and Sauer, D. 2019. Rice-residue management practices of small holder farms in Vietnam and their effects on nutrient fluxes in the soil-plant system. Sustainability 11: 2-15.

Jat, H.S., Madhu Choudhary., Ashim Datta., Yadav, A.K., Meena, M.D. and Sharma, P.C. 2020. Temporal changes in soil microbial properties and nutrient dynamics under climate smart agriculture practices. Soil Tillage Research 199: 104-595.

Kausar, H., Sariah, M., Saud, H.M., Alam, M.Z. and Ismail, M.R. 2010. Development of compatible lignocellulolytic fungal consortium for rapid composting of rice straw. International Bio deterioration and Biodegradation 64(7): 594–600.

Koranda, M., Kaiser, C., Fuchslueger, L., Kitzler, B., Sessitsch, A., Zechmeister–Boltenstern, S. and Richter, A. 2014. Fungal and bacterial utilization of organic substrates depends on substrate complexity and N availability. FEMS Microbiology Ecology 87: 142–152.

Kumar, A., Gaind, S. and Nain, L. 2008. Evaluation of thermophilic fungal consortium for paddy straw composting. Biodegradation 19(3): 395–402.

Kumar, A., Kushwaha, K.K., Singh, S., Shivay, Y.S., Meena, M.C. and Nain, L. 2019. Effect of paddy straw burning on soil microbial dynamics in sandy loam soil of Indo-Gangetic plains. Environmental Technological Innovations16: 100-469.

Ladha, J.K., Mark, P.B, Reddy, M.P., Biswas, J.C., Alan, B., Mangi, L.J. and Timothy, J.K. 2022. Biological nitrogen fixation and prospects for ecological intensification in cereal-based cropping systems. Field Crops Research 283: 108541.

Lal, R. 2005.World crop residues production and implications of its use as a biofuel. Environment International 31: 575–584

Liang, F., Li, J., Yang, X., Huang, S., Cai, Z., Gao, H., Ma, J., Cui, X. and Xu, M. 2016.Three-decade long fertilization-induced soil organic carbon sequestration depends on edaphic characteristics in six typical croplands. Scientific Reports 6: 30350.

Lou, YL., Liang, W.J., Xu, M.G., Xin Hua., Wang, Y.D. and Zhao Kai. 2011. Straw coverage alleviates seasonal variability of the topsoil microbial biomass and activity. Catena 86 (2):117–120.

Mallareddy, M., Ramanjaneyulu, A.V., MadhuBindu, G.S., Prabhakar Reddy, T., Rajaiah, P., Ramya, V., Rajitha, G. and Jagadeeshwar, R. 2021. Crop residue management for augmenting carbon sequestration and agricultural sustainability. Indian Journal of Agronomy 66(5): S69-S81.

Mandal, K.G., Arun, K.M., Kuntal Hai., Kali, K.B., Ghosh, P.K. and Mohanty, M. 2004. Rice residue-management options and effects on soil properties and crop productivity Journal of Food Agriculture and Environment 2(1): 224-231.

Manpreet, K, Dharm Pal, M., Gurdeep Singh, M., Virender, S., Nanthi, S.B, Rattan, L and Siddique, H.M. 2022. Rice residue management in the Indo-Gangetic Plains for climate and food security-a review. Agronomy for Sustainable Development 42(5): 101-126.

Mondal, M., Skalicky, M., Garai, S., Hossain, A., Sarkar, S., Banerjee, H., Kundu, R., Brestic, M., Barutcular, C. and Erman, M. 2020. Supplementing nitrogen in combination with rhizobium inoculation and soil mulch in peanut (Arachis hypogaea L.) production system: Part ii. Effect on phenology, growth, yield attributes, pod quality, profitability and nitrogen use efficiency. Agronomy 10: 1513.

National Policy for Management of Crop Residues (NPMCR). 2014. Department of Agriculture & Cooperation, Ministry of Agriculture, Government of India.

Paterson, E., Osler, G., Dawson, L.A., Gebbing, T., Sim, A. and Ord, B. 2008. Labile and recalcitrant plant fractions are utilized by distinct microbial communities in soil: Independent of the presence of roots and mycorrhizal fungi. Soil Biology and Biochemistry 40: 1103-1113.

Piccoli, I., Sartori, F., Polese, R. and Berti, A.2020. Crop yield after 5 decades of contrasting residue management. Nutrient Cycling in Agroecosystems 117: 231-241.

Pituello, C., Polese, R., Morari, F. and Berti, A. 2016. Outcomes from a long-term Study on Crop residue effects on plant yield and nitrogen use efficiency in contrasting soils. European Journal of Agronomy 77: 179-187.

Poeplau, C.; Reiter, L.; Berti, A. and Kätterer, T. 2017. Qualitative and quantitative response of soil organic carbon to 40 years of crop residue incorporation under contrasting nitrogen fertilization regimes. Soil Research 55: 1-9.

Rajput, R.; Arunachalam, K. and Arunachalam, A. 2017. Role of crop residue management practices on microbial dynamics of soil-a mini review. Indian Journal of Hill Farming 30: 1-6.

Ramulu, C., Pateriya, R.N., Naik, M.A., Dinesh, K.V., Abed, A. and Ahmed, Z.D. 2023. A residue management machine for chopping paddy residues in combine harvested paddy field. Scientific reports 13, 5077.

Riggs, E.C. 2016. Mechanisms driving the soil organic matter decomposition response to nitrogen enrichment in grassland soils. Soil Biology and Biochemistry 99 (2): 54-65.

Rusinamhodzi, L., Corbeels, M. and Giller, K.E. 2016. Diversity in crop residue management across an intensification gradient in Southern Africa: System dynamics and crop productivity. Field Crops Research 185: 79-88

Sahu, A., Manna, M.C., Bhattacharjya, S., Rahman, M.M., Mandal, A., Thakur, J.K., Sahu, K., Bhargav, V.K., Singh, U.B., Sahu, K.P. and Patra, A.K. 2020. Dynamics of maturity and stability indices during decomposition of biodegradable city waste using rapo–compost technology. Applied Soil Ecology 155: 103-670.

Salas, A.M., Elliott, E.T., Westfall, D.G., Cole, C.V. and Six, J. 2003. The role of particulate organic matter in phosphorus cycling. Soil Science Society of America Journal 67: 181-189.

Salinas-Garcia, J.R., Baez-Gonzalez, A.D., Tiscareno-Lopez, M. and Rosales-Robles, E. 2001. Residue removal and tillage interaction effects on soil properties under rain-fed corn production in Central Mexico. Soil and Tillage Research 59: 67-79.

Samui, I., Skalicky, M., Sarkar, S., Brahmachari, K., Sau, S., Ray, K., Hossain, A., Ghosh, A., Nanda, M.K. and Bell, R.W. 2020. Yield Response, nutritional quality and water productivity of tomato (Solanum lycopersicum L.) as influenced by drip irrigation and straw mulch in the coastal saline ecosystem of Ganges Delta, India. Sustainability 12: 67-79.

Santosh, K., Gandhamanagenahalli, A., Rajanna., DharamBir Yadav., Venkatesh, P., Chandra Mohan, M., Prakash Kumar, J., Surendra Singh and Shikha Singh. 2022. Impacts of mechanized crop residue management on rice-wheat cropping system- A review. Sustainability 14: 2-22.

Sarkar, S., Ghosh, A. and Brahmachari, K. 2020. Application of APSIM model for assessing the complexities of rice-based cropping systems of South-Asia. New Delhi Publishers Kolkata, India. pp. 212-233.

Schimel, J.P. and Schaeffer, S.M. 2012. Microbial control over carbon cycling in soil. Frontiers in Microbiology 3: 348.

Schjonning, P., Jensen, J.L., Bruun, S., Jensen, L.S., Christensen, B.T., Munkholm, L.J., Oelofse, M., Baby, S. and Knudsen, L. 2018. The role of soil organic matter for maintaining crop yields: Evidence for a renewed conceptual basis. Advances in Agronomy 150: 35-79.

Sharma, S., Pritpal singh. and Sandeep kumar. 2020.Responses of soil carbon pools, enzymatic activity, and crop yields to nitrogen and straw incorporation in a rice-wheat cropping system in North-Western India. Frontiers in Sustainable Food Systems. 4: 532-704.

Sidhu, B.S. and Beri, V. 2005. Experience with managing rice residue in intensive rice- wheat cropping system in Punjab. In: Abrol IP, Gupta RK and Malik RK (eds) Conservation Agriculture-Status and Prospects. Centre for Advancement of Sustainable Agriculture (CASA), New Delhi, India, pp 55-63.

Singh, B.R.Z. and Bowden, J.W. 2006. Carbon, nitrogen and sulphur cycling following incorporation of canola residue of different sizes into a nutrient-poor sandy soil. Soil Biology and Biochemistry 38: 1591-1597.

Singh, G., Dhakal, M. and Yang, L. 2020. Decomposition and nitrogen release of cover crops in reduced- and no-tillage systems. Agronomy 112(5): 3605-3618.

Singh, V.K., Dwivedi, B.S., Yadvinder-Singh., Singh, S.K., Mishra, R.P., Shukla, A.K., Rathore, S.S., Shekhawat, K., Majumdar, K. and Jat, M.L. 2018. Effect of tillage and crop establishment, residue management and K fertilization on yield, K use efficiency and apparent K balance under rice-maize system in North-Western India. Field Crops Research 224: 1-12.

Smitha, G.R., Basak, B.B., Thondaiman, V. and Saha, A. 2019. Nutrient management through organics, bio-fertilizers and crop residues improves growth, yield and quality of sacred basil (Ocimum sanctum). Industrial Crops and Products 128: 599-606.

Sombrero, A. and DeBenito, A. 2010. Carbon accumulation in soil. Ten-year study of conservation tillage and crop rotation in a Semi-Arid area of Castile-Leon, Spain. Soil and Tillage Research 107: 64-70.

Stawinski, C., Cymerman, J., Witkowska-Walczak, B. and Lamorski, K. 2015. Impact of diverse tillage on soil moisture dynamics. International Agrophysics 26: 301-309.

Swarna, R. and Behera, U.K. 2018. Effect of tillage, crop residues and nitrogen management practices on growth performance and soil microbial parameters in Wheat. International Journal of Current Microbiology and Applied Sciences 7(1): 845-858.

Tarafdar, J.C and Yadav, B.K. 2017. Residue management and Composting: In: Tarafdar, Tripathy, K.P and Kumar (eds) Organic agriculture. Scientific Publisher India, pp 52-71.

Tiemann, L.K., Grandy, A.S., Atkinson, E.E., Marin-Spiotta, E. and McDaniel, M.D. 2015. Crop rotational diversity enhances belowground communities and functions in an agroecosystem. Ecology Letters 18: 761-771.

Walter, S., Kurt, M. and Gunter, L. 2008. Effects of biogas digestion of clover/grass-leys, cover crops and crop residues on nitrogen cycle and crop yield in organic stockless farming systems. European Journal of Agronomy 29: 125-134.

Wang, W.J., Dalal, R.C. and Moody, P.W. 2005. Soil carbon sequestration and density distribution in a vertisol under different farming practices. Soil Research 42: 875-882.

Wang, X., Qi, J., Zhang, X.Z. Li, S., Latif Virk, A., Zhao, X., Xiao, X. and Zhang, H. 2019. Effects of tillage and residue management on soil aggregates and associated carbon storage in a double paddy cropping system. Soil and Tillage Research 194: 104339.

Wang, X.L., Yang, Z.L. and Liu, X. 2020. The composition characteristics of different crop straw types and their multivariate analysis and comparison. Waste Management 110: 87-97.

Wei, W., Yan, Y., Cao, J., Christie, P., Zhang, F. and Fan, M. 2016. Effects of combined application of organic amendments and fertilizers on crop yield and soil organic matter: An integrated analysis of long-term experiments. Agriculture Ecosystems and Environment 225: 86-92.

Whitbread, A., Blair, G., Konboon, Y., Lefroy, R. and Naklang, K. 2003. Managing crop residues, fertilizers and leaf litters to improve soil C, nutrient balances and the grain yield of rice and wheat cropping systems in Thailand and Australia. Agriculture Ecosystem and Environment 100: 251-263.

Yadvinder, B.S. and Timsina, J. 2005. Crop residue management for nutrient cycling and improving soil productivity in rice-based cropping systems in the tropics. Advances in Agronomy 85: 269-407.

Yang, F., Ali, M., Zheng, X., He, Q., Yang, X., Huo, W., Liang, F.C. and Wang, S.M. 2017. Diurnal dynamics of soil respiration and the influencing factors for three land-cover types in the hinterland of the Taklimakan Desert, China. Journal of Arid Land 9: 568-579.

Zhao, H.L., Shar, A.G., Li, S. and Yanlong Chen.2018. Effect of straw return mode on soil aggregation and aggregate carbon content in an annual maize-wheat double cropping system. Soil Tillage Research 175: 178-186.

Zhao, X.L., Yuan, G.Y., Wang, H.Y, Dainjun Lu., Xiaoqin, C. and Jhainmin, Z. 2019. Effects of full straw incorporation on soil fertility and crop yield in rice-wheat rotation for silty clay loamy cropland. Agronomy 9(3): 133.

Zheng, Y., Han, X.R., Li, Y.Y., Jinfeng, Y., Na, L and Ning, A. 2019. Effects of biochar and straw application on the physicochemical and biological properties of paddy soils in northeast China. Scientific Reports 9(1): 16531.

6

Insect Pest Management in Organic Farming: Strategies and Challenges

P. Duraimurugan[1], P.S. Shanmugam[2], T. Srinivasan[2], V. Baskaran[2] and C. Sowmiya[2]

[1]*ICAR-Indian Institute of Oilseeds Research, Hyderabad-500 030, Telangana, India*

[2]*Tamil Nadu Agricultural University, Coimbatore-641 003, Tamil Nadu, India*

Abstract

Insect pests are responsible for 10.8% of crop losses on an average. Their management is crucial in agriculture to protect crops and boost output. In organic farming systems, preventative measures are taken more often than curative. Understanding the impacts that the local climate, topography, soils, and all other facets of the production system are expected to have on crop performance and potential host/pest complexes is essential for developing successful organic crop protection techniques. Crop rotation, soil health management, host plant resistance, field location, monitoring pests and their natural enemies, vegetation management to enhance natural enemies and exert direct effects on pest populations, releases of bio-control agents and use of approved insecticides of biological origin are highlighted in organic farming in order to reduce pest damage. The safe and environmentally friendly methods for managing the pests in organic farming are discussed in this chapter.

Keywords: *Integrated pest management, organic farming*

1. Introduction

The fundamental tenet of organic pest management is the use of environmentally sound measures as outlined by worldwide and national organic agricultural standards. The emphasis of organic farming's pest management strategies is on the employment of numerous and diverse strategies integrated into the cropping system, which ultimately lowers the pest incidence. Pest and disease options mainly dependent on preventive measures in organic farming rather than curative and maintaining healthy ecosystems is priority. The farmer should have a better grasp of the pests and beneficial organisms that exist in the specific crop ecology when practising organic farming. The farmers must follow a monitoring mechanism in which they may able to act on the pest before they reach devastating levels.

2. Components of Pest Management in Organic Farming

The strategies or techniques adopted to manage pests in organic farming was grouped under four phases by Zehnder *et al.* (2007) (Table 1). This classification was based on the model proposed by the Wyss *et al.* (2005).

Table 1. Strategies of pest management in organic farming

I Phase	Cultural practices	Field/site location, soil quality management, crop rotation, habitat manipulation/diversity, use of pest-resistant cultivars
II Phase	Vegetation management	Border crops, trap crops, cover crops and inter crops
III Phase	Biocontrol	Inundative and inoculative release of parasitoids and predators
IV Phase	Approved insecticides	Bio-pesticides, botanicals, mineral origin insecticides, mating disruption chemicals – pheromones, repellents

The above four phase strategies are the major components of Integrated Pest Management (IPM), have been recommended to farming community for the last five decades (Kogan, 1998). The different components of organic farming pest management are discussed below.

2.1. Cultural practices

Cultural practices are among the first methods which are used for pest control. Disrupting the suitable environment for pests in crop ecosystem is the major objective of cultural control. These cultural control methods are more preventive than curative and hence proper planning is prerequisite to harness the benefits (Devi and Ram, 2018). The nutrition and irrigation management also enhance the plant capability to tolerate insect pests. The different tactics followed are detailed below.

2.1.1. Sowing

The time of sowing should be adjusted in such way to avoid peak season attack of the particular pest or disease. Since the pest incidence changes depending on area, plantings should also be based on pest monitoring information. The late sowing and also crops suffering from abiotic stresses are more amenable for insect attack. Thrips incidence is more in early planted crops than in late planted crops. Off-season cultivation of cabbage and cauliflower increases the incidence of diamond back moth (Shivalingaswamy and Satpathy, 2007).

In order to avoid pests, crop maturity is also crucial. Eg. Delayed sowing of wheat reduces the incidence of Hessian fly (Mayetiola destructor). Early sowing reduces shoot fly incidence in sorghum and shoot borer incidence in sugarcane. The importance of time of sowing and pest occurrence as reported by several authors (Dhawan *et al.*, 2008; Devi and Ram, 2018) is given below (Table 2).

Table 2. Effect of sowing time on incidence of insect pests in different crops

Crops	Insect pests	Response
Rice	Leaf folder or leaf roller (*Cnaphalocrocis medinalis*)	Early rice planting reduced population
	Brown plant hopper (*Nilaparvata lugens*)	Planting in late July in *kharif* and early in *rabi* season reduces occurrence in Andhra Pradesh
	Gall midge (*Orseolia oryzae*)	Low incidence in August/October planting
Chickpea	Gram pod borer (*Helicoverpa armigera*)	Increased pod damage due to delay in sowing
Sorghum	Shoot fly (*Atherigona soccata*)	Earlier sowing (September to October) reduced incidence
Cotton	Leafhopper (*Amrasca devastans*)	Increased occurrence in late-sown crops
Chillies	Thrips (*Scirtothrips dorsalis*)	Increased thrips and leaf curl virus incidence in late planted crop
Tomato	Whitefly (*Bemisia tabaci*)	Reduced incidence if planted within July-November

2.1.2. Nutrient management

Organic standards reduce or exclude usage of synthetic inputs and promote maximum usage of local natural resources. Since soil is treated organically, it can act as a buffer to keep plants nutritional and mineral balance at their ideal levels. This in turn influences the performance of insect pests infesting plant. E.g. the low level of nitrogen concentration in foliar tissues or phloem in organic farming recorded lower incidence of whiteflies and aphids than conventional farming (Van Bruggen, 1995).

2.1.3. Water management

Water management plays an important role in pest control, as it can effect pest populations and their impact on crops. Overwatering can lead to humid conditions that encourage the development of certain pests, such as fungal pathogens and some insect pests. By providing plants with the exact amount of water they need and preventing excessive wetness, we can reduce the frequency of pests and diseases. Using drip irrigation or root zone watering methods helps deliver water directly to plant roots, curtailing wet foliage.

Moist foliage can generate a favourable environment for pests and diseases. By keeping foliage dry, we can reduce the possibility of pest infestations and fungal diseases. Consider the pest life cycle when planning irrigation. For instance, some pests, like slugs or snails, are more active at night and attracted to wet environments. You can deter their presence by altering irrigation timing to minimise prolonged moisture periods during their active periods. Proper drainage is critical to avoid waterlogging, which can lead to root diseases and generate habitats for pests like water-loving insects or slugs. Ensuring that excess water drains away from the field or using drainage systems can help keep ideal soil moisture levels and discourage pests.

2.1.4. Host plant resistance

Growing resistant varieties are important management strategies to protect the plants from target insects. Mostly leaf colour, hairiness, glossiness, secretion of chemicals and biochemical characters decide the resistance in plants (Patil *et al.,* 2011). Resistant cultivars develop structural, chemical and protein-based defense, which helps to detect the damage causing insects and restrict them from causing economic damage (Chakraborty *et al.,* 2008). The economic injury levels for the resistant plants are higher than that of susceptible cultivars. The cultivation of resistant varieties, apart from reducing insect pest population also attracts natural enemies in the field by the way of reducing insecticides. Varieties known to produce induced odor emissions attract parasitoids (Degen *et al.,* 2004). Selection of varieties with ability to recover quickly after pest damage is also ideal strategy (Waring and O'Hara, 2005). The time consumed for developing resistant varieties through conventional breeding programme is much more and availability of varieties against critical insect pests are less.

2.1.5. Field sanitation

Removing weeds, crop wastes and alternative host places will benefit to manage certain insect pests and diseases (Duraimurugan and Jagadish, 2004). In many instances, weeds provide habitat for the natural enemies. The natural enemies move to the field from the border, whenever the pest population increases, hence the weed management should be done carefully. The biological properties of soil also influence the capability of plants to overcome infestation by pests and diseases (Altieri and Nicholls, 2003). Low oviposition by the European corn borer (*Ostrinia nubilalis*) on a corn crop grown organically was observed by Phelan *et al.* (1995). The microbial activity in the organically maintained soils affects the performance of phytophagous insects (Zimmer and Topp, 2002). The organically raised tea plantations recorded more natural enemies than conventional tea plantations (Das *et al.,* 2010). Some examples of alternate host of insect pests (Sushil Kumar *et al.*, 2021) are enlisted below (Table 3).

Table 3. Alternate hosts of insect pests in different crop ecosystem

Crops	Insect Pests	Weed or alternate hosts to be removed
Rice	Gall midge (*Orseolia oryzae*)	Wild rice species (*Oryza nivara*)
	Green leaf hopper (*Nephotettix virescens*)	Swamp rice grass (*Leersia hexandra*), Deccan grass (*Echinochloa colona*), Common barnyard grass (*Echinochloa crusgalli*), Bermuda grass (*Cynodon dactylon*)
	White backed planthopper (*Sogatella furcifera*)	Swollen finger grass (*Chloris barbata*)
Groundnut	Thrips (*Caliothrips indicus*)	Devil's horsewhip (*Achyranthes aspera*)

2.2. Vegetation management

Enhancing the activities of natural enemies in the crop ecosystem through habitat management is a crucial method. The farming practices intended to reduce the pest population are integrated component of organic farming. The practices *viz.*, crop rotation, intercropping, border cropping, cover cropping, mulches, patch cropping, etc. are the important methods employed in pest management in organic framing conditions. Planting of combination of crops which can flower at different duration can be a best option to attract natural enemies throughout the main crop period. Habitat diversity around the main crop increase the natural enemy population (Landis, 2017). Enhancement of arthropod biodiversity, prevention of disease and pest outbreaks, and provision of food and shelter for natural enemies and reproduction rates are important functions of intercropping (Huss *et al.*, 2022). Maize crop grown along the border offers food and shelter for number of lepidopteran parasitoids and consequently help as eco-feast crop.

2.2.1. Cover crops

Before the main crop is planted, cover crops have a number of advantages, including a decrease in soil weeds, illnesses, and pests. Additionally, it increases soil fertility and offers a useful supply of organic manure. E.g. the 'Navadanya' – nine pulses were mixed and sown as cover crops. Before flowering these crops are incorporated *in-situ.*

2.2.2. Trap crop

To set up a trap, small plantings of a vulnerable or favoured crop may be made next to a larger one. The insect is eliminated or the infected plant parts or plants may be removed after the pest has been drawn to the crop. For the control of the diamond back moth, 25 rows of cabbage should be followed by two rows of mustard planted as a trap crop. 15 days before planting cabbage, the first crop of mustard is sown, or 20-day-old mustard seedlings are planted. When planted in 1:16 rows, 25-day-old tomato seedlings and 40-day African tall marigolds lessen *H. armigera* damage.

2.2.3. Inter crop

Inter crops are crops grown in between the main crop rows. These crops offer a favourable microclimate on the field for the local natural enemies. Growing numerous distinctly similar plant species, as demonstrated by Rao *et al.* (2001), might negatively impact the habitat by obstructing specialist herbivores visually or chemically. The presence of variety of crops increases the local biodiversity (Shrivastava *et al.,* 2010). The planting time, spacing and extent of trap crop are critical features for the success of trap crop in reducing pest population. In sorghum and maize, lablab or cowpea is grown as inter crops which increase the population of coccinellids and other predators. Intercropping of redgram or cowpea in sorghum reduces earhead bug and stem borer, respectively. Redgram+sorghum intercropping reduces leafhopper and pod borer infestation. Intercropping of sorghum in greengram and groundnut reduces leafhopper infestation. Mustard grown as intercrop in chickpea reduced gram pod borer infestation. Greengram or balckgram grown as intercrop in sugarcane reduces early shoot borer (Balasubramanian *et al*., 1998; Huss *et al*., 2022).

Inter cropping of maize in cotton fields augmented natural enemies *viz*., spiders, predatory coccinellid beetles and green lacewings population as compared to control fields. Maize also served as a trap crop for bollworm (*H. armigera)* lowering the number of eggs and damage to cotton caused by the second generation. Pulses intercropped with cotton decreased the leafhopper population (Rabindra, 1985), whereas lablab bean intercropped with sorghum decreased the incidence of sorghum stem borer. Growing tomato and marigold in 3:1 ratio recorded maximum reduction in tomato fruit borer (*H. armigera).*

2.2.4. Border or strip crop

It is possible to slow the spread of pest spices in the field by planting sections of the field or field borders with spices and kinds that differ from the main crop. Field borders and margins with flowering plants can provide as a habitat and a source of food for beneficial insects, particularly coleopterans and hymenopterans. Predators like coccinellids, syrphids, and other natural enemies are encouraged when cowpea is grown in cotton as a bund crop. Raise of blackgram, greengram, sesame, sunflower, sunhemp, etc., around paddy to reduce the hopper and stem borer population (Shanmugam *et al.,* 2022).

2.2.5. Crop rotation

Soil structure improvement and fertility renewal are the important benefits of crop rotation, which increases the crop ability to tolerate insect pest and pathogen attack (Tiemann *et al.,* 2015). The ability of the pathogen to survive without its host and the pathogen's host range are two elements that determine whether rotation in the case of diseases is successful in preventing disease. The large variety of host pathogens will be harder to manage.

During crop rotation, considerations may be given to the possible unfavourable effect of pest oppressive rotation on crop yield *viz.,* competition and secondary plant compounds. For maximum effect, rotation should be combined with other cultural practises. Insect pests *viz*., soil-born insects, bollworms, whitefly, and nematodes are less common in cotton grown followed by cereals like maize / sorghum. Crop rotation can effectively manage Colorado potato beetle, brinjal shoot and fruit borer, carrot rust fly and onion maggot (Buckland *et al.,* 2013). The rotations *viz*., maize and soybean, potato and legume, and sugarcane and paddy reduced the incidence of maize stem borer, wireworm and white grub, respectively.

2.2.6. Living mulches

Living mulches are a crucial component of organic farming methods and provide several advantages for the agroecosystem. In order to offer ground cover and different ecosystem services, living mulches entail cultivating particular plant species alongside the primary crop. Living mulches promote biodiversity by providing a home for helpful insects. By attracting pollinators and pests' natural enemies, they enhance biological management and ecological harmony in the farming system. The prevalence of insect pests is decreased by living mulches or straw mulches. The spider population in the rice ecosystem will rise if a straw bundle is placed in rice fields (Pauli *et al.,* 2011).

2.3. Biological control

Biological control is the use of live organisms to decrease the population density or impact of a specific pest organism, making it less abundant or less harmful than it would otherwise be (Eilenberg *et al*., 2001). The predators, parasitoids, natural enemies or entomopathogens either naturally happening or introduced in the ecosystem will manage the harmful insects. Highly species-specific, limited side effects and self-perpetuation in the introduced environment are some of the advantages of biological control. For permanent establishment and long-term pest management, classical biological control entails the purposeful introduction of an exotic, typically co-evolved, biological control agent. Exploitation of the environment to augment the endurance and effectiveness of natural enemies are the main entity of classical biological control (Eilenberg *et al.,* 2001; Frank and Gillett-Kaufman, 2006). Bio-agents are parasites, predators and pathogens (virus, bacteria, fungi, protozoan and entomopathogenic nematodes) proffer huge possibility in biological control of insect pests (Duraimurugan and Regupathy, 2005a; Duraimurugan and Regupathy, 2005b; Geetha *et al.*, 2020). The vital approaches in bio-control are: a) *Importation:* Importing biological control agents to eradicate pests of foreign origins is known as the classical method of biological control; b) *Conservation:* This technique involves modifying the environment to safeguard the bio-control agents. The many naturally existing beneficial organisms that are present in the crop should be preserved by the organic farmer. The organic farming practices helps to build up large number of natural enemies or beneficial insects in the field which ultimately manages the insect pest population; c) *Augmentation*: Mass production of microbial bio-agents and natural enemies is the goal of augmentation. This category also includes genetically modifying biocontrol agents to have superior features. Inundative and inoculative release of parasites and predators based on the pest intensity reduces the pest damage in the crops. More than 230 species of natural enemies are marketed for supplemental biological control all over the world (Van Lenteren, 2012).

2.3.1. Parasitoids and predators

Small insects known as parasitoids feed on and survive on hazardous insects, going through their entire life cycle inside the host insect and eventually killing it. Large insects called predators which hunt and eat hazardous insects. Many parasitoids have been known to be extremely successful, including *Trichogramma chilonis* (egg parasitoid of tomato fruit borer, *H. armigera)*, *Isotima javensis* (larval parasitoid of sugarcane top borer), *Epiricarnia melanoleuca* (adult and nymphal parasitoid of *Pyrilla purpusilla*), *Campoletis chlorideae* (larval parasitoid of *H. armigera*) and many more (Hem Saxena and Duraimurugan, 2009; Hem *et al.*, 2012). The predators that are most effective include predatory coccinellid beetles, reduviid bugs, pentomid bugs, carabids, predatory ants and spiders (Duraimurugan and Srinivasan, 2009). A large number of arthropods and insectivorous birds have been also reported as a successful predator (Rabindra *et. al.*, 2005). Some important parasitoids and predators commercially produced are enlisted below (Table 4).

Table 4. Bio-control agents to control insect pests of different crops

Bio-control agents	Insect pests managed	References
I. Parasitoids		
Egg parasitoids		
T. chilonis	Bollworms and Borers pests	Fatouros and Huigens, (2012)
T. brasiliensis	*H. armigera* in tomato	
T. pretiosum	*H. armigera* in tomato	
T. japonicum	Paddy stem borer	
Telenomus remus	Tobacco caterpillar	
Egg larval parasitoid		
Chelonus blackburni	Cotton bollworms	Fatouros and Huigens, (2012)
Larval parasitoids		
Bracon brevicornis, Goniozus nephantidis, Elamus nephantidis	Coconut black headed caterpillar (*Opisina arenosella*)	Walker *et al.* (2004)
Bracon kirkpatrici and Bracon hebetor	Cotton bollworms	
Acerophagus papaya	Papaya mealy bug	
Isotima javensis	*Scipophga excerptalis*	
Stumiposis infernes	*Chilo auricillus*	
Nymphal parasitoids		
Encarsia formosa, Encarsia guadeloupae	Whitefly	Birkett *et al.* (2003) Elango, (2019)
Aphidius sp.	Aphid	
Pupal parasitoids		
Brachymeria spp., *Tetrastichus israeli, Trichospilus pupivora*	Coconut black headed caterpillar	Le and Tran, (2022)
II. Predators		
Chrysoperla carnea	Soft bodied homopteran insect pests	Garzón *et al.* (2015)
Cryptolaemus montrouzieri	Mealy bugs	

2.3.2. Entomopathogens

The microbe's virus, bacteria, fungi, nematodes and protozoans are able to cause different types of diseases in insects and there by reduces the insect population (Table 5). In the same way some of the beneficial pathogens have detrimental effect on the disease causing pathogens. The nuclear polyhedrosis virus (NPV), which is more virulent than granulosis virus (GV) and cytoplasmic virus (CPV), has the most potential among the various types of entomopathogenic viruses. The nuclear polyhedrosis virus (NPV) of pod borer, *H. armigera* is successfully controlling the pod borer in Bengal gram, red gram, cotton, tomato, groundnut and sunflower. The NPV of *Spodoptera litura* is also very effective in cotton and tobacco.

One of the first microbial insecticides to be manufactured commercially was *Bacillus thuringiensis* (Bt). In India, the bacterial bio-insecticide has been used in the management of numerous insect pests *viz*., gram pod borer (*H. armigera*) in Bengalgram and redgram; diamondback moth (*Plutella xylostella*) in vegetables; tobacco caterpillar (*S. litura*) in cabbage and oilseed crops and fruit borer (*Earias* spp.) of okra (Vimala Devi *et al*., 2020; Vineela *et al*., 2020; Vimala Devi *et al*., 2021; Duraimurugan and Vimala Devi, 2021). There are numerous commercial Bt formulations in the market.

Entomopathogenic fungi *viz*., *Beauveria bassiana, Metarhizium anisopliae,* and *Metarhizium rileyi* and *Lecanicillium lecanii* were found promising in managing numerous crop insect pests (Vimala Devi and Duraimurugan, 2013). *M. rileyi* is effective against several lepidopteran pests and occur in epizootic form on *S. litura*, *H. armigera*, *S. exigua* and *S. frugiperda* in field conditions. *Aspergillus flavus* and *Aspergillus niger* are reported from the larvae of *H. armigera* in pulse eco-system and found promising against *H. armigera*.

Protozoans *viz*., *Nosema* sp. and *Vairiomorpha* sp. were reported on *H. armigera.* A huge number of entomopathogenic nematodes (EPN) were found infective on *H. armigera, Liriomyza trifolii* and *Conogethes punctiferalis* (Gayatri and Duraimurugan, 2019; Gayatri *et al*., 2020). EPN species *viz*., *Steinernema masoodi* and *S. seemae* have been reported from the larvae of *H. armigera*, causing upto 67% larval mortality (Ali *et al*., 2010).

Table 5. Entomopathogens to control insect pests of different crops

Entomopathogens		**Pests managed**	**References**
Entomopathogenic Bacteria	*Bacillus thuringiensis*	Lepidoptera, Coleoptera, Diptera	Bravo *et al.* (2017)
Entomopathogenic Virus	NPV, GV, CPV	Lepidoptera, Coleoptera	Kalha *et al.* (2014)
Entomopathogenic fungi	*Beauveria bassiana, Metarhizium anisopliae, Metarhizium rileyi, Lecanicillium lecanii*	Lepidoptera, Hemiptera, Coleoptera	Shah and Pell (2003)
Entomopathogenic nemtaodes	Heterorhabditidae and Steinernematidae	Lepidoptera, Coleoptera, Diptera	Lacey *et al.* (2015)

The disease caused by the fungal pathogens can be effectively managed using *Trichodermma* sp. and *Pseudomonas fluorescens*. The seeds with these biofungicides effectively reduce the wilt disease. *Trichoderma viride* and *Trichoderma harzianum* are effective against wilt and root rot causing fungi *viz., Rhizoctonia solani, Macrophomina phaseolina, Fusarium* spp. in pulses, cotton, oilseeds and vegetable crops (Sennoi *et al.,* 2013). *Pseudomonas fluorescens* was reported to be effective against root rot causing fungi in various crops (Moin *et al.,* 2020).

2.4. Approved biorational insecticides/methods

2.4.1. Behavioural control

The insect communicates between themselves to reproduce, to find members of same species, to locate resources, protect territory, camouflage or mimic other organisms (Richard and Hunt, 2013). Semiochemicals are volatile organic compounds either used for intraspecific communication (pheromones) or interspecific communication (allelochemicals). The specialized receptor presents in antennae, legs etc. detect the semiochemicals which were emitted by the emitter (same species or related species) in the environment.

The semiochemicals are employed to draw the same species to entrap and pull them (Gregg *et al.,* 2018). Pheromones are widely used in pest management programme for monitoring and mass trapping against lepidopteran insects. The highly species specific nature reduces significance in pest management for example in tomato for fruit borer *Helicoverpa armigera* and *Tuta absoluta* both causes extensive damage and the farmers has to instal both traps separately to monitor, which ultimately enhance the production cost.

2.4.2. Pheromones

Pheromones are chemical released by insects which attract other individuals of the same species. They were extensively utilised in organic farming pest management through strategies like monitoring, mass trapping and mating disruption. (a) Monitoring: The adults attracted in the traps were periodically monitored and based on weekly count in the traps the management strategies are decided. Eg. *H. armigera* sex pheromone traps @ 12 ha^{-1} to monitor the adult movement in the field. (b) Mass trapping: Mass trapping refers to the use of a high density of traps in a crop to be protected in order to provide some level of protection through the removal of a sizeable fraction of the population through trapping. E.g. Installation of *Tuta absoluta* sex pheromone traps @ 100 ha^{-1} to mass trap the adults. (c) Mating disruption: The sex pheromones are used to avert male insects finding females and ultimately reduce the number of fertile eggs laid in the ecosystem (Welter *et al.,* 2005). The species specific synthetic chemical lures and food or host attractants are used to mass trap the insects. The number of traps and strength of lures are important factors in this technique. For instance, food baited traps for management of fruit fly in melons and methyl eugenol along with insecticide impregnated in to a suitable substrate reduces the fruit flies (Ravikumar and Viraktamath, 2007). In the following crops pheromones are used (Table 6).

Table 6. Pheromones used to monitor or mass trap insect pests of different crops

Crop	**Insect pest managed**
Sex pheromones	
Tomato	*Helicoverpa armigera*
	Spodoptera litura
Brinjal	*Leucinodes orbonalis*
Bhendi	*Helicoverpa armigera*
	Earias vittella
Cabbage, Cauliflower	*Plutella xylostella*
Sugarcane	*Chilo infuscatellus*
Groundnut	*Aproaerema modicella*
	Spodoptera litura
Paddy	*Scirpophaga incertulas*
Mango	*Bactrocera dorsalis*
Melons	*Bactrocera cucurbitae*
Aggregation Pheromones	
Coconut	*Oryctes rhinoceros*
	Rhynchophorus ferrugineus

2.4.3. Traps

Light traps and sticky traps are widely used in the pest management. Light traps are used to attract adult moths during the evening hours and thereby monitoring and suppression. Sticky traps are used to attract the sucking pests. The whitefly, aphid, thrips and leaf miner population can be effectively monitored using yellow sticky traps, whereas blue sticky traps will be used against thrips (Tang *et al.*, 2016). Bait traps were used to manage shoot fly in sorghum and fruit fly in cucurbits. Funnel type, water pan type and delta trap are the three types of traps used in the pest management depending upon the nature of the pest.

2.4.4. Botanicals

About 2400 plant species distributed in 189 plant families having pesticidal capabilities were reported by Grainge and Ahmed (1988). More than 100 different chemicals have been discovered in neem seeds, with azadirachtin being the most physiologically active. Neem products have biological properties that control insect growth, prevent eating, and prevent oviposition. It has been demonstrated that neem oil at 2% and neem seed kernel extract

(NSKE) at 5% combined with liquid soap at 0.05% are effective against major rice pests, cotton and vegetable sucking bugs, and other pests. Neem cake applied (250 kg ha^{-1}) at the last ploughing before sowing was found effective against cotton stem weevil, plant parasitic nematodes in tomato, shoot and fruit borer in brinjal and soil insect pests of many other crops. There are a variety of commercial neem formulations on the market that range in azadirachtin content from 0.03% to a maximum of 5%. Neem formulations with various brand names are produced in India by more than 50 different companies.

Apart from neem, *Pongamia pinnata* cake and extracts were also widely used in pest management. Soap formulations in neem and pungam were used for the management of sucking pests. There has been evidence of insecticidal action in plants from the Annonaceae, Asteraceae, Fabaceae, Lamiaceae, Rutaceae, and many more families. Essential oils from *Mentha* sp., *Ocimum* sp., *Origanum* sp., *Rosmarinus* sp., *Piper* sp. and *Cinnamomum* sp. were acts as insect's neurotoxins. Patil and Chavan (2009) found that the extracts of *Acacia concianna* were most effective in managing sugarcane woolly aphid *Ceratovacuna lanigera*. The hexane extracts of *Atalantia monophylla* exhibited antifeedant, and larvicidal action against okra fruit borer *Earias vittella*. The Organic Materials Review Institute approved plant-based extracts *viz*., neem oil, pyrethrum, ryanea, rotenone and sabadilla, mineral derived pesticides and mineral oils to apply on need basis in organic farming systems. Botanicals derived from papaya leaves, garlic and wood ash also used to manage insect infestation (Mochiah *et al.,* 2011).

2.5. Importance of organic pest management

The agroecosystem's health, biodiversity, biological cycles and biological activities are improved through organic farming (Santhoshkumar *et al*., 2017). The organic farming pest management completely excludes the usage of synthetic chemical insecticides and transgenics and rely on cultural, physical, biological, semiochemical etc. for the management of pests (El-Shafie, 2019). The pest management in organic ecosystems developed around the frame of integrated pest management (Zehnder *et al.,* 2007). Monitoring of insect pests, exact identification and biology of insect pests, selection of appropriate natural enemies and deploying the natural enemies are the important process involved in organic farming.

- Early detection of pest problem
- Determination of location and density
- Establish growth trends
- Historical record
- Proper application of management strategy & follow up

The preventive strategies *viz.*, cultural, mechanical, natural enemies and resistant cultivars form integral part of organic pest management than curative control (Speiser *et al.,* 2006). The strategy should be followed in organic pest management apart from reducing incidence of insect's pest's incidence should also promote healthy and diversified crop ecosystem (Morgera *et al.,* 2012).

Steps in Organic pest management

Risk assessment both qualitative and quantitative

↓

Analysis of damage by comparing with ETL and management costs

↓

Choosing the right management option

↓

Decision making process (How and When)

3. Conclusion

IPM is a holistic system approach used in organic farming that combines a variety of crop production and protection techniques in order to lessen the financial losses brought on by pests and diseases. It places a strong emphasis on vigilant pest surveillance, the preservation of beneficial species, and need-based critical pest management intervention. The majority of pest management strategies have a strong emphasis on prevention, particularly before the crop is sown. The cost of pest control can be significantly decreased by conducting regular pest monitoring programmes, evaluating ETL, and pest scouting. In order to get the most out of pest and disease control strategies, decision-makers often use analytical interpretation of the population dynamics of pests, defenders, abiotic factors, and farmers' prior experiences.

4. Way Forward

Organic-approved insecticides made from natural sources will continue to exist in the changing organic farming scene as a last resort. Future developments, however, might see a stronger emphasis on precise application methods and the creation of innovative formulas that maximize effectiveness while reducing environmental impact. Strict attention to organic standards, guaranteeing that any pesticide used is licensed for organic farming and applied in compliance with established norms, will be essential to this growth. In the field of organic agriculture, cooperation among researchers, farmers, and legislators will be essential to advancing sustainable insect pest management techniques and promoting productivity and resilience in the face of upcoming obstacles.

5. References

Ali, S.S., Asif, M., Duraimurugan, P., Akhtar, M. H., Pervez, R., Ahaheen, A. and Ahmad, I. 2010. In vivo production and infectivity of Oscheius amsactae on four agriculturally important insect pests. Trends in Biosciences 3(2): 232-233.

Altieri, M.A. and Nicholls, C.I. 2003. Soil fertility management and insect pests: harmonizing soil and plant health in agroecosystems. Soil and Tillage Research 72(2): 203-211.

Balasubramanian, A., Mahadevan, N.R., Venugopal, M.S. and Murali-Baskaran, R.K. 1998. Influence of intercropping on infestation of early season sucking pests of cotton (*Gosspium hirsutum*). Indian Journal of Agricultural Sciences 68: 315-316.

Bravo, A., Pacheco, S., Gomez, I., Garcia-Gomez, B., Onofre, J. and Soberón, M. 2017. Insecticidal proteins from Bacillus thuringiensis and their mechanism of action. In: Bravo A, Pacheco S, Gómez I, Garcia-Gómez B, Onofre J, Soberón M (eds), Bacillus thuringiensis and Lysinibacillus sphaericus: Characterization and Use in the Field of Biocontrol Springer International Publishing, Cham pp 53-66.

Birkett, M.A., chamberlian, K., Guerrieri, E., Pickett, JA., Wadhams, L.J., and Yasuda, T. 2003. Volatiles from whitefly-infested plants elicit a host- location response in the parasitoid, Encarsia formosa. journal of chemical Ecology, 29 (7): 1589-1600.

Buckland, K., Reeve, J. R., Alston, D., Nischwitz, C. and Drost, D. 2013. Effects of nitrogen fertility and crop rotation on onion growth and yield, thrips densities, Iris yellow spot virus and soil properties. Agriculture, Ecosystems and Environment 177: 63-74.

Chakraborty, S., Luck, J., Hollaway, G., Freeman, A., Norton, R. and Garrett, K.A. 2008. Impacts of global change on diseases of agricultural crops and forest trees. CABI Review 3: 1-5.

Das, S., Roy, S. and Mukhopadhyay, A. 2010. Diversity of arthropod natural enemies in the tea plantations of North Bengal with emphasis on their association with tea pests. Current Science 1457-1463.

Degen, T., Dillmann, C., Marion-Poll, F. and Turlings, T.C. 2004. High genetic variability of herbivore-induced volatile emission within a broad range of maize inbred lines. Plant Physiology 135(4): 1928-1938.

Devi S and Ram P. 2018. Effect of dates of sowing on population of sucking insect pests in desi cotton (*Gossypium arboreum* L.). Journal of Entomology and Zoology Studies 6(1): 1041-1044.

Dhawan, A.K., Bharathi, M. and Singh, K. 2008. Impact of castor as a trap crop in relation to major insect pests in cotton agroecosystem. Indian Journal of Ecology 35(1): 70-72.

Duraimurugan, P. and Jagadish, A. 2004. Non chemical management of Scirtothrips dorsalis Hood damaging rose flowers. Journal of Applied Zoological Research 15(2): 163-165.

Duraimurugan, P. and Regupathy, A. 2005a. Push-pull strategy with trap crops, neem and nuclear polyhedrosis virus for insecticide resistance management in Helicoverpa armigera (Hubner) in cotton. American Journal of Applied Sciences 2(6): 1042-1048.

Duraimurugan, P. and Regupathy, A. 2005b. Stimulo-deterrent diversionary strategy with trap crops, neem and Bacillus thuringiensis Berliner for the management of insecticide resistant Helicoverpa armigera (Hubner) in cotton. Journal of Biological Sciences 5(6): 681-686.

Duraimurugan, P. and Srinivasan, T. 2009. Carabid predator of Maruca vitrata in mungbean – a new record. Pulses Newsletter 20(3): 5.

Duraimurugan, P. and Vimala Devi, P.S. 2021. Assessment of Bacillus thuringiensis var. kurstaki (Bt-127) SC formulation as a component of IPM in soybean through farmer's participatory approach. Journal of Oilseeds Research 38(4): 358-361.

Eilenberg, J., Hajek, A. and Lomer, C. 2001. Suggestions for unifying the terminology in biological control. Bio Control 46(4): 387-400.

Elango, K. 2019. Biology, distribution and host range of new invasive pest of India coconut rugose spiralling whitefly Aleurodicus rugioperculatus Martin in Tamil Nadu and the status of its natural enemies. International Journal of Agriculture Sciences 11(9):8423-8426.

El-Shafie, H.A.F. 2019. Insect pest management in organic farming system. In Moudry J, Ferreira Mendes K, Bernas J, Teixeira RdS and de Sousa RN (eds) Multifunctionality and Impacts of Organic and Conventional Agriculture IntechOpen, London. pp. 1-20.

Fatouros, N.E. and Huigens, M.E. 2012. Phoresy in the field: natural occurrence of Trichogramma egg parasitoids on butterflies and moths. BioControl 57(4): 493-502.

Frank, J.H. and Gillett-Kaufman, J.L. 2006. Glossary of expressions in biological control. Extension Data Information Source 15.

Garzón, A., Medina, P., Amor, F., Vinuela, E. and Budia, F. 2015. Toxicity and sublethal effects of six insecticides to last instar larvae and adults of the biocontrol agents Chrysoperla carnea (Stephens) (Neuroptera: Chrysopidae) and Adalia bipunctata (L.) (Coleoptera: Coccinellidae). Chemosphere 132: 87-93.

Gayatri, B. and Duraimurugan, P. 2019. Bio-efficacy of Heterorhabditis bacteriophora (Rhabditida: Heterorhabditidae) against serpentine leafminer, Liriomyza trifolii Burgess (Diptera: Agromyzidae) in oilseed crops. Journal of Biological Control 33(2): 169-172.

Gayatri, B., Duraimurugan, P. and Sarada, C. 2020. Virulence of entomopathogenic nematodes against castor capsule borer Conogethes punctiferalis (Guenee). Indian Journal of Entomology 82(1).

Geetha, B., Venkatachalam, S.R. and Duraimurugan, P. 2020. Evaluation of field efficacy of microbial and chemical insecticides for the management of defoliator pests in castor. Journal of Oilseeds Research 37(3): 323-326.

Grainge, M and Ahmed, S. 1988. Hand book of plants with pest control properties. John Wiley and Sons, New York, p. 470.

Gregg, P.C., Del Socorro, A.P. and Landolt, P.J. 2018. Advances in attract-and-kill for agricultural pests: Beyond pheromones. Annual Review of Entomology, 63(1): 453-470.

Hem, Saxena and Duraimurugan, P. 2009. Record of Bracon hebetor Say (Hymenoptera: Braconidae) on Helicoverpa armigera (Hübner) in chickpea. Journal of Biological Control 23(2): 191-192.

Hem, Saxena, Duraimurugan, P. and Iquebal, M.A. 2012. Seasonal parasitism and biological characteristics of Habrobracon hebetor (Hymenoptera: Braconidae) - a potential larval ectoparasitoid of Helicoverpa armigera (Lepidoptera: Noctuidae) in a chickpea ecosystem. Biocontrol Science and Technology 22: 305-318.

Huss, C.P., Holmes, K.D. and Blubaugh, C.K. 2022. Benefits and risks of intercropping for crop resilience and pest management. Journal of Economic Entomology 115(5): 1350-1362.

Kalha, C.S., Singh, P.P., Kang, S.S., Hunjan, M.S., Gupta, V. and Sharma, R. 2014. Entomopathogenic viruses and bacteria for insect-pest control, In: Abrol D P (eds) Integrated Pest Management: Current Concepts and Ecological Perspective. Academic Press, pp 225-244.

Kogan, M. 1998. Integrated pest management: historical perspectives and contemporary developments. Annual Review of Entomology 43: 243-270.

Lacey, L.A., Grzywacz, D., Shapiro-Ilan, D. I., Frutos, R., Brownbridge, M. and Goettel, M. S. 2015. Insect pathogens as biological control agents: back to the future. Journal of Invertebrate Pathology 132: 1-41.

Landis, D.A. 2017. Designing agricultural landscapes for biodiversity-based ecosystem services. Basic and Applied Ecology 18: 1-12.

Le, K.H. and Tran, D.H. 2022. Biology of Brachymeria kamijoi Habu (Hymenoptera: Chalcididae), a pupal parasitoid of the coconut black-headed caterpillar, Opisina arenosella Walker (Lepidoptera: Xyloryctidae). Research on Crops 23(4): 866-869.

Mochiah, M.B., Banful, B., Fening, K.N., Amoabeng, B.W., Offei Bonsu, K., Ekyem, S., Briamah, H. and Owusu-Akyaw, M. 2011. Botanicals for the management of insect pests in organic vegetable production. Journal of Entomology and Nematology 6(3): 85.

Moin, S., Ali, S.A., Hasan, K.A., Tariq, A., Sultana, V., Ara, J. and Ehteshamul-Haque, S. 2020. Managing the root rot disease of sunflower with endophytic fluorescent Pseudomonas associated with healthy plants. Crop Protection 130: 105066.

Morgera, E., Caro, C.B. and Duran, G.M. 2012. Organic agriculture and the law. FAO legislative study, Food and Agriculture Organization, Rome, Italy, pp 302.

Patil, D.D., Mhaske, D.K., Wadhawa, G.C. 2011. Antidiabetic activity of bark and root of Caesalpinia bonduc. Journal of Pharmaceutical Biology 2(10): 27502752.

Patil, D.S. and Chavan, N.S. 2009. Bioefficacy of some botanicals against the sugarcane woolly aphid, Ceratovacuna lanigera Zehnter. Journal of Biopesticides 2(1): 44-47.

Pauli, N., Barrios, E., Conacher, A.J. and Oberthür, T. 2011. Soil macrofauna in agricultural landscapes dominated by the Quesungual Slash-and-Mulch Agroforestry System, western Honduras. Applied Soil Ecology 47(2): 119-132.

Phelan, P.L., Mason, J.F. and Stinner, B.R. 1995. Soil-fertility management and host preference by European corn borer, Ostrinia nubilalis (Hübner), on Zea mays L.: A comparison of organic and conventional chemical farming. Agriculture, Ecosystems & Environment 56(1): 1-8.

Rabindra, R.J. 1985. Transfer of Plant Protection Technology in Dry crops. In: Jayaraj S (eds) Proceedings of the National Seminar on Integrated Pest and Disease Management, Tamil Nadu Agricultural University, Coimbatore pp 337-383.

Rabindra, R.J., Singh, A., Saxena, H. and Chandish, R.B. 2005. Biological control of insect pests and diseases in food legumes. In: Kharkwal M.C. (ed.) Fourth International Food Legumes Research Conference (October 18-22, 2005), Indian Agricultural Statistics Research Institute, New Delhi, pp 25-26.

Rao, M., Terborgh, J. and Nunez, P. 2001. Increased herbivory in forest isolates: implications for plant community structure and composition. Conservation Biology 15: 624-633.

Ravikumar, P. and Viraktamath, S. 2007. Attraction of female fruit flies to different protein food baits in guava and mango orchards. Karnataka Journal of Agricultural Sciences 20(4): 745.

Richard, F.J. and Hunt, J.H. 2013. Intracolony chemical communication in social insects. Insectes Sociaux 60(3): 275-291.

Santhoshkumar, M., Reddy, G.C. and Sangwan, P.S. 2017. A review on organic Farming-Sustainable agriculture development. International Journal of Pure and Applied Bioscience 5(4): 1277-1282.

Sennoi, R., Singkham, N., Jogloy, S., Boonlue, S., Saksirirat, W., Kesmala, T. and Patanothai, A. 2013. Biological control of southern stem rot caused by Sclerotium rolfsii using Trichoderma harzianum and arbuscular mycorrhizal fungi on Jerusalem artichoke (Helianthus tuberosus L.). Crop Protection 54: 148-153.

Shah, P.A. and Pell, J.K. 2003. Entomopathogenic fungi as biological control agents. Applied Microbiology and Biotechnology 61(5): 413-423.

Shanmugam, P.S., Sangeetha, M., Ayyadurai, P. and Prasad, Y.G. 2022. Demonstration of ecological engineering based pest management in rice Oryza sativa L. through farmer's participatory approach. Agricultural Science Digest-A Research Journal 42(3): 290-295.

Shivalingaswamy, T.M. and Satpathy, S. 2007. Integrated pest management in vegetable crops. In: Jain, PC and Bhargava, MC (eds), Entomology: Novel Approaches. New India Publishing Agency, New Delhi, India, pp 353-375.

Shrivastava, G., Rogers, M., Wszelaki, A., Panthee, D. R. and Chen, F. 2010. Plant volatiles-based insect pest management in organic farming. Critical Reviews in Plant Sciences 29(2): 123-133.

Speiser, B., Wyss, E. and Maurer, V. 2006. Biological control in organic production: First choice or last option? In: An Ecological and Societal Approach to Biological Control, Springer, Dordrecht (2): 27-46.

Sushil Kumar, Malay K Bhowmick and Puja Ray. 2021. Weeds as alternate and alternative hosts of crop pests. Indian Journal of Weed Science 53(1): 14-29.

Tang, L.D., Zhao, H.Y., Fu, B.L., Han, Y., Liu, K. and Wu, J.H. 2016. Colored sticky traps to selectively survey thrips in cowpea ecosystem. Neotropical Entomology 45(1): 96-101.

Tiemann, L.K., Grandy, A.S., Atkinson, E.E., Marin-Spiotta, E. and McDaniel, M.D. 2015. Crop rotational diversity enhances belowground communities and functions in an agroecosystem. Ecology Letters 18(8): 761-771.

Van Bruggen, A.H.C. 1995. Plant disease severity in high-input compared to reduced-input and organic farming systems. Plant Disease 79: 976- 984.

Van Lenteren, J.C. 2012. The state of commercial augmentative biological control: plenty of natural enemies, but a frustrating lack of uptake. BioControl 57(1): 1-20.

Vimala, D., and Duraimurugan, P. 2013. Exploitation of Nomuraea rileyi and Beauveria bassiana for the management of lepidopteran pests. Kavaka 41: 43-67.

Vimala, D., Duraimurugan, P, Chandrika, K.S.V.P., Vineela, V. and Hari, P.P. 2020. Novel formulations of Bacillus thuringiensis var. kurstaki: An eco-friendly approach for management of lepidopteran pests. World Journal of Microbiology and Biotechnology 36: Article No. 78.

Vimala, D., Duraimurugan., P, Chandrika, K.S.V.P. and Vineela, V. 2021. Development of a water dispersible granule (WDG) formulation of Bacillus thuringiensis for the management of Spodoptera litura (Fab.). Biocontrol Science and Technology 31(8): 850-864.

Vineela, V., Vimala D. and Duraimurugan, P. 2020. Effect of novel formulation of Bacillus thuringiensis var. kurstaki (DOR Bt-127) against semilooper (Achaea janata) in castor. Journal of Oilseeds Research 37:188-189.

Walker, G.P., Qureshi, M.S. and Wallace, A.R. 2004. Parasitism of lepidopteran larvae collected from vegetable crops and associated weeds at Pukekohe. New Zealand Plant Protection 57: 1-7.

Waring, K.M., and O'Hara, K.L. 2005. Silvicultural strategies in forest ecosystems affected by introduced pests. Forest Ecology and Management 209: 27-41.

Welter, S., Pickel, C., Millar, J., Cave, F., Van Steenwyk, R. and Dunley, J. 2005. Pheromone mating disruption offers selective management options for key pests. California Agriculture 59(1): 16-22.

Wyss, E., Luka, H., Pfiffner, L., Schlatter, C., Uehlinger, G. and Daniel, C. 2005. Approaches to pest management in organic agriculture: a case study in European apple orchards. Organic Research 33-36.

Zehnder, G., Gurr, G. M., Kuhne, S., Wade, M. R., Wratten, S. D. and Wyss, E. 2007. Arthropod pest management in organic crops. Annual Review of Entomology 52: 57-80.

Zimmer, M. and Topp, W. 2002. The role of coprophagy in nutrient release from faeces of phytophagous insects. Soil Biology and Biochemistry 34(8): 1093-1099.

7

Importance of *Trichoderma* Application in Organic Agriculture

***G. Rajesha*[1], *K. Sakthivel*[2], *S. Madhavan*[3], *K. Sankari Meena*[2] *P. Valarmathi*[4], *I.K. Das*[1] *and R.D. Prasad*[2]**

[1]*ICAR- Indian Institute of Millet Research, Hyderabad-500 030, Telangana, India*

[2]*ICAR- Indian institute of Oilseeds Research, Hyderabad-500 030, Telangana, India*

[3]*ICAR-Directorate of Floricultural Research, Regional Station Vemagiri Rajahmundry-533 125, Andhra Pradesh, India*

[4]*ICAR-Central Institute for Cotton Research (CICR) Regional Station Coimbatore-641 003, Tamil Nadu, India*

Abstract

Trichoderma is fungus is well adapted to different ecosystem and identified > 250 species for agricultural use. It is primarily used as a bio-fungicide because of the variety of secondary metabolites produced and applied mainly for biocontrol and bioremediation and growth promotion activities. Bio-control potential of Trichoderma used for the control of soil borne diseases through various mechanisms responsible for broad range of cell wall degrading enzymes, anti-microbial and organic volatile compounds produced by Trichoderma. During abiotic stress, Trichoderma play an important role in plant growth promotion through mobilization of various nutrients. Presently Trichoderma is used as an effective biocontrol agent against plant parasitic nematodes. It is also used for one of the alternate remedy for improving the quality of the soil by removing the heavy metal in agricultural field, makes them suitable for cultivation.

Keywords: *Trichoderma, bio-fungicide, bioremediation*

1. Introduction

Trichoderma is one of the genus of fungus species which belongs to Ascomycetes (Order Hypocreales) present in diverse soil ecosystems, colonise under various agro ecological niches and most of the prevalent are culturable microorganism. They are primarily used as efficient biocontrol agents against a variety of soil-borne plant diseases. However, they have also shown potential in processes like bioremediation, promoting plant growth, and managing nematode populations.

2. Application of *Trichoderma* as Bio fungicide

It has been observed that *Trichoderma* species exhibit strong biocontrol capabilities, especially in the context of managing soil-borne diseases. Approximately 250 *Trichoderma* species have been identified, and they are commonly found in agricultural and forest soils (Harman *et al.*, 2004; Bissett *et al.*, 2015). Among these species, *T. harzianum, T. hamatum, T. polysporum T. hamatum, T. polysporum, T. atroviride, T. pseudoko- ningii T. koningii, T. viride, T. longibrachiatum,* and *T. reesei* have been extensively studied worldwide for their antagonistic potential. Mycoparasitism, competition for nutrients and space in the rhizosphere, and development of antibiotics that serve to reduce plant diseases are the key mechanisms underpinning the biocontrol potential of *Trichoderma* species. (Manigundan *et al.,* 2016*; Sood et al.*, 2020).

2.1. Mycoparasitism

Mycoparasitism refers to the capacity of one fungus to parasitize and eliminate another fungus, and members of the *Trichoderma* genus are renowned for their mycoparasitic prowess. They can efficiently colonize and break down the cell walls of other fungi by producing a diverse range of CWDE (cell wall degrading enzymes) like chitinases, proteases, glucanases, xylanases, and other secondary metabolites, which enables them to prevent the growth of plant pathogens. The fungal cell wall contains components such as chitin, β-1,3-glucans, proteins, cellulose, and hemicellulose, all of which are targeted and degraded by *Trichoderma's* chitinases, glucanases, proteases, and xylanases as they grow. This ability allows them to inhibit the growth of plant pathogens. Several studies confirmed the broad spectrum potential of *Trichoderma* strains with high levels of CWDEs. Mach *et al.*, 1999 reported that two major chitinase genes, *ech42* (encoding the CHIT42 endochitinase) and *nag1* (encoding the CHIT73 *N*-acetyl-β-D-glucosaminidase), were found responsible for broad spectrum biocontrol potential of *Trichoderma atroviride strain P1.* Chitinase of *T. asperellum* PQ34 strain can inhibit the growth of two strains of *S. rolfsii* and *Colletotrichum* sp (Loc NH 2020). Mercado *et al.* (2015) reported that expression of the β-1,3-glucanase gene (bgn13.1) from *Trichoderma harzianum* in strawberry increases tolerance to crown rot disease and enhance with plant growth. Ting and Chai 2015 reported that the high chitinase and β-1,3-glucanase activities of *Trichoderma harzianum* is responsible for broad spectrum biocontrol potential against pathogenic (*Fusarium oxysporum* f. sp. *cubense* race 4, *Ganoderma boninense*) and non-pathogenic isolates

(*T. viridescens, Serratia marcescens, Streptomyces griseus*).

2.2. Competition of nutrition and space

Competition for space and nutrition has direct role on biocontrol behaviour for any potential antagonists. *Trichoderma* species are known for their competitive ability in the soil ecosystem for nutrients and space with other microorganisms, including plant pathogens. Starvation for essential nutrients and death for microorganisms due to competitive potential enhances biological control of any antagonists. *Trichoderma* has a superior capacity to mobilize and takeup soil nutrients compared to other organisms. *Trichoderma* can secrete siderophores, which are iron-chelating compounds that enable to scavenge iron from the soil and compete with other microorganisms for this essential nutrient. *T. asperellum* Q1 strain was reported to enhance cucumber growth by siderophore production and alleviating negative effect of salinity and available iron deficiency (Qi and Zhao, 2013). Oszust *et al.,* 2020 reported that *Trichoderma* spp. could utilize more D-arabitol, i-erythritol, glycerol, D-mannitol and D-sorbitol than pathogenic fungi *Botrytis* sp., *Colletotrichum* sp., *Verticillium* sp. and *Phytophthora* sp. In addition to its competition for nutrients, *Trichoderma's* rapid growth and colonization behavior further contribute to protecting plants from a variety of soilborne microorganisms.

2.3. Antibiosis

Antibiosis is a mechanism of biological control used by *Trichoderma* species to suppress the growth of plant pathogens. This mechanism encompasses the production of secondary compounds, including volatile organic compounds (VOCs) and antibiotics, which possess antimicrobial properties against other microorganisms.

2.4. Production of antibiotics

Trichoderma species are well known for their ability to produce a range of antibiotics that can inhibit the growth of plant pathogens. Some examples of antibiotics produced by *Trichoderma* include gliotoxin, harzianic acid, and trichokonins. Gliotoxin is a non-ribosomal peptide that has broad-spectrum antimicrobial activity, and can inhibit the growth of fungi, bacteria, and even some higher organisms (Hua *et al.,* 2021). Harzianic acid is a polyketide that has been shown to have strong antifungal activity against a range of plant pathogens (Vinale *et al.,* 2013). Trichokonins are peptaibols that have potent antifungal activity, and have been shown to inhibit the growth of *Fusarium oxysporum*, *Rhizoctonia solani*, and other plant pathogens. Growth of soil-borne pathogens like *R. solani*, *Phytophthora cinnamomi*, *Pythium middletonii*, *Fusarium oxysporum* and *Bipolaris sorokiniana* was observed to be negatively affected in the presence of Koninginin D (Sood *et al.*, 2020). Similarly, viridins obtained from *Trichoderma* spp. like *T. koningii, T. viride*, and *T. virens* contained the spore germination of *Botrytis allii, Colletotrichum lini, Fusarium caeruleum, Penicillium expansum, Aspergillus niger* and *Stachybotrys atra* (Reino *et al.*, 2008).

2.5. Volatile organic compounds (VOCs)

Trichoderma can also produce VOCs that have antimicrobial activity (Manigundan *et al.*, 2016). These volatile organic compounds (VOCs) are organic molecules with the ability to be emitted into the atmosphere and permeate through the soil, thereby inhibiting the proliferation of plant pathogens. Some examples of VOCs produced by *Trichoderma* include 6-pentyl-α-pyrone (6-PP) and 1-octen-3-ol. 6-PP has been shown to have antifungal activity against a range of plant pathogens, including *Botrytis cinerea* and *Fusarium oxysporum*. 1-octen-3-ol has also been shown to have antifungal activity against several plant pathogens, including *Fusarium oxysporum, Sclerotinia sclerotiorum,* and *Phytophthora infestans* (Contreras-Cornejo *et al.*, 2022).

3. Role of *Trichoderma* in Plant Growth Promotion

Trichoderma has been used as a biocontrol agent since the 1930s (Weindling, 1932). Furthermore, *Trichoderma* is recognized for its ability to enhance phosphorus solubilization, promote root growth through the synthesis of indole-3-acetic acid (IAA) and enhance plant resistance to abiotic stress factors, all of which are recognized as factors that stimulate plant development (Harman, 2011).

3.1. Growth promotion through solubilisation and efficient use of nutrients

Among the many ways *Trichoderma* promotes plant growth, improving crop growth and yield through increased nutrient availability and solubilizing nutrients are significant activities. The mechanisms by which plants can more effectively utilise nitrogen are not fully known, but one aspect likely involved is the capacity of fungi to promote deep rooting, which in turn increases the area of soil that plant roots can occupy. For instance, *T. harzianum* T22 has been reported to increase the efficiency of nitrogen utilization in maize, allowing for maximum yields to be produced with 38% less nitrogen than is typically advised per hectare. *Trichoderma*-treated wheat seeds reportedly yielded greater yield following treatment (Harman *et al.*, 2008). *Trichoderma* has the potential to transform complex and less accessible forms of plant nutrients into simpler, more accessible forms, as demonstrated with Fe-sensitive Catharanthus plants

(Harman, 2000). While Singh *et al.* (2014) discovered that *Trichoderma's* solubilization of nutrients raised tomato yield and improved their availability to the plants. The results of an *in vitro* experiment using *Trichoderma* isolates obtained from the Andaman Nicobar Islands showed that tomato and chilli germination percentage, shoot and root growth are significantly improved when treated with *Trichoderma* isolates (Manigundan *et al.*, 2016). *Trichoderma* strains isolated from the Amazon rain forest were discovered to have boosted soybean growth by up to 41% and phosphate uptake by up to 145% (Bononi *et al.*, 2020). It was recently discovered by Yadav *et al.* (2023) that bio-priming of chilli seeds with *T. harzianum* and *T. asperllum* enhances their growth promotion and raises their immunity against *Colletotrichum truncatum.*

3.2. Growth promotion through improved rooting and abiotic stress tolerance

Trichoderma treatment has led to increased root development and enhanced yields in several crops, including poinsettia, soybean, corn, and soybean (Harman, 2000). The *Trichoderma*-colonized improved root system helps to mitigate the effects of drought stress. When plants and fields are treated with various combinations of *T. harzianum* with *Pseudomonas fluorescens* and arbuscular mycorrhiza, Kabdwal *et al.* (2019) discovered that tomato yield is enhanced while reducing plant mortality caused by disease and promoting plant growth. Priming tomato plants with *Trichoderma* strains improves the shoot and root growth of plants by producing IAA and solubilizing phosphorus. The tomato treated with *Trichoderma* strains also shows increased chlorophyll synthesis and a 10-30% reduction in *Fusarium* wilt disease (Bader *et al.*, 2020). *Trichoderma* treatment resulted in improved root growth and plant height in mungbean plants (Inayati *et al.*, 2021). When tomatoes were treated with *T. atroviride*, a substantial increase in yield was observed (Gravel *et al.*, 2007). Even in abiotic and other physiologically stressful conditions, *Trichoderma* is proven to improve crop seed germination. Mastouri *et al.* (2010) discovered that *T. harzianum* treatment increased the germination percentage of tomato cv. Jubilee seeds under osmotic, salt, and freezing temperature stress, along with lower levels of lipid peroxidase concentration. *Trichoderma* might be one of the potential treatments for increasing the salt stress tolerance of crops. Wheat seedlings treated with *Trichoderma longibrachiatum* under a 150 mM NaCl concentration demonstrated significant tolerance through improved shoot and root growth, as well as a significant up-regulation of antioxidant enzymes such as superoxide dismutase (SOD), peroxidase (POD), and catalase (CAT) in wheat seedlings, which were increased by 29, 39, and 19%, respectively (Zhang *et al.*, 2016). Ayyandurai *et al.* (2021) discovered that groundnut seeds treated with *Trichoderma longibrachiatum* exhibited higher germination percentage (93.33) and vigor index (2246.2) compared to untreated seeds. Similarly, apple rootstock seedlings that had been treated with *Trichoderma simmonsii* had 70% more root dry biomass and 16% more shoot length than control plants.

4. *Trichoderma* in Nematode Control

Trichoderma is used as an effective biocontrol agent against plant parasitic nematodes. They directly reduce the damage caused by the nematodes by the mechanism of parasitism, antibiosis and paralysis. By fostering competition for food and space, they indirectly lessen the harm that the nematodes do to the plants. By turning on the hormone-mediated plant defense mechanism, they also develop resistance to the nematodes. Also, by producing various metabolites and enzymes, they provide long-lasting biocontrol methods against the nematodes (Poveda *et al.*, 2020).

Efficacies of various *Trichoderma* species were tested under glasshouse condition against nematodes. Studies revealed that the application of *Trichoderma* strains 10^{10} spores of fungi /g soil reduced root galling and nematode multiplication and promoted plant growth and development. Among the different strains testes, *T. harzianum* outperformed over the other strains (Javeed *et al.*, 2016).

Since autoclaved soil was used in the glass house experiments, their results cannot be directly extended to field environment, where the natural soil has a significant impact on the extent to which biocontrol agents combat nematodes. Hence, field studies are a viable method for determining the bioagent's real effects. In order to evaluate the effectiveness of the fungi against the nematodes, a few case studies involving field tests were analyzed and the results of the reports proved the nematicidal effectiveness of *Trichoderma* spp. both semi-field (D'Errico *et al.*, 2021) and field studies (Fan *et al.*, 2020; Gautam *et al.*, 2021).

Field experiment conducted by Fan *et al.* (2020) in tomato against root knot nematode reported 50% reduction in the gall index of seedlings treated with *T. citrinoviride* over control, which indicated a perceivable decrease of the nematode during the growing seasons. Corresponding to this, soil treatment with *T. harzianum* and *T. viridae* enhanced nematode control effectiveness by reducing nematode galling on tomato by up to 30.8% (Dababat *et al.*, 2006; Goswami *et al.*, 2008). Similarly, antagonistic potential of *T. asperellum* (T-16) and *T. brevicompactum* against *M. incognita* on tomatoes has been proved under the field studies conducted by Affokpon *et al.* (2011). Besides tomato, inoculation of *Trichoderma* spp. led to an increase in the growth and yield of soybean, tobacco, and capsicum in field trials (Goswami *et al.,* 2008). High rhizosphere competency and ease of root colonisation by *Trichoderma* spp. diminishes nematode feeding sites, nematode galls and egg masses from the roots, which prevent the juveniles to penetrate the host roots.

In short-term trials, pre-exposure of nematode-infested soils to T. harzianum preparations led to enhanced development of nematode-infected plants with decreased root-galling index and the number of eggs per g of root. Pre-exposure of *T. harzianum* in long-term trial led to better growth and increased yield in nematode-infected plants but no discernible change in the root galling index was observed (Spiegel and Chet, 1998).

Trichoderma, besides applying individually, their combination with other methods gives more efficient control of the nematodes. Fluensulfone, the chemical nematicide, when applied to guava plant either alone or in conjunction with *T. harzianum*, increased guava productivity by lowering the population of *M. enterolobi* (Benitez *et al*., 2004). Another study conducted by Sharon *et al*. (2011) by combining the fumigant 1,3-D with the commercial formulation of the biocontrol agent, *Trichoderma harzianum* (TH) and an organic fertiliser (OF) on the root-knot nematode, *M. incognita* infesting tomato revealed that the combination of three items had the strongest nematicidal effect. The biggest gains in plant growth and yield came from fumigant combination with TH and OF. Consequently, it was proved that the integrated management of root-knot nematodes using a soil fumigant, a bionematicide (*T. harzianum*), and a source of organic matter was excellent for their reduction.

Synthesis of toxic compounds is the primary trait that distinguishes this organism as a useful biological control agent for *Meloidogyne* spp. This means that *Trichoderma* spp. are the additional option for the management of *Meloidogyne* spp., and among its attributes that make it an effective nematode control agent are its ability to protect plants beforehand, restore the microbial community, and recover the soil structure that has been weakened by intensive agricultural practise, in addition to releasing secondary metabolites that promote plant growth (Sousa *et al*., 2006).

Due to the aforementioned characteristics, *Trichoderma* has been demonstrated as a prudent, consistently efficient, and cost-effective method of treating nematodes. Since combining two (or more) different types of biocontrol agents and chemicals can increase their biocontrol activity, soil treatment with lower amounts of nematicides and *Trichoderma* can be employed in integrated management methods.

5. *Trichoderma* in Bioremediation

Currently, there is a growing concern about the accumulation of heavy metals in agricultural soils due to the indiscriminate use of agrochemicals. Heavy metals are among the most challenging contaminants in agricultural fields as they cannot be easily eliminated. Microorganisms have been recognized for their ability to mitigate the effects of toxic substances through various mechanisms of metal resistance. These mechanisms include the expulsion of metals from cell membranes, intracellular binding by metallothionein proteins, and the production of phytochelatins, which are peptides derived from glutathione (Kamizomo *et al.*, 1989; Presta and Stillman, 1997; Doblera *et al.*, 2000; Tripathi *et al.*, 2007).

The beneficial *Trichoderma* present in soil and belong to the most resistant microorganisms for natural or synthetic chemicals and toxins and are able to reduce some of them. Because of this, *Trichoderma* is an important fungal genus to be explored as a genetic resource to work in process of bioremediation of toxic. *Trichoderma* spp. are play a significantly involved in the breakdown of pesticides, herbicides, insecticides, heavy metals and polycyclic aromatic hydrocarbons by means of biovolatilization, bioaccumulation, biosorption and degradative enzymes/acids (Singh and Khajuria, 2018) and turned into non-toxic chemicals.

5.1. Bio volatilization of pollutants

The process of enzymatic conversation of organic and inorganic compounds of metal (loid)s into their volatile derivatives is called as bio volatilization of pollutants. This was achieved by biomethylation (Mukhopadhyay *et al.*, 2002). Application of chlamydospores of *T. asperellum* SM-12F1 in soil was significantly accelerated the as volatilization from soils and available as content in soils was decreased by 23.7% (Wang, 2015).

5.2. Biosorption

The process of independent ability of biological materials to remove pollutants such as heavy metals from wastewater through metabolically mediated or physico-chemical pathways of uptake is called as biosorption. Dhara Shukla *et al.* (2014) reported that, concentrations of chromium ions at pH 5.5 of chromium solution was drastically noticed the reduction of 97.39% by non-pathogenic species of *Trichoderma*. For removal of Ni(II), Cd(II), and Cr(VI) ions the dried biomass of *Trichoderma* was used (Rahman *et al.*, 2016) and it is one of the option the treatment of toxic heavy metal ions from the industrial effluents as well as natural waters.

5.3. Bioaccumulation

The processes of removal of metals by living cells is called as bioaccumulation, which is achieved from the mechanism of energy dependent metal influx (Ting and Choong, 2009). In this process of mechanism, the isolates of *Trichoderma* have been exploited to accumulation of several heavy metals under *in-vitro* conditions (Zeng *et al*., 2010). Studies have been conducted on application of *T. viride* with copper and recorded the bioaccumulation as a mechanism of copper tolerance during growth. Adding of *T. viride* at 3.4 g L^{-1} of the biomass was effectively removed the 81% of Cu(II) in 72 h (Anand *et al.,* 2006).

Bioaccumulation capacity of isolate *Trichoderma* SP2F1 carried out from the samples of Penchala River which is heavily contaminated from industrial effluents. The viable cells of SP2F1 has effectively removed 19.60 mg g^{-1} of Cu (II) after 168 h inoculation (Ting and Choong, 2009). Which implies that, *Trichoderma* has good potential microbe to remove the metals from the sewage.

Table 1. Mode of treatment with *Trichoderma* and its Purpose

S. No.	Name of the crop	Mode of treatment with Trichoderma	Purpose (Disease/pest controlled/others)	Reference
1	Pigeon pea	Seed treatment (20 g kg^{-1} seed) and soil amendment (20 g per 9 m^2)	Pigeon pea wilt (*Fusarium udum*)	Prasad *et al.,* 2002
2	Cacao	Suspension of *T. harzianum* and *T. virens*	Pod Rot (*Phytophthora palmivora*)	Rina Sriwati *et al.,* 2019
3	Tomato	Soil application (2.15×10^6 spores ml^{-1})	Root knot nematode, *Meloidogyne incognita*	Fan *et al.,* 2020
4	Tomato	Soil pretreatment with Trichoderma peat-bran preparations	Root knot nematode, *Meloidogyne javanica*	Sharon *et al.,* 2001
5	Green gram	Soil application	Root knot nematode, *Meloidogyne javanica*	Mukhtar *et al.,* 2021
6	Pea	Soil application along with botanical powders (3% of soil weight)	Root knot nematode, *Meloidogyne incognita*	Nagdi *et al.,* 2019
7	Tomato	Soil application of culture suspension (10^6 spores ml^{-1})	Root knot nematode, *Meloidogyne incognita*	Khan *et al.,* 2018

6. Conclusion

The interaction between *Trichoderma*, plants, and pathogens is a highly dynamic and interdependent system. *Trichoderma* employs various intricate methods for biocontrol, targeting a broad range of biotic threats including fungi, bacteria, insects, and nematodes, as well as abiotic stresses like adverse environmental conditions. Moreover, it holds the potential to serve as a foundation for novel phytoremediation technologies due to its resilience against various toxic substances, whether organic or inorganic. It can also enhance a plant's ability to withstand stress factors in the presence of pollutants. Importantly, these solutions align with the principles of sustainable agriculture.

7. Way Forward

Future research endeavours should focus on the identification of versatile *Trichoderma* strains suitable for effective incorporation into diverse agricultural settings and in conjunction with other microbial bioagents. Additionally, investigations should explore into how *Trichoderma* can support plants in coping with the challenges posed by climate change, including issues like heat stress, drought, and changing disease patterns. Further exploration is warranted into the means of delivering *Trichoderma* strains and their metabolites through nano formulations and biostimulants within plant ecosystems. Further, the research has to be aimed at the accurate and quick estimation of the risk of using BCAs based on Trichoderma, their toxicity and ecotoxicity not only in in vitro conditions, but above all in natural growing conditions in vitro and in situ, before being introduced to the market as biocontrols, biostimulations or bioremediation preparations.

8. References

Affokpon, A., Coyne, D.L., Htay, C.C., Agbede, R.D., Lawouin, L. and Coosemans, J. 2011. Biocontrol potential of native Trichoderma isolates against root-knot nematodes in West African vegetable production systems. Soil Biology and Biochemistry 43: 600- 608.

Anand, P., Isar, J., Saran, S., & Saxena, R.K. (2006). Bioaccumulation of copper by Trichoderma viride. Bioresource technology 97(8): 1018-1025.

Ayyandurai, M. R., Akhila, K., Manonmani, M., Theradimani, S and Vellaikumar. 2021. Phytostimulation and growth promotion activities of Trichoderma spp. on groundnut (Arachis hypogaea L.) crop. Journal of Applied and Natural Science 13(4): 1172-1179.

Bader, A.N., Salerno, G.L., Covacevich, F. and Consolo, V.F. 2020. Native Trichoderma harzianum strains from Argentina produce indole-3 acetic acid and phosphorus solubilization, promote growth and control wilt disease on tomato (Solanum lycopersicum L.) Journal of King Saud University Science 32: 867-873.

Benitez, T., Rincon, A.M., Limon, M.C. and Codon. A.C. 2004. Biocontrol mechanisms of Trichoderma strains. International Microbiology 7(4): 249-260.

Bissett, J., Gams, W., Jaklitsch, W. and Samuels, G.J. 2015. Accepted Trichoderma names in the year 2015. IMA Fungus 6: 263–295.

Bononi, L., Chiaramonte, J. B., Pansa, C.C., Moitinho, M. A. and Melo, I. S. 2020. Phosphorus-solubilizing Trichoderma spp. from Amazon soils improve soybean plant growth. Scientific Reports 10: 2858.

Contreras-Cornejo H.A., Orozco-Granados, O., Ramírez-Ordorica, A., García-Juárez, P., López-Bucio, J. and Macías-Rodríguez, L. 2022. Light and mycelial injury influences the volatile and non-volatile metabolites and the biocontrol properties of Trichoderma atroviride. Rhizosphere 22: 100511.

D'Errico, G., Mormile, P., Malinconico, M., Bolletti Censi, S., Lanzuise, S., Crasto, A., Woo, S.L., Marra, R., Lorito, M. and Vinale, F. 2021. Trichoderma spp. and a carob (Ceratonia siliqua) galactomannan to control the root-knot nematode Meloidogyne incognita on tomato plants. Canadian Journal of Plant Pathology 43: 267 - 274.

Dababat, A.A, Sikora, R.A. and Hauschild, R. 2006. Use of Trichoderma harzianum and Trichoderma viride for the biological control of Meloidogyne incognita on tomato. Communications in agricultural and applied biological sciences 71(3): 953 – 961.

Dhara Shukla, Padma S. Vankar. 2014. Role of Trichoderma Species in Bioremediation Process: Biosorption Studies on Hexavalent Chromium, Editor(s): Vijai K. Gupta, Monika Schmoll, Alfredo Herrera-Estrella, R.S. Upadhyay, Irina Druzhinina, Maria G. Tuohy, Biotechnology and Biology of Trichoderma, Elsevier 405-412.

Doblera R, Sanera M. and Bachofena R. 2000. Population changes of soil microbial communities induced by hydrocarbon and heavy metal contamination. Bioremediation Journal 4:41–56.

Fan, H., Yao, H., Zhao, D., Zhu, X., Wang, Y., Liu, X., Duan, Y. and Chen, L. 2020. Isolation and effect of Trichoderma citrinoviride Snef 1910 for the biological control of root-knot nematode, Meloidogyne incognita. BMC Microbiology 20: 299.

Fan, H., Yao, M., Wang, H., Zhao, D., Zhu, X., Wang, Y., Liu, X., Duan, Y. and Chen L. 2020. Isolation and effect of Trichoderma citrinoviride Snef1910 for the biological control of root-knot nematode, Meloidogyne incognita. BMC Microbiology 20: 299.

Gautam, S.B., Chand, R., Singh, S.K., Chandra, S. and Maurya, M.K. 2021. Efficacy of Trichoderma viride and T. harzianum culture filtrate on egg hatching and larval mortality of root-knot nematode, Meloidogyne incognita. Annals of Plant Protection Sciences 29: 241 - 245.

Goswami, J., Pandey, R.K., Tewari, J. and Goswami, B. 2008. Management of root knot nematode on tomato through application of fungal antagonists, Acremonium strictum and Trichoderma harzianum. Journal of Environmental Science and Health 43(3): 237 – 240.

Gravel, V., Antoun, H and Tweddell, R.J. 2007. Growth stimulation and fruit yield improvement of greenhouse tomato plants by inoculation with Pseudomonas putida or Trichoderma atroviride: Possible role of indole acetic acid (IAA). Soil Biology & Biochemistry 39: 1968-1977.

Harman, G.E. 2000. Myths and dogmas of biocontrol. Changes in perceptions derived from research on Trichoderma harzianum T-22. Plant Disease 84: 377-393.

Harman, G.E. 2011. Trichoderma-not just for biocontrol anymore. Phytoparasitica 39: 103-108.

Harman, G.E., Bjorkman, T., Ondik, K and Shoresh, M. 2008. Changing paradigms on the mode of action and uses of Trichoderma spp. for biocontrol. Outlooks Pest Management 19: 24-29.

Harman, G.E., Howell, C. R., Viterbo, A., Chet, I. and Lorito, M. 2004. Trichoderma species—opportunistic, avirulent plant symbionts. Nature Reviews Microbiology 2:43.

Hua L, Zeng H, He L, Jiang Q, Ye P, Liu Y, Sun X. and Zhang M. 2021. Gliotoxin Is an Important Secondary Metabolite Involved in Suppression of Sclerotium rolfsii of Trichoderma virens T23. Phytopathology 111(10): 1720-5.

Inayati, A., Setyowati, L., Aini, L. Q. and Yusnawan, E. 2021. Plant growth promoter produced by Trichoderma virens and its effect on mungbean (Vigna radiata (L.) Wilczek) seedling. IOP Conf. Series: Earth and Environmental Science. 803. 012013.

Javeed, M.T., Al-Hazmi, A.S. and Molan, Y.Y. 2016. Antagonistic effects of some indigenous isolates of Trichoderma spp. against Meloidogyne javanica. Pakistan Journal of Nematology 34(2): 183-191.

Kabdwal, B.C., Sharma, R., Tewari, R. et al. 2019. Field efficacy of different combinations of Trichoderma harzianum, Pseudomonas fluorescens, and arbuscular mycorrhiza fungus against the major diseases of tomato in Uttarakhand (India). Egyptian Journal of Biological Pest Control 29. 1.

Kamizomo, A., Nishizawa, M., Teranishi, Y., Murata, K. and Kimura, A. 1989. Identification of a gene conferring resistance to zinc and cadmium in the yeast Saccharomyces cerevisiae. Molecular Genetics and Genomics 219: 161–167.

Khan, M.R., Ahmad, I. and Ahamad, F. 2018. Effect of pure culture and culture filtrates of Trichoderma species on root-knot nematode, Meloidogyne incognita infesting tomato. Indian Phytopathology 71: 265–274.

Loc, N.H., Huy, N.D., Quang, H.T., Lan, T.T. and Thu Ha, T.T. 2020. Characterisation and antifungal activity of extracellular chitinase from a biocontrol fungus, Trichoderma asperellum PQ34. Mycology 11(1): 38-48.

Mach, R.L., Peterbauer, C.K., Payer, K., Jaksits, S., Woo, S.L., Zeilinger, S., Kullnig, C.M., Lorito, M. and Kubicek, C. P. 1999. Expression of two major chitinase genes of Trichoderma atroviride (T. harzianum P1) is triggered by different regulatory signals. Applied and environmental microbiology 65(5): 1858–1863.

Manigundan, K., Sakthivel, K., Gautam, R.K., Krishna Kumar, Anantharaj, A., Velmurugan, A., Singh, P. K and Dam Roy, S. 2016. Characterization of species from vegetable and spice rhizospheres of Andaman Islands for broad spectrum antagonism and plant growth promotion. Journal of Environmental Biology 37: 341-347.

Mastouri, F., Bjorkman, T. and Harman, G.E. 2010. Seed Treatment with Trichoderma harzianum alleviates biotic, abiotic, and physiological stresses in germinating seeds and seedlings. Phytopathology 100 (11): 1213-1221.

Mercado, J.A., Barceló, M., Pliego, C., Rey, M., Caballero, J.L., Muñoz-Blanco, J., Ruano-Rosa, D., López-Herrera, C., de Los, Santos B., Romero-Muñoz, F., Pliego-Alfaro, F. 2015. Expression of the β-1, 3-glucanase gene bgn13. 1 from Trichoderma harzianum in strawberry increases tolerance to crown rot diseases but interferes with plant growth. Transgenic research 24:979-89.

Mukhopadhyay, R., Rosen, B.P., Phung, L.T., Silver, S. 2002. Microbial arsenic: from geocycles to genes and enzymes. FEMS Microbiology Reviews 26:311-325.

Mukhtar, T., Khan, M. and Aslam, M.N. 2021. Bioefficacy of Trichoderma species against javanese root-knot nematode, Meloidogyne javanica, in green gram. Gesunde Pflanzen 73: 265-272.

Nagdi, W.M.A., Youssef, M.M.A., El-Khair, H.A. and Elgawad, M.M. 2019. Effect of certain organic amendments and Trichoderma species on the root-knot nematode, Meloidogyne incognita, infecting pea (Pisum sativum L.) plants. Egyptian Journal of Biological Pest Control 29: 75.

Oszust, K., Cybulska, J. and Frąc, M. 2020. How do Trichoderma genus fungi win a nutritional competition battle against soft fruit pathogens? A report on niche overlap nutritional potentiates. International Journal of Molecular Sciences 21(12): 4235.

Poveda., J. 2020. Trichoderma parareesei favors the tolerance of rapeseed (Brassica napus L.) to salinity and drought due to a chorismate mutase. Agronomy 10:118.

Prasad R.D., R. Rangeshwaran, S.V. Hegde and C.P. Anuroop. 2022. Effect of soil and seed application of Trichoderma harzianum on pigeonpea wilt caused by Fusarium udum under field conditions. Crop Protection 21(4): 293-297.

Presta, A., Stillman and M.J. 1997. Incorporation of copper into the yeast Saccharomyces cerevisiae. Identification of Cu (I)- Methallo thionein in intact yeast cells. Journal of Inorganic Biochemistry 66: 231–240.

Qi, W. and Zhao, L. 2013. Study of the siderophore□producing Trichoderma asperellum Q1 on cucumber growth promotion under salt stress. Journal of basic microbiology 53(4): 355-64.

Rahman, N.A, Mohammad, Shahadat, Fatehah Mohd Omar, Ann Won Chew and Mohd Omar Ab Kadir. 2016. Dry Trichoderma biomass: biosorption behavior for the treatment of toxic heavy metal ions. Desalination and Water Treatment 57(28): 13106-13112.

Reino, J.L., Guerrero, R.F., Hernández-Galán, R. and Collado, I.G. 2008. Secondary metabolites from species of the biocontrol agent Trichoderma. Phytochemistry Reviews 7: 89-123.

Rina Sriwati, Tjut Chamzurni, Loekas Soesanto and Munazhirah. 2019. Field Application of Trichoderma suspension to control Cacao Pod Rot (Phytophthora palmivora). Journal of Journal of Agricultural Science 41(1): 175-182

Shalini, Singh. and Robinka, Khajuria. 2018. Trichoderma spp. in bioremediation current status and scope. Microbial Cell Factories. CRC Press, 1st Edition. pp20.

Sharon, E., Bareyal, M., Chet, I., Herreraestrella, A., Kleifeld, O. and Spiegel, Y. 2001. Biological control of the root-knot nematode Meloidogyne javanica by Trichoderma harzianum. Phytopathology 91: 687–693.

Sharon, E., Chet, I. and Spiegel, Y. 2011. Trichoderma as a biological control agent. In: Biological Control of Plant-Parasitic Nematodes: Building Coherence Between Microbial Ecology and Molecular Mechanisms, Progress in Biological Control; Davies, K., Spiegel, Y. (eds.). Springer: Dordrecht, The Netherlands, pp 183 - 201.

Singh, S.P., Singh, H.B., Singh, D.K. and Amitava, R. 2014. Effects of Trichoderma on tomato growth and disease resistance. Egyptian Journal of Biology 16: 29-38.

Sood, M., Kapoor, D., Kumar, V., Sheteiwy, M.S., Ramakrishnan, M., Landi, M., Araniti, F. and Sharma, A. 2020. Trichoderma: The "secrets" of a multitalented biocontrol agent. Plants 9(6): 762.

Sousa, R.M.; Nogueira, M.S., Lima, I.M., Melarato, M. and Dolinski, C.M. 2006. Management of the guava root-knot nematode in São João da Barra, Brazil, and report of new hosts. Nematologia Brasileira, Brasilia 30(2): 165-169.

Spiegel, Y. and Chet, I. 1998. Evaluation of Trichoderma spp. as a biocontrol agent against soilborne fungi and plant-parasitic nematodes in Israel. Integrated Pest Management Rev 3: 169 - 175.

Ting, A.S. and Chai, J. 2015. Chitinase and β-1, 3-glucanase activities of Trichoderma harzianum in response towards pathogenic and non-pathogenic isolates: Early indications of compatibility in consortium. Biocatalysis and agricultural biotechnology 4(1):109-13.

Ting, A.S.Y. and Choong, C.C. 2009. Bioaccumulation and biosorption efficacy of Trichoderma isolate SP2F1 in removing copper (Cu(II)) from aqueous solutions. World Journal of Microbiology and Biotechnology 25: 1431–1437.

Tripathi, R.D., Srivastava, S., Mishra, S., Singh, N., Tuli, R., Gupta, D.K. and Maathuis, F.J.M. 2007.Arsenic hazards: strategies for tolerance and remediation by plants. Trends in Biotechnology 25:158-165.

Vinale, F., Nigro, M., Sivasithamparam, K., Flematti, G., Ghisalberti, E.L., Ruocco, M., Varlese, R., Marra, R., Lanzuise, S., Eid, A. and Woo, S.L. 2013. Harzianic acid: a novel siderophore from Trichoderma harzianum. FEMS microbiology letters (2):123-9.

Wang, X.R., Su, S.M., Zeng, X.B., Bai, L.Y., Li, L.F., Ran, D.U.A.N., Wang, Y.N. and Wu, C.X. 2015. Inoculation with chlamydospores of Trichoderma asperellum SM-12F1 accelerated arsenic volatilization and influenced arsenic availability in soils. Journal of Integrative Agriculture 14(2): 389-397.

Weindling, R. 1932. Trichoderma lignorum as a parasite of other soil fungi. Phytopathology 22: 837-845.

Yadav, M., Divyanshu, K., Dubey, M.K. et al. 2023. Plant growth promotion and differential expression of defense genes in chilli pepper against Colletotrichum truncatum induced by Trichoderma asperellum and T. harzianum. BMC Microbiology 23: 54.

Zeng X, Su S, Jiang X, Li L, Bai L and Zhang Y. 2010. Capability of pentavalent arsenic bioaccumulation and biovolatilization of three fungal strains under laboratory conditions. Clean: Soil, Air Water 38: 238–241.

Zhang, S., Gan, Y. and Xu, B. 2016. Application of plant-growth- promoting fungi Trichoderma longibrachiatum T6 enhances tolerance of wheat to salt stress through improvement of antioxidative defense system and gene expression. Frontiers in Plant Science 7: 1405.

8

Bio-Rational Approaches for Pest Management in Organic Farming

R.S. Ramya[1], R.R. Rachana[1], K.T. Shivakumara[1], B.S. Gotyal[1] R. Varshney[1] and M. Ranjith[2]

[1]*ICAR-National Bureau of Agricultural Insect Resources, Bengaluru-560 024 Karnataka, India*

[2]*Central Integrated Pest Management Centre, Jaivik Bhavan, Bengaluru-560 067 Karnataka, India*

Abstract

The concept of insect pest management and the strategies employed have evolved over the time. With the increased emphasis on safe agricultural practices and elevated demand for safe and healthy food, the various solutions for pest management have shifted more towards use of biocontrol agents. Bio- rational pest management methods have now evolved as important pillar of the integrated pest management to replace the synthetic insecticides and also other associated advantages. They include the use of botanical pesticides, insect growth regulators, pheromones and allelochemicals. The major positive attributes associated with all of them are their low mammalian toxicity, safety to beneficial and non-target organisms, higher specificity, low risk of development of resistance and low environmental persistence. Though bio-rational insecticides are considered to be safe, in an organic ecosystem, it should be done with caution because over reliance on them for longer periods may facilitate the target insect pests to develop some kind of resistance against them. So, proper deliberations on the positive and negative attributes of bio rational pest management strategies should be done so as to facilitate their long term usage in organic farming.

Keywords: *Bio-rational, botanicals, growth regulators, pheromones*

1. Introduction

Sustainable management of insect pests poses a major challenge to the farmers who adopt organic agricultural practices. Insects are notorious for easily finding ways to overcome the various pest control tactics. Most of the conventional chemical pesticides which are used for the insect pest management have broad toxicity and have found to have deleterious effects on non-target organisms including beneficial insects, soil arthropods and water dwelling organisms. However, there are instances when the cultural practices will not be sufficient to prevent or control the crop pests. In such a scenario, the utilisation of biological, botanical or mineral inputs have been allowed by many of the agencies associated with organic farming like the United States Department of Agriculture (USDA) National Organic Program. Hence, the use of non-synthetic and safe alternatives to chemical pesticides is permitted in the organic agriculture. The pest management practices in organic agriculture is focused mainly on sustainably managing the insect population and not on their complete elimination.

The pest management scenario has undergone drastic changes since the commercialization of first synthetic insecticidal compounds in the 1940s. The current world situation demands safer and sustainable pest management tactics especially in the organic farming systems. One of the major pillars of such safe and sustainable biocontrol tactics is the use of biorational methods of pest management.

"Bio rational" as a term was first used by Djerassi *et al.* (1974). The term when it was first used meant the insect pest management using insect pheromones, hormones and hormone anatagonists. They were associated with high specificity, lethal at low doses and also considered safe to beneficial as well as other non-target organisms. Later on, Ware (1989) included chemical analogues of natural biochemical compounds such as pheromones and insect growth regulators as "bio rational". Uri (1998) opined that pest management options which are effective in Integrated Pest Management (IPM) and have no adverse effects on non-target organisms and also environment can be termed as "bio-rationals". The chemicals allowed to be used for pest management in organic cultivation are strictly regulated by the NPOP (National Programme for Organic Production) for India. Similarly, specialized organisations exist in other countries to regulate the chemicals used in organic farming. Hence, the bio-rational strategies of pest management include the use of botanical products, insect growth regulators, pheromones and allelochemicals. Each of them are discussed in detail bellow.

2. Components of Bio-rational Strategies

2.1. Use of plant-based products

Botanical pesticides are chemicals with insecticidal properties extracted from plants and are mostly slow-acting and considered environmentally safe. Many botanical pesticides are being used in organic farming (Table 1).

The isolation, structure, biological activity, toxicology and agricultural applications of natural products and synthetic compounds were extensively reviewed by Duke (1995). The low residual toxicity and the ecological safety of crude extracts and commercial formulations from plants like tobacco, pongamia and neem make them ideal candidates for the management of insect pests in organic farming. In addition to this, options like homemade herbal teas and other plant extracts which have some pest repellent properties are also approved to be used in organic agriculture. Mehta *et al.* (2005) evaluated the crude extracts of drupes of *Melia azedarach*, leaves of *Lantana camara* and *Artimisia brevifolia* and roots of *Rumex nepalensis* and found them to be efficient against the lepidopterous larval complex in cabbage. Non synthetic extracts and derivatives of neem can be resorted to for pest management in an organic farming plan. The neem products are mostly prepared from the seeds of neem tree, *Azadirachta indica*. These products are generally formulated as Emulsifiable Concentrates. Neem products are known to inhibit metamorphosis, act as a repellent, antifeedant and also as a stomach poison.

Padmanabam (1997) recorded high per cent mortality (44.80%) of the white grub, *Leucopholis burmeisteri* after treatment with Karanj cake. The extract which was made by taking equal quantity of *Parthenium* and neem leaves was found to be effective in reducing the *H. armigera* damage in chilli when sprayed at 20 ml/10 litres of water (Kumar and Yadav, 2007). Studies done by Haldhar *et al.* (2014) proved that the organic IPM system which consisted of spraying neem oil at 20 DAS, installation of pheromone trap (10 ha^{-1}) at 42 DAS, spray of tumba fruit extract (TFE 5%) at 50 DAS and spray of spinosad 46 SC at 60 DAS was the most effective and economical approach for the management of *Aphis gossypii*, *Diaphania indica*, *Epilachna vigintiopunctata* and *Bactrocera cucurbitae* in musk melon.

However, while using these plant based products, some things have to be kept in mind as they may result in some deleterious effects on some stages of beneficial insects like predators and parasitoids. For example, even though neem products are found to be safe to adults of many beneficial insects, eggs of coccinellids and also to spiders, nymphal and larval instars were observed to be sensitive under laboratory conditions. However, the side effects of neem on the immature stages of beneficial insects are found to be mostly non-significant in field conditions. Smitha (2002) evaluated the safety of plant extracts and bioagent formulations against coccinellids by comparing the natural activity of predators and have found them to be reasonably safe to coccinellids. Similar evaluation studies were done by Varghese (2003) who confirmed the safety of several organics and botanicals to coccinellids and predatory mites. Several indigenous plant products and cow urine were tested by Kumar (2004) for their safety against natural enemies in chilli ecosystem and found the indigenous sprays to be safe to coccinellids and *Chrysoperla* spp.

Table 1. Botanical biopesticides, target pests and their mode of action

Sl. No.	Botanicals	Target pests	Mode of action
1.	Pyrethrum	Aphids, leafhoppers, spider mites etc.	Disrupts sodium and potassium ion exchange and interrupts the transmission of nerve impulses
2.	Neem	Cutworms, Armyworms etc.	Acts on digestive, nervous and hormonal systems
3.	Rotenone	Aphids, spider mites etc.	Inhibits cellular respiration
4.	Ryania	Codling moth, potato aphids, thrips etc.	Prevents muscles from contracting, causes Paralysis
5.	Sabadilla	Grasshoppers, codling moth, armyworms, aphids etc.	Affects nerve cell membrane action, causes loss of nerve function, paralysis
6.	Garlic oil	Leafhoppers, whiteflies, aphids	Repellent or toxic and antifeedant on insects
7.	Essential oils	All types of insects, bacteria, fungi and nematodes	Destroy cell membrane

(*Source:* Costa, 2016)

2.2. Use of insect growth regulators

These include compounds which do not directly cause mortality or prove lethal to insects, but affect the development of insects. The mechanism can be either by inhibition of the chitin synthesis or by affecting the secretion and action of insect growth hormones (Table 2). Dhaliwal and Arora (2003) has categorized the insect growth regulators into two classes based on their mode of action.

2.2.1. Insect cuticle based insecticides

Insect cuticle is mainly composed of chitin which is made up of many subunits of N-Acetyl-D-glucosamine. Insecticides which can interfere with the biosynthesis and biodegradation of cuticle and also the cuticular sclerotization can be very good candidates for pest management in organic agriculture as they are target specific and safer to humans and other non-target organisms (Dhaliwal and Arora, 2003). Many natural molecules or their analogues which have this mode of action have been commercialized and will be ideal for use in organic farming. Insect cuticle based insect growth regulator insecticides can be further categorised into:

2.2.1.1. Chitin synthesis inhibitors

Chitin synthesis inhibitors (CSI) usually interfere with the polymerisation step of chitin synthesis which is catalysed by the enzyme, chitin synthase. CSIs interfere with the permeability of the membrane and block the availability of chitin at the active site of chitin synthase (Dhaliwal and Arora, 2003). When chitin synthesis is inhibited, it affects the development of new exoskeleton in larval instars during moulting stage. Insects without normal exoskeleton or with abnormal exoskeleton cannot survive for long and will die of dehydration or desiccation. Compounds which are having the chitin synthesis attributes and are commonly used for insect pest control are benzoyl ureas, triazine/pyramidine derivatives, buprofezin and plumbagin. When these CSIs are applied on foliage with the intention of killing insect pests, they are ingested by only the insect pests and not by the parasitoids or bees. Hence, the CSIs are considered safe to the non-target organisms. Buprofezin is the most common CSI which is recommended and used against several insect pests. It affects the moulting process of larvae, inhibits ovisposition in adults and also cause sterility. It is found to be effective against sucking pests and mites in rice, cotton, fruits, vegetables and many other field crops and ornamental plants (Sontakke *et al.*, 2013). Benzoyl phenyl urea CSIs like lufenuron, triflumuron, diflubenzuron, chlorfluazuron and and teflubenzuron are also being used against several ecocnomically important pest species (Gogi *et al.*, 2006; Saeed *et al.*, 2012).

2.2.1.2. Cuticle sclerotization inhibitors

In insects, different types of proteins and phenolic compounds are covalently linked by a process known as sclerotization which is regulated by the hormone, Bursicon (Pedigo, 2003). Either the inhibition or the acceleration of the sclerotization process can cause improper hardening of the cuticle which will expose the insects to dehydration or desiccation and ultimate death. Compounds having sclerotization inhibition characters which can be employed for pest management are MON-0585 (only commercialized sclerotization inhibitor developed by Monsanto), α-methyl DOPA and Cryomazine (Dhaliwal and Arora, 2003).

2.2.2 Insect growth hormone based insecticides

The growth and development of insects is governed by several hormones which are secreted from the glands of endocrine system. These are Brain Hormone, Moulting Hormone, Juvenile Hormone, Eclosion Hormone and Tanning Hormone (Pedigo, 2003). These hormones are ideal candidates for the development of biorational insecticides which will interfere with the active sites of these enzymes or their synthesis and eventually result in abnormalities in the growth and development of the insects and hence mortality (Ishaaya and Horowitz, 2004).

2.2.2.1. Insecticides based on moulting hormone

The insect Moulting Hormones (MH) are secreted by the prothoracic glands and they regulate various mechanisms of molting, reproduction, growth and maturation. These hormones constitute of steroidal peptides like ecdysone, ecdysteroids and ecdysterone (Dhaliwal and Arora, 2003). Phytoecdysteroids which are steroids extracted from plants can create the moulting hormone imbalance in insects and thus abnormal growth and development which will ultimately cause mortality. In addition to Phytoecdysteroids, there are molecules that act as moulting hormone agonists which can disrupt the ecdysteroid receptor and can cause abnormal growth in insects (Harada *et al.*, 2011). These MH agonists mimic the moulting hormone and compel the insects to undergo immature moulting which will lead to cessation of feeding and eventually death. They can also cause increased mortality of eggs, reduce fecundity and reproduction rate. The commercially available moulting hormone agonists are Tebufenozide, Chromafenozide, Halofenozide and Methoxyfenozide which are Bisacylhydrazines (BSH). There are also moulting hormone antagonists which can inhibit the production of prothoracicotropic hormone and thus, disrupt the action of ecdysone. There are two moulting hormone antagonists based insecticides available commercially, i.e., Diofenolan and Azadirachtin (Dhadialla *et al.*, 1998).

2.2.2.2. Insecticides based on juvenile hormone

Juvenile hormones regulate several physiological processes in insects like diapause, development, embryogenesis, metamorphosis, molting, reproduction, etc. (Li *et al.*, 2007; Parthasarathy *et al.*, 2008). When compared to the natural juvenile hormones, the stability of juvenile hormone analogues is more. There are many such JH analogues that are commercialized. Examples include methoprene, hydroprene, kinoprene, triprene, fenoxycarb, pyriproxyfen, epofenonane and dayoutong. There are also some plant products which mimic the juvenile hormone and have similar activity. Examples are juvabione and its analogue, dehydrojuvabione which are isolated from balsam fir tree, *Abies balsamea* (Dhaliwal and Arora, 2003). The most commonly used Juvenile hormone mimics in agroecosystems for plant protection against insect pests include fenoxycarb and pyriproxyfen.

There are also other compounds which inhibit the natural JH production, cause the degradation of JH or destroy the corpus allatum in insects. These chemicals are known as precocenes or JH antagonists. The early examples are precocene-I, prococene-II and precocene-III which were isolated from *Ageratum houstonianum* and *A. conyzoides.* When these precocenes enter the body of insects, they can interrupt the function of corpora allata, cause premature metamorphosis, cause sterility in females, facilitate diapause and can also disrupt the synthesis of sex pheromones (Dhaliwal and Arora, 2003).

Table 2. List of insect growth regulators, active ingredient and target insects

Category	Major group	Active ingredient	Mode of action/effects	Target insects	References
Chitin Synthesis Inhibitor	Benzoylphenyl Urea	Diflubenzuron,Triflumuron, Teflubenzuron, Hexaflumuron, Chlorfluazuron,Chlorbenzuron, Dichlorbenzuron, Flucycloxuron, Flufenoxuron, Bistrifluron, Lufenuron, Novaluron, Noviflumuron, Penfluron, Flucycloxuron	Stomach toxicant/ ovicidal action/chitin synthesis inhibition/ malformed cuticle/larval and pupal abnormalities	Coleoptera, Diptera, Lepidoptera Eg. *Trichoplusia ni*, *Spodoptera litura*, eggs of *Helicoverpa armigera* and *S. litura*, *Cydia pomonella*	Khater (2003), Dhaliwal and Arora (2003)
	Triazine derivatives	Cyromazine	Growth inhibition, expansion of the body walls, abnormal internal growth and development	Dipterans, leaf miners	O'Brien and Fahey (1991)
	Pyrimidine derivative	Buprofezin	Inhibition of biosynthesis of chitin	Most of the sucking insects especially hemipterans	Dhaliwal and Arora (2003)
		Plumbagin (Derived from roots of *Plumbago capensis*)	Inhibits ecdysis, ovicidal effect, inhibits ecdysteroid biosynthesis, disrupt mating and reproduction	Lepidopterous pests, like *Pectinophora gossypiella*, *Helicoverpa armigera* etc.	
Sclerotization inhibitors		Ditertiary butyl alcohol (MON-0585)	Reduction in phenoloxidase activity, malformations in the exoskeleton	Mosquito and other Dipterans	Dhaliwal and Arora (2003)

MH agonists	Ecdysteroid agonist	Tebufenozide	Bind to the ecdysteroid receptors, thereby accelerating the molting process	Lepidopteran insects, especially, *Spodoptera* sp., *Helicoverpa zea* and *Manduca sexta* in cotton, cereal, maize, rice and vegetables	Smagghe and Degheele (1992); Palli *et al.* (1996)
		Chromafenozide	Contact and stomach Action, interaction with ecdysteroid receptor proteins, creates Hyperecdysonism, synthesis and apolysis of cuticle but no sclerotization		Retnakaran *et al.* (1997)
		Halofenozide	Act as MACs (Molt Accelerating Compounds); Mimicking action of ecdysone, premature molting, cessation of feeding and ultimately death of insects	Lepidoptera, Coleoptera and some Hemiptera	Potter (1998)
		Methoxyfenozide	Accelerating the molting process, abnormal growth and development	Lepidopteran insects but more effective bollworms and diamondback moth	Smagghe and Degheele (1998)

MH antagonists	Ecdysone antagonist	Diofenolan	Delay in larval-larval and larval–pupal ecdysis, deformed pupae, inhibition of adult emergence	Lepidopteran pests and scale insect's in fruit, field crops etc.	Streibert *et al.* (1994); Singh and Kumar (2011)
		Azadirachtin	Inhibition of PTTH Production, antifeeding and antioviposition effects, inhibit the development of egg, larvae and pupae, induce sterility in adults	Whiteflies, aphids, thrips, beetles, mealybugs, leaf miners etc.	
JHAs	Juvenile hormone agonists	Hydroprene Methoprene	Juvenile hormone mimic, abnormal growth, development and reproduction	*Culex pipiens,* whiteflies, aphids, mealybugs and scales	Hamaidia and Soltani (2014)
		Fenoxycarb Pyriproxyfen	Ecdysis inhibition, inhibits hatching, mimics juvenile hormone action on molting and reproduction, larvicide, ovicide	Cockroaches, fleas, stored product pests, ants, mosquito larvae, scale insects etc.	Richardson and Lagos (2007)

AJHAs	Juvenile hormone antagonists	Precocene-I, Prococene-II Precocene-III	Inhibits juvenile hormone biosynthesis and ovarian activation; disrupt vitellogenesis, delays moulting, inhibits melanation and sclerotization	Majority of insects belonging to Hemiptera, Orthoptera, Dictyoptera, Coleoptera, Lepidoptera etc.	Eid *et al.* (1988)

2.3. Use of pheromones and allelochemicals

Insects mostly rely on chemical communication for inter and intra specific dialogues. The chemicals facilitating inter and intra specific communication are known as semiochemicals. They cause some kind of behavioural responses between same species or two different species so as to accomplish various purposes like finding food and mates, avoiding natural enemies and competition (Ayasse, 2010). Semiochemicals are utilised for designing of several biorational approaches which can cause sustainable management of insect pests (Meyer, 2006). The semiochemicals cause two types of changes in the receiving insect's behaviour. It can be short term and reversible or long term and irreversible. The insect pest management approaches based on semiochemicals utilise mostly the short term reversible changes caused by them. The important reversible changes are attraction, aggregation and repellence (Pedigo, 2003; Chapman, 2013). Semiochemicals are classified into two groups, intraspecific semiochemicals (pheromones) and interspecific semiochemicals (allelochemicals). Pheromones are of different types based on the function they perform, for eg. sex pheromones, alarm pheromones, trail pheromones, epideictic pheromones and aggregation pheromones.

Sex pheromones are mainly employed to interrupt the pest ecology so as to reduce the pest population in a sustainable way. They can be either used for mating disruption or for mass trapping. Mating disruption is achieved by high ambient concentration of the pheromone so as to confuse the males or to mask the trails of calling females (Stelinski *et al.,* 2009; Vassiliou, 2009). Mass trapping using pheromones cause attraction and aggregation of insects and trapping them in specific traps (Landolt *et al.,* 2012; Chen *et al.,* 2014; Hampton *et al.*, 2014). One of the best examples where mating disruption has been successfully used for insect pest management is in the case of Codling moth, *Cydia pomonella* (Stelinski *et al.,* 2009). In attract and kill technique, the pheromone is co-administered along with an insecticide. The insects attracted to the pheromone will be killed because of the insecticide. Male-Annihilation technique is used when male attracting pheromone is employed which is widely used for the control of fruit flies (Pedigo, 2003). Alarm pheromones also have great potential to be used in organic farming systems for sustainable and eco-friendly management of insect pests of fruits, vegetables, ornamentals, etc. (Verheggen *et al.,* 2010).

Pheromones are being largely employed for detection and monitoring of pink bollworm (*Pectinophora gossypiella*), American bollworm (*Helicoverpa armigera*) and fruit flies (*Bactrocera zonata*, *B. dorsalis*, *B. cucurbitae*). Use of pheromones for pest management have an edge over synthetic insecticides as pheromones are species specific and will not cause any adverse effects on the non-target organisms. They are also environmentally sustainable as they are non-persistent and don't leave any harmful residues, possess low mammalian toxicity, and required in less quantity and there are less chance of the insect pest developing resistance against them.

Allelochemicals are plant metabolites that will facilitate chemical communication between plants and insects. These allelochemicals have either antixenotic or antibiotic effects which will help in imparting resistance to the plants against the infesting insect pests (Pedigo, 2003). Allelochemicals can act as kairomones, allomones or synomones depending on the reaction that they elicit in the organism involved. Allomones can be utilised for repelling the insect pest species so as to feed or oviposit on the crops and thus reducing crop damage. Kairomones can be utilised for attracting the insect pests into trap crops or food baits. Synomones can also be employed in insect pest management by facilitating the natural enemies in their search for hosts.

3. Conclusion

It is a fact beyond argument that biorational insecticides are safer, target specific, cause low residues and are considered to be not causing any deleterious effects on the non-target organisms. But their use in organic agriculture also should be judicious and need dependent because even though the chances of them causing harmful effects on beneficial insects like predators, natural enemies etc. are less, there is a necessity to use and recommend them with caution. We should also use the biorational insecticides with restraint as over reliance on them for longer periods may facilitate the target insect pests to develop some kind of resistance against them. In order to ensure the long-term use of biorational insecticides, it is important to carefully consider and strike a balanced approach between their advantages and disadvantages.

4. Way Forward

Biorational pesticides, which include biopesticides, semiochemicals, and other environmentally friendly pest control products, offer several advantages over conventional chemical pesticides. To further advance the use of biorational pesticides, research and development efforts should focus on enhancing their efficacy, expanding their range of target pests, and improving their compatibility with integrated pest management (IPM) approaches. Additionally, regulatory frameworks should be adapted to facilitate the registration and adoption of these products, ensuring that they meet stringent safety standards while reducing barriers to market entry. Education and training programs for farmers, agronomists, and other stakeholders should also be prioritized to promote awareness and understanding of the benefits and proper use of biorational pesticides. By embracing biorational pesticides and integrating them into comprehensive pest management strategies, we can enhance agricultural sustainability, reduce chemical residues, protect biodiversity, and safeguard human health.

5. References

Ayasse, M. 2010. Chemical ecology in deceptive orchids. Chemoecol 20: 171-178.

Chapman, R.F. 2013. The Insect: Structure and Function. 5 (eds), Cambridge University Press, New York, USA, pp 932.

Chen, R., Klein, M.G., Li, Q. and Li, Y., 2014. Mass trapping Popillia quadriguttata using Popillia japonica (Coleoptera: Scarabaeidae) pheromone and floral lures in northeastern China. Environmental Entomology 43(3): 774-781.

Costa, C.I. 2016. Integrated pest management and the (un)sustainable use of pesticides. www. repository.utl.in.

Dhaliwal, G.S. and Arora, R. 2003. Principles of Insect Pest Management. 2nd edition, Kalyani Publishers, Ludhiana, India, pp 374.

Djerassi, C., Shih-Coleman, C. and Diekman, J. 1974. Insect control of the future: operational and policy aspects. Science 186: 596-607.

Duke, S.O. 1995. Agrochemicals from natural products. In: Godfrey C.R.A. (ed) Agriculture, Ecosystems & Environment, Marcel Dekker, New York, pp 65-66.

Eid, M.A.A., Salem, M.S. and Taha, G.Z. 1988. Effects of precocene II on morphogenesis of the desert locust Schistocerca gregaria. Biochemical Systematics and Ecology 16: 515-520.

Gogi, M.D., Sarfraz, R.M., Dosdall, L.M., Arif, M.J., Keddie, A.B. and Ashfaq, M. 2006. Effectiveness of two insect growth regulators against Bemisia tabaci (Gennadius) (Homoptera: Aleyrodidae) and Helicoverpa armegera (Hubner) (Lepidoptera: Noctuidae) and their impact on population bensities of arthropod predators in cotton in Pakistan. Pest Management Science 62: 982-990.

Haldhar, S.M., Choudhary, B.R., Bhargava, R. and Sharma, S.K. 2014. Development of an organic integrated pest management (IPM) module against insect-pests of muskmelon in arid region of Rajasthan, India. Journal of Experimental Biology and Agricultural Sciences 2: 19-24.

Hamaidia, K. and Soltani, N. 2014. Laboratory evaluation of a biorational insecticide, kinoprene, against Culex pipiens larvae: Effects on growth and development. Annual Research and Review in Biology 4: 2263-2273.

Hampton, E., Koski, C., Barsoian, O., Faubert, H., Cowles, R.S. and Alm, S.R., 2014. Use of early ripening cultivars to avoid infestation and mass trapping to manage Drosophila suzukii (Diptera: Drosophilidae) in Vaccinium corymbosum (Ericales: Ericaceae). Journal of the Entomological Society 107(5): 1849-1857.

Harada, T., Nakagawa, Y., Ogura, T., Yamada, Y., Ohe, T. and Miyagawa, H. 2011. Virtual screening for ligands of the insect molting hormone receptor. Journal of Chemical Information and Modeling 51: 296-305.

Ishaaya, I. and Horowitz, A.R. 2004. Biorational insecticides mechanisms, selectivity and importance. In: Horowitz, A.R., Ishaaya, I. (eds) Insect Pest Management. Springer Berlin Heidelberg, New York, pp 1-28.

Khater, H.F. 2003. Biocontrol of some insects. PhD Dissertation, Zagazig University, Benha Branch, Egypt.

Kumar, S. and Yadav, B.P. 2007. Efficacy of fungicides and phyto extracts on Colletotrichum spp. Physiology and Molecular Plant Pathology 37(2): 363-364.

Kumar, R. 2004. Evaluation of organics and indigenous products for the management of Helicoverpa armigera in chilli. University of Agricultural Sciences, Dharwad, India.

Landolt, P., Adams, T. and Rogg, H., 2012. Trapping spotted wing drosophila, Drosophila suzukii (Matsumura) (Diptera: Drosophilidae), with combinations of vinegar and wine, and acetic acid and ethanol. Journal of Applied Entomology 136: 148-154.

Li, Y., Zhang, Z., Robinson, G.E. and Palli, S.R. 2007. Identification and characterization of a juvenile hormone response element and its binding proteins. Journal of Biological Chemistry 282: 37605-37617.

Mehta, P.K., Sharma, P.C. and Sood, A.K. 2005. Evaluation of some botanicals against lepidopterous larval complex of cabbage in Himachal Pradesh. First Congress on Insect Science, Punjab Agricultural University, Ludhiana, pp 110-11.

Meyer, J.R. 2006. Insect Communication. https://genent.cals.ncsu.edu.

O'Brien, D.J. and Fahey G. 1991. Control of fly strike in sheep by means of a pouron formulation of cyromazine. Macedonian Veterinary Review, 129: 351-353.

Padmanabam. 1997. Evolution of oil cake against white grub. Annals of Agricultural Research 24(3): 461-465.

Parthasarathy, R., Tan, A. and Palli, S.R. 2008. HLH-PAS family transcription factor methoprene-tolerant plays a key role in JH action in preventing the premature development of adult structures during larval-pupal metamorphosis. Mechanisms of Development 125: 601-616.

Pedigo, L.P. 2003. Plant resistance to insects. In: Pedigo LP (ed) Entomology and Pest Management 4th Edition. Prentice hall of India Private Limited, New Delhi, pp 413-424.

Potter, D.A. 1998. Destructive Turfgrass Insects: Biology, Diagnosis, and Control. Ann. Arbor Press, Chelsea, USA. pp 400.

Retnakaran, A., MacDonald, A., Tomkins, W.L., Davis, C.N., Brownwright, A.J. and Palli, S.R. 1997. Ultrastructural effects of a non-steroidal ecdysone agonist, RH-5992, on the sixth instar larva of the spruce budworm, Choristoneura fumiferana. Journal of Insect Physiology 43: 55-68.

Richardson, M.L. and Lagos, D.M. 2007. Effects of a juvenile hormone analogue, pyriproxyfen, on the apterous form of soybean aphid (Aphis glycines). Journal of Applied Entomology 131: 297-302.

Saeed, Q., Saleem, M.A. and Ahmad, M. 2012. Toxicity of some commonly used synthetic insecticides against Spodoptera exigua (Fab.) (Lepidoptera: Noctuidae). Pakistan Journal of Zoology 44: 1197-1201.

Singh, S. and Kumar, K. 2011. Diofenolan: a novel insect growth regulator in common citrus butterfly, Papilio demoleus. Phytoparasitica 39: 205-213.

Smagghe, G. and Degheele, D. 1992. Effects of RH 5849, the first nonsteroidal ecdysteroid agonist, on larvae of Spodoptera littoralis (Boisd.) (Lepidoptera: Noctuidae). Archives of Insect Biochemistry and Physiology 21:119-128.

Smitha, M.S. 2002. Management of yellow mite, Polyphagotarsonemus latus (Banks) [Acari; Tarsonemidae] on chilli. University of Agricultural Sciences, Dharwad, India.

Sontakke, B.K., Mohapatra, L.N. and Swain, L.K. 2013. Comparative bioefficacy of buprofezin 25 EC against sucking pests of cotton and its safety to natural enemies. Indian Journal of Entomology 75: 325-329.

Stelinski, L.L., Il'ichev, A.L. and Gut, L.J. 2009. Efficacy and release rate of reservoir pheromone dispensers for simultaneous mating disruption of codling moth and oriental fruit moth (Lepidoptera: Tortricidae). Journal of Economic Entomology 102(1): 315-323.

Streibert, H.P., Frischknecht, M.L. and Karrer F. 1994. Diofenolan - a new insect growth regulator for the control of scale insects and important lepidopterous pests in deciduous fruits and citrus. Proceedings / Brighton Crop Protection Conference, Pests and Diseases, 1: 23-30.

Uri, N.D. 1998. Government policy and the development and use of biopesticides. Futures, 30: 409-423.

Varghese, T.S. 2003. Management of thrips, Scirtothrips dorsalis Hood and mite, Polyphagotarsonemus latus Banks on chilli using biorationals. University of Agricultural Sciences, Dharwad, India.

Vassiliou, V.A. 2009. Control of Lobesia botrana (Lepidoptera: Tortricidae) in vineyards in cyprus using the mating disruption technique. Journal of Crop Protection 28(2): 145-150.

Verheggen, F.J., Haubruge, E. and Mescher, M.C. 2010. Alarm pheromones. In: Litwack, G. (ed.) Pheromones. Amsterdam, The Netherlands: Elsevier 83:215-39.

Ware, G.W. 1989. The Pesticide Book (3rd Edn.). Thomas Publications, Fresno, USA, p. 320-322.

9

Role of Botanicals in Organic Farming and Their Future Prospectus and Potential Challenges

M. Rajashekhar*, B. Keerthana, N. Priyanka, Ch. Asritha K. Bhargavi, G. Akhil, R. Viswakarma and A.V. Ramanjaneyulu

Professor Jayashankaar Telangana Agricultural University, Rajendranagar Hyderabad-500 030, Telangana, India

Abstract

Since many years, there has been an extensive use of synthetic pesticides against which has caused pest resistance, resurgence, residues environmental problems and toxicity to non-target organisms. Therefore, attention is given towards the alternate methods which are ecologically, economically sustainable. Botanical pesticides that control insects by using plant oils and extracts have proven to be the most effective. As they have promising role in organic farming, they help in reduction of pest load and improve crop quality and thereby increase crop yield without disturbing the environment and helps in maintaining the balance of natural ecosystems. This book chapter, discuss about the active functional ingredients of each botanical pesticide and mode of action, its importance in organic farming, their commercial formulations, major insect pests suppressed or controlled, their challenges in usage in agriculture and horticulture and its future prospects.

Keywords: *Organic farming, neem, pyrethrum, nicotine, rotenone, essential oils, sabadilla*

1. Introduction

Synthetic pesticide use has caused pest resistance, environmental problems and toxicity to non-target organisms. It is essential to follow alternate methods because these synthetic pesticides have led to acute and chronic poisoning in farm workers, applicators, and even customers. Employing botanical pesticides, is the most effective way to replace the widespread use of synthetic pesticides, is one of the important alternative strategies. Among them, plant-based bio-pesticides that control insects by using plant oils and extracts have proven to be the most effective. Botanicals being natural and non-toxic to humans, animals, and the environment, have a significant role in organic farming as they help in promoting plant growth, controlling pests and diseases, and enhancing soil health. The use of botanicals is an essential component of organic farming as it helps in maintaining the balance of natural ecosystems, improves crop quality, and increases yield without the negative impacts of synthetic chemicals. As the use of synthetic pesticides and fertilizers is prohibited in organic agriculture. Organic farmers rely on natural methods to protect their crops from pests, diseases, and weeds. The beneficial insects, crop rotation, and natural remedies such as neem oil or garlic sprays are used to protect the crops. The use of natural methods to manage pests and diseases promotes biodiversity, maintains soil health, and minimizes the environmental impact of farming. Thus, plant protection is crucial in organic farming to ensure sustainable and responsible agriculture practices.

Botanicals are also an important source of bioactive compounds that can be used as natural pesticides and fungicides in organic farming. For example, a study by Kurtz *et al.* (2017) found that botanical extracts from rosemary, thyme, and oregano had potent antifungal activity against several plant pathogens, including *Fusarium oxysporum* and *Alternaria alternata.* Similarly, a study by Rajakumar *et al.* (2014) showed that botanical-based formulations from *Azadirachta indica*, *Trichoderma harzianum*, and *Pseudomonas aeruginosa* suppressed the growth of soil-borne fungal pathogens in tomato crops. Several studies have demonstrated the benefits of using botanicals in organic farming. For instance, a study conducted by Shaik *et al.* (2019) revealed that the application of botanical-based extracts from neem, turmeric, and garlic significantly reduced pest infestations in tomato crops without causing any adverse effects on the environment. Another study by Singh *et al.* (2017) reported that the use of botanical-based formulations from marigold and cow urine significantly improved the soil quality and increased the yield of okra crops in organic farming systems. Organic farming plays a crucial role in India's sustainable development by improving soil health, ensuring environmental sustainability, and enhancing public health, providing economic benefits, and preserving biodiversity. There is less doubt that biopesticides have great potential to control insects and vector-borne diseases, bacterial and fungal pathogens, and weeds. An expansion of biopesticides over conventional pesticides is observed globally, with the global biopesticides market size exceeding USD 4 billion at the outset of this decade, expecting to double by the year 2025, with bioinsecticides being approximately fifty per cent of the total biopesticide share. However, practicing and promotion of organic farming in India is still at infancy stage despite the fact of having great scope for eco-friendly, economically sustainable, sociologically acceptable and culturally feasible methods. Hence, here summarized the various plant-based bio-pesticides and their importance, formulations used for management of the insect pests and their future prospects.

1.1. Neem (*Azadirachta indica*) importance in organic farming

Neem (*Azadirachta indica*) belongs to the Meliaceae family has gained as a highly potent bio-pesticide. *Azadirachtin* is the key ingredient in seeds, neem oil, leaves, flowers, and fruits which possess insecticidal qualities. Seeds comprises of 40% of oil with *azadirachtin* as the active ingredient, which is responsible for neem's insecticidal properties (Isman *et al.,* 1990). The mevalonic acid route isolates the most significant active principle with 16 chiral carbon centres which is a complex tetranortriterpene. Due to its repelling qualities, neem leaves have been used for millennia to combat pests that attack stored grains. While neem oil acts as an antifertility agent and neem leaf extract has anti-inflammatory properties. There haven't been any reports of insects becoming resistant to neem formulations till now. The main factor behind this advantage is that neem formulations are nearly crude products with many additional related components in the neem kernel or oil, and these related compounds to *azadirachtin* may have various mode of actions. All these properties of *azadirachtin* makes it an ideal bio-pesticide, because it doesn't have non-specific toxicity for mammals and is biodegradable.

Table 1. Commercial *azadirachtin*-based products available worldwide

Products	**Manufacturer**	***Azadirachtin* Percent (%)**
Agroneem plus	Agro logistic systems Inc	0.15
AzaGuard	BioSafe systems	3.0
Azamax	E.I.D parry Ltd	1.2
AzaPRO	CANN-CARE company	1.2
Azasol	ARBORJET Inc	6.0
Debug TRES	AGRO Logistic systems	3.0
EcoZin plus	AMVAC chemical Corp.	1.2
Fortune aza	Fortune biotech Ltd	3.0
MOLT-X	BIOWORKS Inc	3.0
Neem Azal TS	Trifolio-M GmbH	1.0
Neemarin	AZA-Direct Gowan company LLC	0.15
Neemfol	Gassin pierre	5.0
Neemix	Certis	0.25
Ornazin	SEPRO corporation	3.0

1.1.1. Mode of action

Several modes of action, including sterility in mature female insects, deterrent of feeding and oviposition, interference with reproduction, or growth and development (ecdysteroid synthesis and release) have been identified.

1.1.2. Formulations

The technical powder containing 12-26% *azadirachtin* is versatile for making a wide range of formulations. These technical powders themselves have acceptable storage stability, losing only 3-5% of the *azadirachtin* content over 1 year when stored at room temperature and low humidity. This facilitates shipping and handling. The creation of storage-stable formulations, particularly Emulisifiable Concentrates (ECs), has however presented a challenge. Currently, there are number of patent with procedures and compositions that claim to confer unique stability to *azadirachtin* in ECs. In general, the higher the *azadirachtin* concentration in the final formulation, the greater the instability. Manufacture of *azadirachtin* formulations with good stability and dispersion characteristics is essential for grower acceptance and usage, especially in high value markets. Formulations with poor stability profiles generally provide unpredictable and ineffective insect control.

Table.2. *Azadirachtin* formulations registered as per CIBRC as on 01-06-2023

S.No	Type of formulation
1	*Azadirachtin* 0.15% EC w/w Min. Neem Seed Kernel Based
2	*Azadirachtin* 0.15% EC
3	*Azadirachtin* 0.30% EC (3000 PPM) Min. Neem Seed Kernel Based
4	*Azadirachtin* 1.00% EC Min. Neem Based
5	*Azadirachtin* 1.00% EC (10000 PPM) Min. Neem Based
6	*Azadirachtin* 0.03% EC Min. Neem Oil Based
7	*Azadirachtin* 0.03% WSP (300 PPM) Neem Oil Based
8	*Azadirachtin* 5.00% w/w Min. Neem Extract Concentrates

1.1.3. Pests controlled or suppressed using the azadirachtin

Azadirachtin interferes with the synthesis of yolk protein and/or its uptake into oocytes, which results in reduced fertility and fecundity in *Spodoptera litura, Dysdercus cingulatus, Galleria mellonella, Drosophila melanogaster*, *Tuta absoluta*, and *Helicoverpa armigera*. The two applications of *azadirachtin* 5ppm at root system against greenhouse whitefly (*Trialeurodes vapororium*) reduced the number from 70-80% in tomato (Pavela and Silva, 2006). The oil obtained from seed has been reported to have larvicidal activity on mosquitoes (Dua *et al.*, 2009). Two sprays of neem seed oil and *azadirachtin* were effective in causing the nymphal mortality of *Aphis glycines* (77% by neem oil and 80% by *azadirachtin*) (Kraiss and Cullen, 2008). Inhibition of the oviposition in *Nezara viridula was also* reported with pure *azadirachtin* (Riba *et al.*, 2003). Neem cake @ 200 g m^{-2} with arbuscular mychorrhiza fungus was effective on increasing the plant height in okra, increased the phosphorous content in the field and was effective in controlling the root knot nematode in okra (Mohapatra *et al.*, 2020).

Table 3. The efficacy of neem pesticides against various agricultural pests

S. No.	Crops	Products used	Pests	References
1.	Rice	Handi Ausadha pot mixture (5:l fermented cow urine + 1 kg fresh cow dung + 1 kg karanj leaves + 1 kg neem leaves + 1 kg calotropis leaves and 50g Gaur @ ml l^{-1})	Reduce the incidence of pest by yellow stem borer (73.13%), Green leaf hopper (75.12%), Gall midge (69.93%), Dead heart and white ear head (69.26%), thrips (79.73%), and leaf folder (85.57%).	Mohapatra, 2018
		Nimbecidine @5ml l^{-1}, 5 ml l^{-1} of Neem oil	Brown plant hopper (*Nilaparvata lugens*)	Choudhary *et al.*, 2017
		Multineem 300ppm @ 2.5 l ha^{-1}	Brown plant hopper, Yellow Stem borer (*Nilaparvata lugens, Scirpophaga incertulas*)	Dash *et al.*, 2019
2.	Maize	Local Neem	*Maize weevil (Sitophilus zeamais)*	Khanal *et al.*, 2019
		Aqueous Neem extract @300 l ha^{-1}	Corn ear worm (*Heliothis armigera*)	Udo and Ibanga, 2019
		Neem leaf extract @ 2 ml l^{-1}	Maize aphid (*Rhophalosiphum maidis*)	Alam *et al.*, 2019
		Neem oil and seed cake	Fall army worm (*Spodoptera furgiperda*)	Shaiba *et al.*, 2019
3.	Wheat	Neem Seed Kernel Extract	Wheat aphid (*Raphalosi phumpadi*)	Matharu and Tanwar, 2019
		Indoneem 1500ppm @ 1200 ml ha^{-1}, 3% Neem oil	Wheat aphid (*Raphalosi phumpadi*)	Bushra *et al.*, 2014
4.	Chick pea	5% Neem seed kernel extract	*Chickpea pod borer Helicoverpa armigera*	Kumar *et al.*, 2019
5	Green gram	Neem leaf powder	*Pulse bruchid (Callosobruchus maculatus)*	Gupta *et al.*, 2015

6.	Cow pea	Neem seed Extract	Cow pea bruchid *(Callosobruchus maculatus)*	Lale and Mustapha, 2000
		Neem + *Metarhizium anisopliae*	Cowpea thrips (*Megolurothipss jostedti*)	Raoul *et al.*, 2019
7.	Pea	Neem seed kernel extract	Pea aphid (*Acyrthosiphon pisum)*	Melesse, 2012
8.	Sorghum	2 % Neem oil	Sorghum shoot fly (*Atherigona soccata)*	Joshi *et al.*, 2016

1.1.4. Challenges

The main causes of the inconsistent efficacy results were the low persistence in the environment, rapid degradation by sunlight and short shelf life of *azadirachtin* formulation is unfavourable from an agribusiness perspective, since they result in reduced efficiency and require more applications. (Pasquoto-Stigliani *et al.*, 2017). The availability of raw neem material, processing (time required; method), storing, and marketing of neem products are all challenges to *azadirachtin*'s wide-spread use. In order to address these problems and encourage the use of neem products for natural pest management, a government or nongovernmental organisation should help a village-based business process and market neem seed. The registration and regulation of bio-pesticides, including products containing *azadirachtin*, can be complex and time-consuming, limiting their availability and use in some regions. Continuous and more use of *azadirachtin* can lead to the development of resistance in target pests over time, making it less effective as a long-term pest management solution. The potency of *azadirachtin* can vary significantly among different neem sources and extraction methods. Standardization of *azadirachtin* content in commercial formulations can be challenging, leading to inconsistent efficacy.

1.1.5. Future prospects of azadirachtin

Implementing nanotechnology is a means of overcoming the limitations of quick photo-degradability by developing more stable formulations of *azadirachtin*. Recent research has focused on the creation of polymeric encapsulated formulations of botanical pesticides with controlled releases (Pasquoto-Stigliani *et al.*, 2017; Das *et al.*, 2014). *Azadirachtin* is protected from photodegradation by natural polymers that include kraft lignin and alginate, which may also be applied to increase its stability and transport to the site of action. These innovative techniques for encapsulating botanical pesticides have a number of advantages, including gradual release, greater stability, usage of small doses, and odour masking (Choudhary *et al.,* 2017).

To develop combinations that can successfully control pests, more research on the synergistic activity of *azadirachtin* and other insecticides is required. Biotechnological approaches might be utilized to improve the production of *azadirachtin* in plants or microbes. Genetically modified crops that naturally produce higher levels of *azadirachtin* could provide built-in pest resistance and reduce the need for external applications.

1.2. Pyrethrum (*Chrysanthemum cinerariifolium*)

In organic farming, management of pests is not about wiping out the insects, it is about keeping biodiversity and reduce insect pests while having minimal impact on beneficial insects. Products containing natural pyrethrins are one of the major insecticide groups in organic farming and should widely used in all countries, especially developing countries. Pyrethrum (*Chrysanthemum cinerariifolium*) which is a part of the chrysanthemum genus in the Asteraceae family. It is known to have originated from the Balkan and Eastern coasts of the Adriatic Sea in the Dalmatia region of current Croatia. The term pyrethrum refers to the dried and powdered flower heads of a white-flowered, daisy-like plant belonging to the Chrysanthemum genus. Pyrethrum's insecticidal properties were recognized in the middle of the 19th century. The white Chrysanthemum flower, pyrethrum, is a mentioned in early Chinese history and it is believed to have passed into Europe along the silk routes. There are more than 40 species within chrysanthemum genus and to large extent, they are found in East-Asia, particularly in China. *C. cinerariifolium* is the most popular one.

Pyrethrins are a set of monoterpene esters with a close similar structure and they result from esterification of two monoterpic acids (pyrethric acid and chrysanthemic acid) with three ketone alcohols (cinerolone, jasmolone and pyrethrolone) and existence of six different active components like, Cinerin–I, II, Pyrethrin–I, II and Jasmolin–I, II which are major insecticidal constituents of pyrethrins. They are especially found in oil glands onto the seed inside the thoroughly packed flower head. In other parts of the plant, Pyrethrin can be also found in small concentrations. On a dry weight basis, a pyrethrum inflorescence contains most the plant's active compounds, reported in the range 1-2% pyrethrin. Almost 80% of natural insecticides produced globally are extracted from Pyrethrum.Pyrethrins are natural products and permitted in organic agriculture because they are degraded, by sunlight and changes in pH, into non-toxic products which are immobile in soil. Pyrethrum which is a natural insecticide has many properties, but the most important are rapid action, insect flushing, excitation, quick knockdown, very low toxicity for mammalian, lack of insect immunity, broad of activity, non-persistence, degraded quickly by UV in sunlight and very effective insect repellent. The low residual effect is one of the advantages of using pyrethrins in organic farming. During postharvest applications, for instance, on grains, it has been revealed that in a period of three to four months of storage, 50% or more of the applied pyrethrins disappear (Gallo *et al.*, 2017). Pyrethrin is environmentally safe and highly biodegradable due to its instability on light, high temperature, oxygen, and water. It is recognized and approved in the European Union regulation (EC) No. 834/2007 on organic production and labelling of organic products. Consequently, it is one of the leading and used insecticides in organic farming systems reported by (Grdiša *et al.*, 2009). Insect mortality is increased by mixing pyrethrins with a synergist, such as piperonyl butoxide (PBO) to create pyrethrum. PBO, however, is not certified to be used on organic crops. Growers, who want to meet organic certification standards, should consult with the Colorado Department of Agriculture before using of products. Pyrethroids are synthetic materials designed to imitate natural pyrethrum. They can be used on houseplants, but because they are synthetic, they are not registered to use on organic crops. Pyrethrum is non-toxic to most mammals, making it among the safest insecticides in use. The Environmental Protection Agency has approved it for more uses than any other insecticide. It is sold under trade names that include Pyrenon, Red Arrow (pyrethrins/rotenone) and Pyrellin (Pyrethrins/rotenone).

1.2.1. Mode of action

Pyrethrin-containing insecticides cause neurotoxicity in insects by blocking voltage-gated sodium channels in nerve axons. When pyrethrins get through the cuticle and into an insect's nervous system, they cause toxicity. The sodium channels located all the way throughout nerve cells are exactly pyrethrins bind to. Because they allow the flow of sodium ions, sodium channels are in charge of transmitting nerve signals all the way along the length of the nerve cell. The normal operation of sodium channels is hindered when pyrethrins attach to them, leading to hyperexcitation of the nerve cell and ultimately a loss of function of the nerve cell. Pyrethroids have an impact on the insect's peripheral and central nervous systems. Pyrethroids have a significantly stronger stimulating impact than DDT. When insects are exposed to pyrethrins, their nervous systems often shut down and they die. Pyrethrins have a moderate toxicity to mammals, however commercial formulations have a much lower toxicity. They lack photostability, are nonpersistent, and are toxic to fish.

1.2.2. Formulations

Pyganic 5% EC, Evergreen 6% EC, Pyrethrum 2% Liquid extract, Diatect.

1.2.3. Pests controlled or suppressed

Pyrethrum is moderately to highly effective against aphids, apple maggot, European apple sawfly, caterpillars, leafhoppers, whiteflies, Lepidoptera larvae (including codling moth), mealybugs, pear psylla, plum curculio, many of the true bugs (Hemiptera), Mexican bean beetle (*Epilachna varivestis*) on soybeans, and flower thrips. Within these groups, pests may have a greater or lesser susceptibility to pyrethrum products. Some insects may be able to recover after the initial knockdown if the dose is too low. Frequent repeat applications of pyrethrum are required because of its rapid photo-degradation and short residual activity in the field. Pyrethrins are highly irritating to insects; as a result, they might be used as "flushing agents," causing insects to come out of hiding, a desirable circumstance when you need to identify an insect that is hiding in the turf grass such as grubs or sod webworms.

1.2.4. Application methods

Pyrethrins (the active chemicals) are rapidly broken-down by sunlight. Therefore, it is recommended that pyrethrum be applied before dawn or in late evening when the target insect pests are active and present in the field, and UV light is minimal. Use of UV-inhibiting adjuvants may allow for longer duration of residual activity.

1.2.5. Prospects and challenges

The industrial development and production of new botanical pesticides are challenging and, e.g., require (i) Availability of pesticidal plant resources in a sustainable and large-scale manner (ii) Standardization in the processing of phytochemical substrates and performance of quality assurance protocols (iii) Authorization approval, which implies extended record data on the environmental fate of the bioactive substrates and possible toxic effects on non-target organisms. Moreover, the combined use of bio-pesticides and bio-fertilizers can improve soil health status and prevent environmental pollution, thereby promoting sustainable agriculture. Yadav *et al.* reported that the adoption of organic farming practices through combined application of bio-fertilizers and bio-pesticides increases sustainability in plant cultivation. Pyrethrum and *azadirachtin* are the main bio-pesticidal compounds in the global bio-pesticide market, with pyrethrum representing 80% of botanical insecticides worldwide. Kenya is the leading country in pyrethrum production, providing 70% of the global share.

1.3. Nicotine (*Nicotiana tabacum*)

One of the first known plant-derived insecticides is nicotine. Around three hundred years ago, *Nicotiana tabacum L.* was being used as a dust or water extract to suppress phytophagous insects, long before anyone was aware that tobacco contains the nicotine alkaloid. A non-systemic pesticide is present. It can be acquired from powdered tobacco leaves, tobacco plant extracts, and tobacco industry trash (Gudeta *et al.*, 2021). A WHO report from 2008 stated that poor countries consumed 25% of the chemical pesticide production worldwide. Chemical pesticides have an impact on nontarget species such as pollinators, domestic and wild animals, and natural insect competitors. Alternatives are required because chemical pesticides have detrimental effects on species that are not intended targets.

The most dangerous natural pesticide, nicotine: is an alkaloid that is isolated from tobacco plants, primarily *Nicotiana tabacum* and *Nicotiana rustica* (Solanaceae). It functions as a contact poison and is a quick-acting nerve toxin. Other alkaloids discovered in commercial tobacco include nornicotine, which is primarily found in *Nicotiana glutinosa* and *Nicotiana sylvestris*, anabasine, which is primarily found in *Nicotiana glauca*, and a variety of other "minor" related alkaloids. Nicotine is active against insect orders *viz:* coleoptera (Beetles), hemiptera (Aphids, whitefly) and thysanoptera (Thrips) etc. and it works best on mites and insects with soft bodies, such as aphids, thrips, leafhoppers, and spider mites.

1.3.1. Mode of action

Neonicotinoids, like nicotine, bind to nicotinic acetylcholine receptors of a cell and trigger a response by that cell. In mammals, nicotinic acetylcholine receptors are located in cells of both the central nervous system and peripheral nervous systems. In insects these receptors are limited to the central nervous system. Nicotinic acetylcholine receptors are activated by the neurotransmitter acetylcholine. While low to moderate activation of these receptors causes nervous stimulation, high levels over stimulate and block the receptors, causing paralysis and death. Acetyl cholinesterase breaks down acetylcholine to terminate signals from these receptors. However, acetyl cholinesterase cannot break down neonicotinoids and their binding is irreversible. Related to nicotine in action and partly in structure, the highly insecticidal neonicotinoids originated from screening novel synthetic chemicals.

1.3.2. Formulations

1.3.2.1. Nicotine sulphate

It contains 40% alkaloid, is safer and is more convenient to use and the free alkaloid is liberated by the addition of soap lime or ammonium hydroxide to the spray solution. Dust formulation of nicotine sulphate releases nicotine in the presence of moisture. It is also used in aerosols. It contains Nicotine Alkaloid 400 gm/1000gm i.e. 40%(w/w) Vegetable Oils, Tar, etc. 20 gm /1000 gm i.e. 2% (w/w) Water and other ingredients 580 gm/1000 gm i.e. 58% (w/w). It is used to kill aphids, bugs, worms, leafhoppers and similar sucking insects which attack and destroy fruit, vegetables, crops and even flowers. It is also effective against Lice, Mites and Ticks which are a menace to livestock

1.3.2.2. Tobacco decoction

This is useful for controlling aphids, Thrips etc. can be prepared by boiling 1kg of tobacco waste in 10 lts of water for 30 minutes or steep it in cold water for a day. Then make it up to 30 litres and add about 90 gm of soap. Addition of soap improves wetting, spreading and killing properties.

1.3.2.3. Nicotine tartrate

This is a powder that is often used to control pests like fleas and ticks on pets.

1.3.2.4. Nicotine fumigant

This is a gas that is used to fumigate spaces like warehouses or storage areas to kill pests like grain beetles and other insects.

1.3.3. Pests controlled or suppressed

Reuben *et al.* (2019) studied the effect of the application of tobacco leaf powder snuff at 3% *w/w* against cowpea weevil (*Callosobruchus maculatus* L.) on stored cowpea (*Vigna unguiculata* L.) grains, finding that this treatment protected the stored grains better than other tested botanicals. Dieng *et al.* (2013) studied the effect of using a pesticide solution manufactured using cigarette butts against *Aedes aegypti* and observed insecticidal activity against its larvae. Henry (2011) studied the effect of nicotine on Colorado potato beetle and concluded that a mixture of phenols with known pesticidal properties work synergistically in controlling the beetles. On the cabbage aphid, the emulsified concentrated extract of nicotine performed best at a ratio of 1: 100 (emulsified concentrated to solvent). Sekar *et al.* (2021) studied that (*Nicotiana tabacum*) leaf extracts as an alternative to synthetic insecticides in the management of rice weevils. The results revealed that that the chloroform and acetone extracts of *N. tabacum* leaves exhibited the highest adulticidal activity against *S. oryzae*.

1.3.4. Challenges

1.3.4.1. Toxicity

Nicotine is highly toxic to mammals, including humans, and can cause serious health problems if ingested or inhaled.

1.3.4.1.2. Persistence

Nicotine is persistent in the environment and can remain in soil or water for extended periods, posing risks to non-target species.

1.3.4.1.3. Non-selectivity

Nicotine is non-selective and can harm both beneficial and harmful insects, leading to disruption of natural ecosystems.

1.3.4.1.4. Resistance

Overuse of nicotine can lead to the development of resistance in insects, reducing the effectiveness of the pesticide.

1.3.4.1.5. Regulations

Nicotine is a controlled substance and requires special permits and regulations for use as a pesticide. This can make it difficult to obtain and use in an effective manner.

1.3.4.1.6. Cost

Nicotine is expensive to produce, making it a less attractive option compared to other chemical pesticides or biological control agents.

1.3.5. Future prospects of nicotine

Development of new nicotine-based pesticides: There is ongoing research into the development of new pesticides based on nicotine, which could be more effective and environmentally friendly than existing alternatives. Widespread adoption in organic farming: The use of nicotine as a pesticide is already allowed in organic farming, and as demand for organic products increases, the use of nicotine could become more widespread. Regulatory challenges: The use of nicotine as a pesticide is currently regulated, and there may be pressure to tighten regulations in the future to ensure that it is used safely and responsibly.

1.4. Rotenone (*Derris elliptica* and *D. malaccensis*)

Rotenone is a broad-spectrum nonselective botanical insecticide that is derived from the roots and stems of tropical legumes such as *Derris elliptica* and *Derris malaccensis,* which contain approximately 4% (w/w) to 5% (w/w) rotenone while *Lonchocarpus utilis* and *Lonchocarpus urucu* contain approximately

8% (w/w) to 10% (w/w) rotenone in dried roots and rotenone is also extracted from *Tephrosia virginiana* (Zubairi *et al.*, 2016). In addition, these plants contain number of other isoflavonoids compound such as deguelin, rotenolone, tephrosin, elliptone, sumatrol, toxicarol and malaccol. Rotenone was first isolated from *Lonchocarpus nicou* in 1895 and given the name "nicouline" by Geoffroy but in 1902, Nagai gave the name rotenone after using the roots of *Derris elliptica* (roten in Japanese) (Manda *et al.*, 2020). Rotenone is extracted by using organic solvents as extractants (green solvents and ionic liquids) in non-conventional or conventional methods. Furthermore, rotenone is safe to humans, non-toxic to plant and unstable to light and heat losing almost all its toxicity after two to three days (Zubairi *et al.*, 2016).

In modern organic agriculture, rotenone is employed either alone or in combination with other substances like pyrethroids, synergistic (piperonyl butoxide), sulphur or copper. Numerous arthropod pests, such as aphids, thrips, moths, beetles and spider mites are controlled and the Colorado potato beetle show high pesticide resistance to rotenone (*Boboescu*, 2020). Though its use is reduced in standard agriculture crops but it is being used in organic farming in California for lettuce and tomato crops. Commercially, it is available as a dusting powder with 1-5% active ingredient for domestic use and liquid formulation for organic farming with about 8% active rotenone. Several different types of insects were tested with rotenone in an aqueous suspension and in dust form, the carrier being diatomaceous earth intimately mixed with precipitated rotenone.

1.4.1. Mode of Action

Rotenone is a naturally occurring chemical with insecticidal, acaricidal and fish-killing properties. WHO classifies rotenone as a moderately hazardous Class II pesticide. Rotenone can enter the insect body through the alimentary canal, tracheae or integument. It appears to kill insects by specific inactivation of the respiratory enzyme, glutamic acid oxidase resulting in death through oxygen (O_2) starvation. In insects, rotenone is both a contact and systemic insecticide that works by inhibiting the transfer of electrons from Fe-S centers in Complex I to ubiquinone in the electron transport chain. This prevents NADH from being converted into usable cellular energy, *i.e.,* adenosine triphosphate (ATP). In mammals and fish, rotenone inhibits the oxidation of reduced nicotinamide adenine dinucleotide (NADH) to nicotinamide adenine dinucleotide (NAD), thereby blocking the oxidation by NAD of substrates such as glutamate, α-ketoglutarate and pyruvate. Rotenone is a strong inhibitor of mitochondrial electron transport. It inhibits the mitochondrial respiratory chain between diphosphopyridine nucleotide and flavine (Gupta, 2019).

1.4.2. Formulations

Formulations include crystalline preparations (95-97% pure), emulsified solutions (1-50 % pure) and dusts (0.75to 8 % pure).

Dusts of ground *Derris* roots are mixed with 3-7 parts of carrier such as talc, clay, gypsum, Sulphur and tobacco or walnut shells. Furthermore, the impregnated dusts are also used and produced by mixing the extract of ground *Derris* roots in a volatile solvent with an adsorptive carrier. Dusts containing 0.5% (w/w) to 1% (w/w) rotenone and 1.8% (w/w) to 3.5% (w/w) total extractives are effective against most insect pests.

Dispersible powders may be made from finely ground *Derris* roots. Two to five pounds (0.9 kg to 2.3 kg) of *Derris* powder with 2 pounds (0.9 kg) of neutral soap or the equivalent of the sulfonated oil make 380 L of spray solution.

The extract of *Derris* is also widely used as dried resin which contains 25% (w/w) to 35% (w/w) rotenone. This dried resin is usually used as emulsifiable concentrates (EC) or known as spray oils. Because of the limited solubility of rotenone in spray oils which is approximately 0.05% (w/w), mutual solvents are generally employed to increase the solubility (Zubairi *et al*., 2016).

1.4.3. Pests controlled or suppressed

The aqueous suspensions were highly toxic to aphids, thrips, white fly larvae, leaf hoppers, larvae of beetles, tent caterpillars and culicine mosquitoes. Adult beetles required a much higher concentration. Poor results were obtained against squash bugs, red spiders and mealy bugs; this might have been remedied by the addition of a penetrating and sticking agent, as there is good evidence that the active ingredients of derris are toxic to these insects under optimum conditions of application. The dusts were effective against chicken lice, roaches and cabbage worms, but results against soft bodied sucking insects were not so good (Davidson, 1930). Rotenone, an insecticide approved for certified organic production, reduced flea beetle damage more than any other material. Rotenone also reduced aphid infestation but did not reduce leafhopper damage in New Jersey (Maletta *et al*., 2004). Saiful Irwan Zubairi and co-workers studied the toxicity and effectiveness of rotenone against *Spodoptera litura* and *Plutella xyllostella* in field and laboratory conditions (Zubairi *et al*., 2016). the effect of rotenone on golden snail, rice skippers, BPH in rice and DBM in vegetables found effective than other treatments (Hien *et al*., 2003).

1.4.4. Challenges

Rotenone has a non-selective impact and may harm beneficial insects like natural predators and pollinators. This may upset the equilibrium of the ecosystem and have unintended effects on species that are not the target. If mistakenly introduced into water bodies during treatment for pest control in aquatic situations, it may be hazardous to aquatic species, particularly fish, and Parkinson's disease in humans. Rotenone use that is prolonged and frequent can cause populations of the targeted pests to become resistant. Because of its short persistence, the long-term environmental effects are reduced, but it also may necessitate repeated applications to maintain pest control effectiveness.

1.4.5. Future prospects of rotenone

Rotenone continues to find applications in sustainable and organic agriculture systems with a demand for natural and environmentally friendly pest control methods. Rotenone plays a role in integrated pest management approaches, which emphasize the use of multiple strategies to control pests effectively while minimizing the reliance on pesticides. Combining rotenone with other pest control methods, such as biological controls, cultural practices and resistant crop varieties, may mitigate its potential negative impacts. Advances in formulation technology could lead to more efficient and targeted delivery of rotenone. This could help reduce the quantity needed for effective pest control, limiting its impact on the environment and non-target species.

1.5. Essential oils

Essential oils are characterized as any volatile oils with potent aromatic components that give a plant a unique fragrance, flavour, or scent. These are secondary metabolites of volatile plants, which are the by-products of plant metabolism. In the leaves, stems, bark, flowers, roots, and/or fruits of some plants, essential oils can be found as liquid droplets in glandular hairs or secretary cavities. Due to these compounds' secure, bioactive, biodegradable, ecologically responsible, and commercially viable qualities, they represent potentially excellent options for biocontrol agents reported by Yuru chang *et al.,* (2022). Based on their antibacterial and antifungal activities, they have a long history of usage in the preservation of stored items (Samar, 2018). Certain chemical components of volatile oils affect how insects' octopaminergic nervous systems function. The main lipophilic chemicals found in essential oils operate as poisons, feeding deterrents, and oviposition deterrents to a range of insect pests. The main components of the essential oils of many different kinds of plants and flowers are terpenes and terpenoids.

A broad range of plants create terpenes, a huge and diverse class of chemical molecules. 1. They are often strong-smelling. They may protect the plants that produce them by deterring herbivores and by attracting predators and parasites of herbivores. 2. Many terpenes are aromatic hydrocarbons and thus may have had a protective function. 3. The difference between terpenes and terpenoids is that terpenes are hydrocarbons, whereas terpenoids contain additional functional groups. Direct toxicity, feeding deterrence and oviposition, attraction or repellency would result from interaction with the insect nervous system, either by antagonism of the octopamine receptors or acetylcholinesterase inhibition. The oils te complex mixtures (hydrocarbon or oxygenated mono and sesquiterpenes; aromatic saliphatics; *etc.)* with some major constituents such as 1,8-cineole (eucalyptus, rosemary), pulegone (pennyroyal), menthol (mints), eugenol (clove), *etc.* Most come from highly aromatic family, Lamiaceae (lavender, thyme, basil, sage, mints, salvia, rosemary, *etc.),* Myrtaceae (myrtle, clove), Rutaceae (lime, lemon, amyris), *etc.,* whose plant species also known for flavorings and spices and many considered to have medicinal use. Several applications to crop protection (stored product pests), control of domestic pests (cockroaches, ants, fleas, *etc.),* mosquito repellence (citronella oil), Varroa mite control, as aphicides and acaricides (cinnamon oil) and urban insect control (eugenol-based products from clove or basil) have been reported. A range of insecticide products may be also blends of plant oils, such as thyme, clove, peanut oil, lemon grass, cinnamon or pennyroyal as components.

1.5.1. Mode of action

Essential oils inhibit some physiological functions of Gamma-Amino Butyric Acid (GABA) receptors, which is the primary inhibitory neurotransmitter of CNS of insects.

1.5.2. Formulations

Microcapsules containing the most promising, oils (*Rosmarinus officinalis* and *Thymus herba-barona*). It was indicated that nanopermethrin has more larvicidal effect against *Culex quinquefasciatus* than the bulk form of permethrin. Solid lipid nanoparticle-loading geranium oil that was used as mosquito repellent. While in other studies, polyethylene glycol (PEG) nanoparticles were used to incorporate garlic and geranium essential oils and were tested against *Tribolium castaneum* and *Rhyzopertha dominica*. Geranium essential oil alone and in the form of nano formulation for management of potato tuber moth. Citrus peel essential oil as an emulsion and loading into polyethylene glycol nanoparticles for management of *Tuta absoluta.*

1.5.3. Pests controlled or suppressed

One of the essential oils that excels in controlling a wide range of insects of the Curculionidae family is eugenol, which is extracted from cloves, nutmeg, and cinnamon. Limonene is an essential oil that has also been studied for pest control. It comes from citrus species such as *Citrus sinensis, Citrus lemon,*

Citrus aurantifolia, and *Citrus reticulata*. The results against spraying of essential oils against stored pest. *S. oryzae* revealed that it was controlled by extractions of Syzygium *aromaticum* (Myrtaceae) and *Lavandula officinalis* (Lamiaceae) (LD_{50} values 0.04 and 0.07 mg/cm^2, respectively) (Abdelgaleil *et al* 2016). *Atalantia guillauminii* (Rutaceae) and *Eucalyptus procera* (Myrtaceae) can be used against *T. castaneum* (Yang *et al*., 2010). Among Anobiidae species, research studies mainly involved *Lasioderma serricorne* (Coleoptera: Anobiidae) and *Callosobruchus* spp. Among the tested EOs, *L. serricorne* showed higher susceptibility to *Perilla frutescens* (You *et al*., 2014). EOs presented remarkable toxic activity against *Ploida interpunctella* adults (Jesser *et al.,* 2017); good mortality rates were reported for contact toxicity with *Cymbopogon martinii* (Poaceae). *Ephestia kuehniella* and *Plodia interpunctella* (Lepidoptera: Pyralidae) are strongly susceptible to *Satureja thymbra* (Lamiaceae) (Ayvaz *et al.,* 2009) Among Psocoptera, *bostrychophila* is the only psocid species studied for EO contact toxicity. Promising insecticidal activities were recorded for EOs extracted from *Laggera pterodonta* (Asteraceae). *Eucalyptus floribundi* (Myrtaceae) When applied to mature *Rhyzopertha dominica* and *O. surinamensis*, EO reduced their consumption rates in a dose-dependent manner, significantly slowing their growth rates (El-Bakry *et al.,* 2016).

1.5.4. Challenges

1.5.4.1. Efficacy

The effectiveness of essential oils for pest management may vary depending on the pest species, life stage, behaviour, and environmental conditions.

1.5.4.2. Stability

Essential oils may degrade quickly when exposed to air, light, heat, and moisture, reducing their potency and shelf life.

1.5.4.3. Formulation

Essential oils may need to be formulated into appropriate products such as sprays, baits, or traps to increase their delivery and efficacy.

1.5.4.4. Safety

Essential oils may be toxic to non-target organisms such as humans, pets, and beneficial insects if used improperly or in high concentrations.

1.5.4.5. Cost

Essential oils may be more expensive than conventional pesticides, particularly if they are sourced from rare or exotic plant species.

1.5.4.6. Availability

Essential oils may not be readily available in some regions or may require specialized equipment and expertise to extract and formulate.

1.5.4.7. Regulation

The use of essential oils for pest management may be regulated or restricted by local or national authorities, limiting their availability and use.

1.5.5. Future prospects of essential oils

One of the major challenges in pest management is the development of resistance by pests to chemical pesticides. By using essential oils, which often have complex chemical compositions, the likelihood of pests developing resistance may be reduced. Essential oils often contain multiple bioactive compounds that can act on different biological targets in pests, making it harder for them to adapt and develop resistance. Plant essential oils can exhibit a broad spectrum of bioactivity against various pests, including insects, mites, nematodes, and fungi. This versatility makes them valuable in integrated pest management (IPM) programs, where multiple pests can be targeted simultaneously with a single natural product. Essential oils can sometimes have synergistic effects when combined with other essential oils or conventional pesticides. These synergistic interactions can enhance the. overall efficacy of pest control while allowing for reduced use of chemical pesticides. The interest in plant essential oils as potential alternatives to synthetic pesticides has led to increased research and development in this field. As scientists and researchers discover new plant species with potent bioactive compounds, the effectiveness and range of essential oils for pest management are likely to expand. With the growing interest in using essential oils for pest management, there may be increased efforts to develop regulations and standards to ensure their proper use and safety. This can lead to better guidelines on formulation, application, and residue limits, further supporting their integration into pest management strategies.

1.6. Sabadilla (*Schoenocaulon officinale*)

Sabadilla is a botanical pesticide derived from ripe seeds of the South American lily *Schoenocaulon officinale* also known as cevadilla or caustic barley and European *Veratrum album*. Traditional used as a pesticide in Central and South America. The veratrine alkaloids make up about 0.3% of the weight of mature sabadilla seeds; cevadine and veratridine are the most effective insecticides among these alkaloids. The seeds and insecticidal extracts also contain the alkaloids sabadinine, sabadiline, and sabadine. The active components, cevandine-type alkaloids are poisonous to animals when present in their purest form. However, commercial formulations normally include less than 1 % active ingredient, providing a margin of safety. Despite having no structural similarity, these alkaloids are quite similar to those of the pyrethrins. Sabadilla is mostly utilised by organic farmers; in California, about one hundred kg is used every year, mainly on avocado and citrus crops. (Isman *et al.,* 2006).

1.6.1. Mode of action

Alkaloids in sabadilla disrupt the function of nerve cell membranes in insects by interfering with Na^+ and K^+ ion movement in nerve axons and leading to a loss of nerve function, paralysis, and death. Veratridine lengthens the duration of action potential, results in recurrent firing, and depolarizes the nerve membrane potential. Some insect species are killed by sabadilla immediately, while others may remain paralysed for several days before dying.

1.6.2. Formulations

"Red Devil" sabadilla dust prepared by aged, heated, or alkali treated seeds to activate the insecticidal alkaloids. The activated seeds are ground and used as crude insecticidal dusts (Weinzierl 2000). In addition, the active alkaloids have been extracted in kerosene and used in spray emulsions or applied to inert carrier dusts.

1.6.3. Pests controlled or suppressed

Sabadilla is effective against "true bugs" (Hemiptera) such as squash bug, *Anasa tristis*; the chinch bug, *Blissus leucopterus* and the stink bugs including the Harlequin bug, *Murgantia histrionica.* Its action against these and other insects can result from both contact exposure and ingestion. Veratridine was more toxic than cevadine to houseflies, *Musca domestica* but cevadine was more toxic than veratridine both to the large milkweed bug, *Oncopeltus fasciatus* and the redlegged grasshopper, *Melanoplus femurrubrum* (Allen *et al.,* 1944).

1.6.4. Challenges

The insecticidal alkaloids in sabadilla degrade rapidly in air and sunlight, resulting in only very short residual control after application. Sabadilla is highly toxic to honeybees; this is particularly noteworthy where dust formulations might be applied to blooming cucurbits where pollinators are active at the same time the most common target pest, the squash bug, causes crop damage.

2. Discussion

Regulation stringency in using chemical pesticides, environmental and safety considerations, and new technological achievements is among the factors boosting biopesticide consumption and global market share. Conversely, the wide use of biopesticides is still restricted compared to synthetic pesticides, due to lower acute activity and higher degradation rate, higher production complexity, poor investment performance, restricted formulation approaches, and past reputations for poor efficiency. Moreover, regulations are still a drawback for the promotion of biopesticides in many countries. Thus, a rapid and straight forward registration process and effective communication within registrants and authorities through easiness in exchanging knowledge and information and improving organization can facilitate the expansion of the global biopesticides market. Enforcement of policies promoting natural biological products through legislative acts can play a pivotal role in boosting the biopesticides market. Regarding farmers, the effectiveness and reliability of biopesticides as compared to conventional agrochemicals are among the major criteria for consideration.

Isman and Grieneisen (2014), reported that the major obstacles for a broad adoption of plant-derived pesticidal products are (i) the lack of scale-up paradigms in international literature, (ii) the limited number of low-cost effective plant pesticidal products, (iii) the strict legislative framework for authorization, (iv) the short persistence of many botanical pesticides during application due to either evaporation or microbial degradation, and (v) the variation in phytochemical composition of plant extracts and mixtures due also to diverse climatic conditions that influence the extraction process, a fact that can result in the production of pesticidal products of weaker activity.

3. Conclusion

Botanicals play a crucial role in organic farming as they provide an effective and natural alternative to synthetic chemicals for promoting plant growth, controlling pests and diseases, and enhancing soil health. The use of botanicals in organic farming is supported by numerous studies that have demonstrated their efficacy and environmental safety. Therefore, the integration of botanical-based products in organic farming systems can help in achieving sustainable and eco-friendly agriculture practices.

4. Way Forward

An integrated plan to inform growers of bio-pesticides availability, uses and advantages should be issued by public authorities and private companies. Thus, collaboration and interaction among farmers, SMEs, key pesticide companies, universities, consumers, and authorities are needed to enhance building capacity in the field of bio-pesticides development and to ensure the scale-up and quality assurance of the novel biological products. Conversely, the wide biodiversity of pesticidal plant species in Africa, which are commonly used by low-income farmers to manage pests, suggests the existence of a new market segment that caters to both resource-poor farmers and organic producers in developed countries. As an example, world pyrethrum production mainly relies on plant species commercially cultivated in Africa and specifically in Kenya, which covers 80% of global production. Enough pesticidal plants at consistent availability should be guaranteed for scale-up development of botanical insecticides. In addition, the recent achievements in chemical synthesis and extraction, biochemical engineering, biotechnology, and molecular biology (e.g., DNA recombinant technologies) can boost the development and commercialization of novel plant pesticidal products.

5. References

Abdelgaleil, S.A.M., Mohamed, M.I.E., Shawir, M.S. and Abou-Taleb, H.A. 2016. "Chemical composition, insecticidal and biochemical effects of essential oils of different plant species from Northern Egypt on the rice weevil, Sitophilus oryzae L. Journal of Pest Science 89(1): 219–229.

Alam, M.J., Ahmed, K.S., Hoque, M., Mansura, A., Rony, M.N.H. and Haque, M.S. 2019. Bio-efficacy of some bio-pesticides against maize aphid, Rhopalosiphum maidis; a threatening pest of maize. Journal of Science, Technology and Environment Informatics 8(01): 563-572.

Allen, T.C., Dicke, R.J. and Harris, H.H. 1994. Journal of Economic Entomology 37:400.

Ayvaz, A., Karaborklu, S. and Sagdic, O. 2009. "Fumigant toxicity of five essential oils against the eggs of Ephestia kuehniella zeller and Plodia interpunctella (Hübner) (Lepidoptera: Pyralidae)," Asian Journal of Chemistry 21(1): 596–604.

Boboescu, N.T. 2020. Perspectives on the use of biopesticides in pest control. Annales of West University of Timisoara, Series of Biology 23(2): 135-146.

Bushra, S., Tariq, M., Naeem, M. and Ashfaq, M. 2014. Efficacy of neem oil and turmeric powder against Sitobion avenae and Rhopalosiphum padi. International Journal of Bioscience 5(12): 439-448.

Choudhary, R., Chandrakar, G., Bhardwaj, J.R., Khan, H.H. and Sahu, R. 2017. Assessment of the efficacy of neem-based insecticides for the management of yellow stem borer, Scirpophaga incertulas Walk in paddy field. Journal of Pharmacognosy and Phytochemistry 6(5): 1446-1449.

Das, R.K., Sarma, S., Brar, S.K. and Verma, M. 2014) Nanoformulation of insecticides: novel products. Journal of Biofertilizers Biopesticides 5: e120.

Dash, S., Mohapatra, L.N., Swain, S. and Swain, D.K. 2019. Bio-efficacy of newer insecticides in combination with neem product against plant hoppers of rice. Journal of Entomology and Zoology Studies 7(1): 1152-1155.

Davidson, W.M. 1930. Rotenone as a contact insecticide, Journal of Economic Entomology 23(5): 868-874.

Dieng , H., Sudha, R. and Hassan, A. 2013. Turning cigarette butt waste into an alternative control tool against an insecticide-resistant mosquito vector, 128(3): 584-590.

Dua, V. K., Pandey, A. C., Raghavendra, K., Gupta, A., Sharma, T. and Dash, A. P. 2009. Larvicidal activity of neem oil (Azadirachta indica) formulation against mosquitoes. Malaria Journal 8:124.

El-Bakry, A.M., Abdel-Aziz, N.F., Sammour, E.A. and Abdelgaleil, S.A.M. 2016. Insecticidal activity of natural plant essential oils against some stored product insects and their side effects on wheat seed germination. Egyptian Journal of Biological Pest Control 26(1): 83–88.

Gallo, M., Formato. A., Ianniello. D., Andolfi. A., Conte, E., Ciaravolo. M., Varchetta. V. and Naviglio. D. 2017. Supercritical fluid extraction of pyrethrins from pyrethrum flowers (Chrysanthemum cinerariifolium) compared to traditional maceration and cyclic pressurization extraction. The journal of supercritical fluid 119:104-112.

Grdiša, M., Klaudija. C., Ivan. K. and Zlatko. S. 2009. Morphological and Biochemical Diversity of Dalmatian Pyrethrum (Tanacetum cinerariifolium (Trevir.) Sch. Bip.). 74(2): 73-80.

Gudeta, B., Solmon, K. and Venkata Ratnam, M. 2021. Bio insecticide production from cigarette wastes. International Journal of Chemical Engineering 1-15.

Gupta, R.C., Miller Mukherjee, I.R., Malik, J.K., Doss, R.B., Dettbarn, W.D. and Milatovic, D. 2019. Insecticides: Biomarkers in Toxicology 455–475.

Gupta, S.C., Prasad, S., Tyagi, A.K., Kunnumakkara, A.B. and Aggarwal, B.B. 2015. Neem (Azadirachta indica): an indian traditional panacea with modern molecular basis. Phytomedicine 34(1): 14-20.

Henry, F. 2011. Tobacco Bio-oil Kills Agricultural Pests. Forum 119(1): 1-8.

Hien, P.P., Gortnizka, H. and Kraemer, R. 2003. Rotenone-potential and prospect for sustainable agriculture, Omonrice 11: 83-92.

Isman M.B. and Grieneisen M.L. 2014. Botanical insecticide research: Many publications, limited useful data. Trends Plant Sciences 19: 140-145

Isman M.B., Koul O., Luczynski A. and Kaminski J. 1990. Insecticidal and antifeedant bioactivities of Neem oils and their relationship to azadirachtin content. Journal of Agricultural Food Chemistry 38(1): 1406-1411.

Isman, M.B. 2006. Botanical insecticides, deterrents, and repellents in modern agriculture and an increasingly regulated world. Annual Review Entomology 51:45-66.

Jesser, E.N., Werdin-González, J.O., Murray, A.P. and Ferrero, A.A. 2017. Efficacy of essential oils to control the Indian meal moth, Plodia interpunctella (Hübner) (Lepidoptera: Pyralidae), Journal of Asia-Pacific Entomology 20(4):1122-1129.

Joshi, S., Hussain, T., Kirar, V. S. and Nagar, R. 2016. Management of sorghum shoot fly, Atherigona soccata Rondani (Diptera: Muscidae) through botanicals. Journal of Biopesticides 9(1): 23-26.

Khanal, S., Adhikari, A., Tiwari, A., Singh, N. B. and Subedi, R. 2019. Efficiency of botanical extract against maize weevil Sitophilus zeamais (Motschulsky, 1855) (Coleoptera: Curculionidae). World News of Natural Sciences 24: 1-8

Kraiss, H. and Cullen, E.M. 2008. Insect growth regulator effects of azadirachtin and neem oil on survivorship, development and fecundity of Aphis glycines (Homoptera: Aphididae) and its predator, Harmonia axyridis (Coleoptera: Coccinellidae). Pest Management Science: formerly Pesticide Science 64(6): 660-668.

Kumar, A., Tripathi, M. K., Chandra, U. and Veer, R. 2019. Efficacy of botanicals and bio-pesticide against Helicoverpa armigera in chickpea. Journal of Entomology and Zoology Studies 7(1): 54-57.

Kurtz, B., Goggins, J. and Junker, R. 2017. Bioactivity of botanical extracts and their use as natural alternatives to synthetic pesticides in organic farming. Journal of Agricultural and Food Chemistry 65(48): 10370-10387.

Lale, N. E. and Mustapha, A. 2000. Potential of combining neem (Azadirachta indica A. Juss) seed oil with varietal resistance for the management of the cowpea bruchid, Callosobruchus maculatus (F). Journal of Stored Products Research 36(3): 215-22.

Maletta, M., Tietjen, W., Ghidiu, G., Holmstrom, K. and Cowgill, W. 2002. Evaluation of controls for flea beetle on eggplant in an organic production system. In XXVI

International Horticultural Congress: Sustainability of Horticultural Systems in the 21st Century 6: 341-346.

Manda, R.R., Addanki, V.A. and Srivastava, S.E. 2020. Microbial bio-pesticides and botanicals as an alternative to synthetic pesticides in the sustainable agricultural production. Plant Cell Biotechnology Molecular Biology 21: 31-48.

Matharu, K.S. and Tanwar, P.S. 2019. Efficacy of different insecticides and biopesticide against wheat aphid. Journal of Entomology and Zoology 7(3): 521-524.

Melesse, T. and Singh, S.K. 2012. Effect of climatic factors on pea aphid, Acyrthosiphon pisum Harris (Homoptera: Aphididae) population and its management through planting dates and biopesticides in field pea (Pisum sativum L.). Journal of Agricultural Technology 8(1): 125-132.

Mohapatra, A., Mahalik, J.K., Dash, B.K. and Samal, B.K. 2020. Eco-friendly management of Meloidogyne incognita in okra through AM fungus (Glomus fasciculatum) and neem oil cake. Annals of Plant Protection Sciences 28(1): 76-79.

Mohapatra, L.N. 2018. Field evaluation of ITKs against insect pests of rice. Indian Journal of Traditional Knowledge 17(2): 360-364.

Pasquoto-Stigliani, T., Campos, E. V. R., Oliveira, J.L., Silva, C.M.G., BileskyJosé, N. and Guilger, M. 2017. Nanocapsules containing neem (Azadirachta Indica) oil: development, characterization, and toxicity evaluation. Scientific Reports 7:5929.

Pavela R. and Silva, T.D.J.A. 2006. New control technologies against pests based on Azadirachtin. Advances and Topical Issues 3(1): 563-566.

Rajakumar, G., Rahuman, A. A., Ambikapathy, V. and Chung, I.M. 2014. Coconut oil and its derived fatty acids as effective alternative biocontrol agents against plant pathogens. Crop Protection 54:1-9.

Raoul, B. B., Albert, N., Manuele, T. and Nchiwan, N. E. 2019. The incidence of aqueous neem leaves (Azadirachta indica A. Juss) extract and Metarhizium anisopliae Metch. on cowpea thrips (Megolurothips sjostedti Trybom) and yield in Ngaoundéré (Adamaoua-Cameroun). Journal of Zoology and Entomology Studies 7(5): 333-338.

Reuben, S., Minza, M., Makundi, R. and Robert, N. 2019. Control of Cowpea Weevil (Callosobruchus maculatus L.) in Stored Cowpea (Vigna unguiculates L.) Grains using Botanicals. Asian Journal of Plant Sciences 5(1): 91-97.

Riba, M., Marti, J. and Sans, A. 2003. Influence of azadirachtin on development and reproduction of Nezara viridula L, (Het., Pentatomidae). Journal of Applied Entomology 127: 37–41.

Samar S.I. 2018. Essential Oil Nano formulations as a Novel Method for Insect Pest Control in Horticulture, pp 1-20.

Sekar, K., lokesh, K. And Rajasingh, R. 2021. Toxicity of tobacco nicotiana tabacum linnaeus (solanaceae) leaf extracts to the rice weevil sitophilus oryzae linnaeus 1763 (coleoptera: curculionidae). The Journal of Basic and Applied Zoology 82(10): 1-5.

Shaiba, Z., Amoore, B., Amoore, I. and Renne, E. 2019. Assessing the impact of neem on fall armyworm damage to maize crops: a field-based study in Nabdam District, UER, Ghana. Journal of Agriculture and Sustainability 12(2):185-201

Shaik, A.S., Mishra, V.K. and Sundaramoorthy, P. 2019. Effectiveness of botanical-based extract on pest infestation of tomato crop in organic farming. Journal of Cleaner Production 239(11): 80-95.

Singh, R., Singh, K., Gautam, N., and Sinha, P.K. 2017. Effect of botanical formulations on soil health, productivity and nutrient uptake of Okra (Abelmoschus esculentus L.) under organic farming system. Journal of Plant Nutrition 40(10): 1426-1434.

Udo, I. and Ibanga, P.T. 2019. Comparative efficacy of aqueous extract of three botanicals on corn earworm (Heliothis armigera). Asian Journal of Agriculture and Food Scences 7(1): 15-19.

Weinzierl, R.A. 2000. Botanical insecticides, soaps, and oils. In Biological and Biotechnological Control of Insect Pests, ed. JE Rechcigl, NA Rechcigl,. Boca Raton, FL: CRC Press, pp 101–21

Yang, F.L., Zhu, F., and Lei, C.L. 2010. Garlic essential oil and its major component as fumigants for controlling Tribolium castaneum (Herbst) in chambers filled with stored grain, Journal of Pest Science 83(3): 311-317.

You, C.X., Yang, K. and Wu, Y. 2014. Chemical composition and insecticidal activities of the essential oil of Perilla frutescens (L.) Britt. aerial parts against two stored product insects, European Food Research and Technology 239(3): 481-490.

Yuru chang, Philip, F. H., Danielle, T.D. and Daniel, C. 2022. Biocontrol Potential of Essential Oils in Organic Horticulture Systems: From Farm to Fork Frontiers 8: 1-5.

Zubairi, S.I., Othman, Z.S., Sarmidi, M.R. and Aziz, R.A. 2016. Environmental friendly bio-pesticide Rotenone extracted from Derris sp.: A review on the extraction method, toxicity and field effectiveness. Journal Teknologi 78: 47-69.

10

Organic Farming in Oilseed Crops

G. Suresh and R.K. Mathur

ICAR-Indian Institute of Oilseeds Research, Rajendranagar, Hyderabad-500 030 Telangana, India

Abstract

Globally, organic oilseeds constitute an important element in organic production. Their products are important feedstuffs (oilcake) and also relevant for human consumption (soya products, oils), and demand for these is continually growing. Almost half of the global organic oilseed area is occupied by soybeans, and another almost 20% is for sunflower seeds and sesame. Rape and turnip rape constitute about 6% of the organic oilseed area. Oilseeds are of pivotal significance in the Indian agricultural economy supporting income of millions of farmers practicing mainly rainfed cultivation. Nine annual oilseed crops are cultivated in India that includes groundnut, rapeseed-mustard, soybean, sunflower, sesame, niger and safflower that are edible and non-edible oilseed crops include castor and linseed. Organic agriculture is one among the broad spectrum of production methods that are supportive of the environment. The principal method of organic farming includes crop rotation, green manures and composts, biological pest control and reduced tillage. The performance of annual edible oilseed crops and development of package in relation to organic management for sustainable agriculture are discussed in this chapter.

Keywords: *Organic farming, groundnut, soybean, sunflower, rapeseed-mustard, linseed*

1. Introduction

In India, both annual and perennial edible oilseed crops are cultivated. Annual edible oilseed crops include groundnut, rapeseed-mustard, soybean, sunflower, sesame, niger and safflower. India, the 4th largest vegetable oil economy in the world, stands first in the production of castor, sesame, safflower, and niger oils, whereas second in groundnut, third in rapeseed mustard, fourth in linseed and fifth in soybean oil. About 14 million farmers are engaged in oilseeds production and another million in their processing. Whereas, the perennial edible oil seed crops include oil palm and coconut. Edible oils are also produced domestically from secondary sources such as rice bran, cotton seed, corn, and other tree-borne oilseeds (TBOs), besides some minor oil producing species of forest origin. Exclusive organic farming is possible only in niche areas and commodities where there are clear opportunities for realizing higher premiums through exports and where there is a strong justification for replacing chemical inputs with organic or bio- inputs. Several expert reviews in the country pointed out that rainfed areas where most of the oilseed crops are ideally suitable for organic farming in view of the existing low use of chemical inputs and less drastic yield declines when farms are converted from chemical to organic management. The scope of niche areas of various commodities with demand for organic production in rainfed areas are listed in Table 1.

Table 1. Selected list of commodities with potential for organic production in rainfed regions

Commodity	Scope/Opportunity	Potential Area
Sesame	Demand for organic sesame seed for medicinal and confectionary uses	Gujarat, Rajasthan
Niger	Demand for niger seeds produced organically for bird feed in Europe	Tribal areas of different states, particular in Orissa and Chhattisgarh
Safflower	Growing market for safflower petals as natural food dye and herbal products	Maharashtra, Telangana
Groundnut	To produce residue/toxin free table varieties	Gujarat
Soybean	Demand for organically produced de-oiled cake (DOC) for livestock feed	Madhya Pradesh
Linseed	Demand for organic linseed seed for medicinal and confectionary uses	Uttar Pradesh, Madhya Pradesh, Maharashtra and NEH region
Cotton	Demand for organically produced lint. To cut down on chemical use	Maharashtra, Andhra Pradesh, Karnataka and Gujarat

(*Source*: Venkateswarlu, 2008)

Oilseeds are of pivotal significance in the agricultural economy of India supporting income of millions of farmers practicing mainly rainfed cultivation. The research carried out by ICAR Institutes/SAUs on cultivation of important annual oilseed crops under organic management are discussed below.

2. Soybean

The soybean [*Glycine max* (L.) Merr.] is a legume with nitrogen fixing ability through the symbiotic bacteria called Rhizobia within the nodules of their root systems. Soybean is a leading oilseed crop in the country with 20% oil and 40% protein. Crop has several food grade uses like vegetable, kuntz trypsin inhibitor (KTI) free, high oleic acid, tofu, soya milk etc. Apart from food uses soya meal can be used and exported for poultry and fig industry as their feed. Besides, soybean relatively easy to cultivate organically as legume crop fix atmospheric nitrogen (60-100 kg ha^{-1}) and incorporate around 30-40 kg N ha^{-1} in soil. Consequently, soybean contributes nitrogen fertility to subsequent crops under soybean based cropping systems. Additionally, soybean consume nutrients moderately compared to other crops and has greater scope under organic farming system.

Soybean is popular oilseed crop predominately grown under rainfed condition during monsoon season. Growing of soybean crop without using synthetic pesticides or fertilizers and relaying on wide range of cultural practices and alternative inputs believed to safer for both consumer and environment (Kuepper, 2003). Fortunately, Madhya Pradesh, Maharashtra, Rajasthan and Karnataka are leading states of soybean production and these states are also having largest area under organic certification. Further, in recent years' soybean is in demand for its meal export for poultry and piggery industry, consumed as vegetable/edamame and other diversified food uses. In addition, availability of vegetable and speciality type soybean varieties with high industrial demand shows, the soybean cultivation through organic farming has enormous scope in India. However, energy input use, including fuels for farm machinery, fertilizers, seeds, and herbicides, is about 30% less under the organic systems. Net returns to the organic rotations, without price premiums, are similar to those for conventional rotations, but are less when the costs of transition years are included. Conversely, the organic price premium required to equalize returns is only 10% above the conventional price, much lower-than-normal price premiums for organic grains. Therefore, identification and development of suitable organic cultivation methods for soybean is the need of the hour.

2.1. Crop and nutrient management

The crop management practices such as seed rate (60-65 kg ha^{-1}), spacing (45 x 10cm) and date of sowing (June 10th to July 5th) are same as that of conventional soybean. However, soybean nutrients requirement can be met through farmyard manure (FYM) and other organic sources. A well decomposed FYM @ 10 t ha^{-1} must be applied/incorporated to soil 20-25 days before sowing. Being a legume crop needs less amount of nitrogen, but requires more phosphorus (P).

Therefore, additional P requirement can be supplied through rock phosphate (16% P_2O_5) @ 375 kg ha^{-1}. Apart from this soybean seed can also be inoculated with bio fertilizers such as rhizobium culture, phosphate solubilizing micro-organism and vesicular arbuscular mycorrhizal (VAM) fungi @ 5-10 g kg^{-1}seed before sowing.

2.2. Tillage and weed management

Tillage is an important and essential operation in organic agriculture system to control weeds and insects-pests. Deep ploughing is essential during summer, after harvesting the *rabi* crop. This facilitates exposing the hibernating insects to extreme heat and predatory birds as well as movement of nutrients and infiltration of soil water. One deep ploughing once in 3-4 years, otherwise one normal ploughing in summer followed by 2 criss-cross harrowings or cultivation for breaking of soil clods will make ideal seed bed for a good crop of soybean cultivation under organic system. As such there are no specific weed control methods to control weeds in organic soybean production system. However, two hand weeding at 20 and 40 DAS or hoeing of field with bullocks during 15-25 DAS can check the weeds under ideal field condition with no continues rains.

2.3. Insect pest and disease management

Management of soybean insects and diseases through organic methods are challenging. However, fungal diseases can be control through seed treatment with *Trichoderma viride* @ 5 gkg^{-1} seed, *Pseudomonas syringae* @ 5 gkg^{-1} seed. *Pseudomonas syringae* is also used as foliar spray @ 10 g l^{-1} of water at initiation of fungal diseases. Insect pest control can be achieved through bio-pesticides/biocontrol agents' spray such as *Bacillus thuringiensis* (Dipel) @ 1.0 lha^{-1}, *Nomuriya rileyi* @ 1 $kgha^{-1}$, *Beauveria bassiana* @ 1$kgha^{-1}$ and nuclear polyhedrosis virus (NPV) @ 250 larval equivalents ha^{-1}. However, stem fly, girdle beetle and pod borer can be controlled by parasitoid naturally with trap/border crops such as Suva (*Anethum graveolens*), Marigold (*Tagetes patula*), cowpea (*Vigna unguiculata*), Dhaincha (*Sesbania rostrata*). Physical method also available to control some of the insect pests includes yellow sticky trap for white fly and pheromone traps for tobacco cater pillar and pod borer @ 5-10 ha^{-1} for monitoring purpose, 20-25 ha^{-1} for mass trapping and 80-100 ha^{-1} for mating disruption.

2.4. Response of soybean to organics

A field experiment was conducted at the research farm of ICAR-Indian Institute of Soil Science, Bhopal to study the effect of different combinations of organic inputs on performance of soybean (cv. JS 335). The results of the study revealed that, the treatments with organic manures either alone or in combination with *Panchagavya* 3% spray at 30 and 45 days after sowing (DAS) and/or biodynamic application of BD 500 (cow horn manure) as soil application @75 g ha^{-1} + BD 501(cow horn silica) as foliar application @2.5 g ha^{-1}) at 30 and 45 DAS improved the performance of soybean crop. The yield of soybean was found comparable and even better (higher 5-13%) with those obtained under Recommended Dose of Fertilizer (RDF), which enhanced nutrient removal by soybean crop by 5–13% with organic combinations. Similarly, the application of organic manures (Cattle dung manure was applied on the N equivalent basis) significantly improved the available nutrient status of soil and showed 19–22% and 28–33% higher available N and P, respectively than RDF. Soil enzyme activities, *viz.* dehydrogenase and alkaline phosphatase increased by 62–72% and 27–35%, respectively under the treatments receiving organic source of nutrients over RDF.

The study revealed that, the application of organic sources of nutrients significantly increased the productivity of soybean, enhanced nutrient uptake and improved soil nutrient status and accelerate soil enzyme activities. For soybean, to achieve maximum performance with sustenance of soil health, application of cattle dung manure (CDM) on N equivalent basis along with foliar application of *Panchagavya* @3% was found beneficial under organic management (Aher *et al.*, 2019).

3. Rapeseed-Mustard

India is the largest rapeseed-mustard growing country in the world, occupying second position in area and fourth position in production. Rapeseed-Mustard is a *rabi* annual, drought hardy crop, capable to grow under diverse agro-climatic conditions and different types of soils. Oil content in rapeseed-mustard varies from 33–46% amongst different species and varieties. Mustard oil is a great affordable alternative of palm oil. Organic farming of mustard is gaining importance in the country in recent years due to increasing soil, human and environmental concerns. The use of chemical fertilizers and pesticides is very low which can be substituted through organic sources.

Bundelkhand, a region lying in the Central and North India split between the states of Uttar Pradesh (U.P.) and Madhya Pradesh cultivates mustard. The soil in this region is categorised as red soil and black soil with poor organic content except black cotton soil. The use of chemical fertilizers and pesticides is very low which can be substituted through organic sources. Therefore, growing of crops under organic management is most appropriate to improve soil health and crop production in Bundelkhand.

Mustard being hardy crop is the most suitable oilseed crop for Bundelkhand region. Field studies conducted at Organic Research Farm, Karguaji, Institute of Agricultural Sciences, Bundelkhand University, Jhansi, Uttar Pradesh during 2017–2021, the production technologies were developed (Gangwar and Deepmala, 2023) for organic mustard are discussed below.

3.1. Ensure early sowing

The most optimum time for sowing is mid-October to first week of November. Timely sowing of crop is important to reduce the incidence of diseases. Studies conducted at organic farm of Bundelkhand University revealed that sowing of mustard during first week of November (4th November) was significantly better than sowing in 3rd week of November (18th November). Short duration varieties like Varuna and KMH-721 were better suited for sowing under delayed conditions.

3.2. Seed treatment and sowing methods

Seed treatment with *Trichoderma viridae* was found beneficial to avoid incidence of seed borne diseases. Normal seed rate of 4–5 kg ha^{-1} is sufficient. Seeds should be sown 3–4 cm deep in a moist soil to ensure good germination. Line sowing with a spacing of (45 x 15 cm) ensures good germination, optimum plant population and ease in intercultural operations which ultimately improves growth and yield.

3.3. Nutrient management through organic sources

Recommended dose of fertiliser for mustard crop in U.P. is 60:40:40 NPK kg ha^{-1} and 12–15 t ha^{-1} of FYM, which can be applied through the use of FYM and vermi-compost in the ratio of 1:1 at the rate of 6 t ha^{-1} of FYM and 3 t ha^{-1} of vermi-compost. FYM and vermi compost are the most commonly and easily available sources of organic manures. FYM is incorporated in the soil at the time of land preparation and vermi-compost is applied 2–3 cm below seeds during sowing in the moist soil. Biofertilizers like PSB and *Azotobacter* can be used to increase economical yield. Studies conducted at organic farms reveal that the phosphorus and sulphur needs can be met out through the use of 20 kg P_2O_5 ha^{-1} through rock phosphate at the rate of 110–160 kg ha^{-1} with bio-fertilisers.

3.4. Choice of suitable varieties

Certified varieties, free from foreign material to be selected to get good germination. Varieties *viz.,* Varuna, Pusa Bold, Giriraj, Super Gold and JM-3 were found most suitable in organic farm of Bundelkhand University, Jhansi. Other suitable varieties were found to be Kranti, Krishna, NarendraRai- 8501.

The highest seed yield (27.5 q ha^{-1}) and stover yield (40.3 q ha^{-1}) was realized in Varuna cultivar followed by Pusa Bold.

3.5. Weed management

One inter-cultivation and hand weeding at 40 - 45 DAS is necessary to avoid initial completion between weeds and crops.

3.6. Mulching

Organic mulches like rice straw, crop residues, stubbles @ 10 t ha^{-1} should be spread evenly to cover the inter row spaces, this checks the water losses through evaporation and at the same time help suppress weeds.

3.7. Integrated approach for pest management

Major pests of mustard in Bundelkhand are mustard aphid, mustard saw fly, painted bug, diamond backmoth, leaf miner, etc., while major diseases are Alternaria blight, white rust, Downey mildew, etc. The following cultural practices were found to reduce the infestation of pests and incidence of diseases:

- Deep ploughing to expose the soil borne pathogens and hibernating stage of defoliators.
- Timely sowing
- Adoption of proper drainage to avoid water stagnation
- Destruction of plant debris.
- Crop rotation to disturb the cycle of obligate pathogens.

3.8. Control measures for important insects and diseases

3.8.1. Aphids

Aphids are most important insects and generally attack from January to March. Therefore, early sowing of crop is desirable. If the infestation reaches at economic threshold level, i.e. 20% plant infestation and 13–14 aphids' $plant^{-1}$, spray 2% neem oil (20 ml oil + 1 ml liquid soap) at flower stage or 5% neem seed kernel extract (NSKE). Spray bio-agent *Verticillium lecanii* 1 kg ha^{-1} in 5000 l ha^{-1} water or use bio-predators like-ladybird beetles as they can feed up to 10–15 adult aphids day^{-1}.

3.8.2. Diamond back moth

Picking and destroying the larvae of pests. If the infestation is severe, installing pheromone traps @ 4 $acre^{-1}$ to monitor the adult activity. Application of 4% NSKE is very effective.

3.8.3. *Mustard painted bug*

Irrigate the crop at 30 DAS to reduce pest attack. Burn the remains of mustard crop so that the stage of insect does not reach the next year crop. Kill the bugs by removing from leaves and stems and sinking them in kerosene water.

3.8.4. *Mustard sawfly*

Collect and destroy larvae of sawfly in morning and evening. Use bitter gourd seed oil emulsion as an anti-feedant.

3.8.5. *Alternaria blight*

Spray of cow urine on affected plants. Proper sanitation of field is desirable. Adopt crop rotation to break the cycle of pathogen.

3.8.6. *White rust/Downy mildew*

Treat seeds with 4 g kg^{-1} seed with *Trichoderma viridae* before sowing. Treat the seeds with freshly prepared garlic bulb extract @2% (w/v) is also recommended. Ensure timely sowing of crop to reduce disease load. Avoid over irrigation and water stagnation. With the adoption of the recommended crop production and protection technologies in Bundelkhand region, and other areas representing similar agro-climatic conditions a seed yield of 25-30 q ha^{-1} of mustard with better quality of oil can be achieved under organic management.

Application of green manure, organic material, bulky organic manures were evaluated in Bharatpur, Rajasthan. Bulky organic manures are applied to improve overall soil health and reduce evaporation losses. Depending upon the availability of raw material and land use conditions various organic sources *viz.,* clusterbean (green manure), Sesbania (green manure), mustard straw @ 3t ha^{-1} and Vermicompost (2.5 7.5 t ha^{-1}) have been evaluated at Bharatpur. Green manure with Sesbania gave significantly higher mustard seed yield at Bharatpur and Bawal. Sesbania green manuring has shown higher mustard yield and improved soil environment. Many bio-stimulants also encourage higher production. At Hisar, foliar spray of Bioforce (an organic formulation) 2ml l^{-1} at the flowering and siliqua formation stage enhanced mustard seed yield (2059 kg ha^{-1}) (AICRP-RM, 2006)

4. Sesame

Sesame (*Sesamum indicum* L.) is an important oilseed crop grown in the states of Gujarat, Rajasthan, Madhya Pradesh, Tamil Nadu, Andhra Pradesh and Chhattisgarh. It is grown mostly under rainfed conditions during *kharif* but *rabi* and summer crop is supported with irrigation. India ranks first in area (29%) and production (26%) of sesame in the world. Sesame seeds are rich source of food, nutrition, edible oil, health care and bio-medicines. International demand and export of sesame are steadily increasing. India's share in world's exports of hulled sesame is around 60%. India exports about 2.5 lakh tonnes of seed valued at Rs.1000 crore. Since it is essentially used in food and medicine in developed countries, the importers are insisting for certified organic produce.

Globally sesamum is exported in the form of seed. Seed with uniform size, lustrous, white, bold, sweet, free from flash, low free fatty acids (<2%), and free from pesticide residues is preferred in the global markets. However, scattered production with varied management practices is resulting in seed of non-homogenous quality. The solution lies in identifying appropriate production zones, standardization of production technologies and adoption of organic production methods through contract farming to produce the desirable products.

A field experiment was conducted during the rainy season of 2005-2009 at CRIDA, Hyderabad to assess the yield and quality of sesame, and soil properties under organic vis-à-vis conventional production systems. The experiment was laid out in 3 blocks *i.e.* control, organic and conventional with a 5-meter buffer zone between the blocks with pearl millet as the barrier crop. Sesame (var. 'Swetha Til') was sown at a seeding rate of 6.5 kg ha^{-1} and a row spacing of 30 cm. The crop was sown during second fortnight of June to first fortnight of July depending upon the onset of monsoon in each year (Gopinath *et al.*, 2011).

The two treatments *i.e.*, conventional (chemical) and organic were designed based on the available data and package of practices evolved for the crop and using permitted inputs in case of organic treatment. The conventional system included application of recommended dose of chemical fertilizers (40:26.5:33.3 kg NPK ha^{-1}) and recommended pest management module *viz.*, seed treatment with thiram 3 g kg $^{-1}$ seed; spray of dimethoate 2 ml l^{-1} water for control of thrips/red spider mites; dimethoate/ chlorpyriphos 2 ml l^{-1} for podfly; monochrotophos 1.6 ml l^{-1} for pod borer; COC 3 g l^{-1} for stem/root rot, mancozeb 1g l^{-1} for Alternaria leaf spot, removal of plants for phyllody and sulfur dust 3 g l^{-1} for control of powdery mildew.

The organic system included application of FYM 3.7 t ha^{-1} + neem cake 900 kg ha^{-1}+ ash 75 kg ha^{-1}+ bone meal 75 kg ha^{-1} + elemental sulfur (ELS) 20 kg ha^{-1} + PSB 5 kg ha^{-1} (soil application) + *Azotobactor* 5 kg ha^{-1}, and pest management with *Trichoderma* (0.4%) seed treatment + neem oil spray thrice at 15, 30 and 45 DAS, and *Azadirachtin* (0.03%) spray at 30 DAS. The organic manures were applied two weeks before sowing and chemical fertilizers were applied at the time of sowing. Two hand weedings were carried out at 20 and 40 days after sowing in all the plots. The results revealed that seed yields were higher under conventional farming than under organic farming and control in all the years. During the initial three years, organic sesame yield was about 20% less compared to conventional farming. However, in the fourth and fifth year, the seed yield reduction under organic farming was 7.7 and 7%, respectively compared with conventional farming indicating that the yield levels under organic farming may further improve over the years. Lower seed yields in the plots under organic management may have been associated with the less readily available nutrients in the initial years of transition as nutrient cycling processes in first-year organic systems change from inorganic N fertilization to organic amendments (Reider *et al.,* 2000) and slower release rates of organic materials (Liebhardt *et al.,* 1989).

4.1. Quality of sesame oil

Sesame is an important oilseed crop having an ideal balance of different fatty acids present in its oil and is practically free from any undesirable constituent making it traditionally suitable oil for frying as well as table purpose. Organically grown sesame seeds had marginally higher oil (%) content compared with conventional and control treatments. No significant variation was observed in the fatty acid composition across different organic and conventional treatments over different years.

4.2. Soil Fertility

Organic cultivation of sesame after five years resulted in the build-up of organic carbon content (0.71%) in soils as compared to conventional (0.63%) and control (0.53%) treatments. The available P and K contents were also higher in the plots under organic management than those under conventional and control treatments.

4.3. Economics of organic sesame production

Economic evaluation of organic sesame cultivation was also done by assuming different price premiums (0-40%) for the produce to assess whether sesame can be profitably grown under organic farming conditions in comparison with conventional practice. The cost of cultivation of sesame was Rs. 10800 and 14050 per hectare under conventional and organic farming, respectively. Higher cost of cultivation under organic management was mainly due to more input costs particularly for purchase of neem cake and bone meal. The net return from organic sesame was about 60% less compared with conventional management. Furthermore, the reduction in net return from organic management was 46, 29, and 13% at 10, 20 and 30% price premium, respectively compared with conventional management. However, at 40% price premium for organic sesame the net returns from organic sesame was comparable with that of conventional sesame. Therefore, at least 40% price premium for organic sesame may be required to offset the higher cost of cultivation and low yields under organic production system compared with conventional production system particularly during initial years (Gopinath *et al.,* 2011)

Field studies conducted at ICAR-Indian Institute of Oilseeds Research, Hyderabad on organic sesame revealed in soybean-sesame cropping system, application of either castor cake (658 kg) or combined application of FYM (680 kg), vermicompost (637 kg) and goat manure (400 kg) (on equal N basis) produced comparable sesame yield as that of nutrient management through inorganic sources. Application of organic sources either FYM/Vermicompost and goat manure; castor cake or combined application of on equal N basis produced comparable sesame yield through inorganic sources. Sesame seeds stored in high density polyethylene (HDPE) bag or Deluxe Lens Pouch (DLP) bag with or without sweet flag @ 10 g kg^{-1} effective against red flour beetle & recorded significantly least seed damage, weight loss and higher germination (%) at 6 months after storage over conventional cloth and jute gunny bags (Annual Report, 2022).

5. Sunflower

Sunflower (*Helianthus annuus* L.) is an oilseed crop with multiple food and industrial uses, such as production of edible oil, the biofuel industry, the manufacture of soap and fodder production. Sunflower is the most important honey plant, considering the large areas cultivated with sunflower and the quality of nectar. Due to its ability to use soil moisture and soil nitrogen reserve better than other species, and to good productivity and the high quality of its by-products, sunflower crop is representing a good option for organic farming systems.

Field experiments were conducted for two consecutive *rabi* seasons at Tirupati, Southern plateau and Hills zone of India on red sandy soils with fourteen treatments comprised of six different sources of nitrogen *viz.,* FA manure, vermicompost, neem leaf, poultry manure, pig manure and fertilizer to supply recommended dose of nitrogen on equivalent nitrogen basis and one absolute control were tried with and without the foliar application of *Panchagavya* @ 3%. All the growth and yield attributes, yield (seed as well as stalk) harvest index, nitrogen uptake and gross returns as well as net returns of sunflower were at their best with recommended dose of fertilizer either with or without 3% *Panchagavya* spray. Reasonably higher yield resulting in higher gross returns and relatively cheaper cost of poultry manure has resulted in higher B-C ratio (Rao *et al.,* 2013). All the five organic manures tried on sunflower crop have resulted in equal performance in terms of yield and economic returns. The results have revealed that sunflower crop could respond to various organic manures in similar way and hence, one can go for the choice of organic sources to sunflower, depending upon the abundant availability locally and cheaper cost. The highest oil content of the seed was recorded with neem leaf manure in combination with *Panchagavya*, which was however, comparable with all the other four organic sources tried in combination with *Panchagavya*, but significantly higher than with the fertilizer either with or without the use of *Panchagavya*. The increase in oil content with the use of *Panchagavya* ranged from 5.2 to 8.0% with different manures. Further, *Panchagavya* is known to contain certain plant growth stimulants, which can improve the biological efficiency and quality of certain crops. Thus, the oil content of sunflower seed might have been improved due to beneficial effect on biosynthesis of oil. These results are in accordance with those of Somasundaram (2003). The highest phosphorus uptake of sunflower was recorded with poultry manure either with or without the spray of 3% *Panchagavya*, while the potassium uptake was the highest with vermicompost either with or without the spray of *Panchagavya*. Irrespective of the source of nutrient supply, foliar application of *Panchagavya* resulted in higher nutrient uptake of sunflower than with the use of respective organic manures alone without the use of *Panchagavya*. Among the organic sources tried, the highest net returns and benefit-cost ratio of sunflower were realized with poultry manure in combination with 3% *Panchagavya*. It was clearly found from this study that foliar application of *Panchagavya* alone to maize and sunflower crops could not exert any beneficial effect either on the productivity or quality and this was just comparable with non-manuring to the two crops through any source. This amply indicates that foliar application of *Panchagavya* @ 3% alone to crops cannot meet the nutritional requirement of the crops and *Panchagavya* can be used as a supplement or additive, whenever organic sources are applied to the crops. Further, *Panchagavya* acts as a top dressing to supplement the nutrient requirement of crop. Hence, it can be tried out higher levels of organic sources and high carbon contents of soils also

6. Linseed

Among the oilseed crops raised during *rabi*, linseed is next in importance to rapeseed-mustard in area as well as in production. In technical oil production, it ranks first in the country. Organic products in India is still a small niche market hence; at present there is not a sufficient number of producers to meet the growing demand in the cities. With recent development of super markets in several cities of India and few showrooms of Organic India, Eco farms, FabIndia and Navdanya etc. the marketing of processed organic produce got accelerated (Singh *et.al,* 2018).

6.1. FabIndia's three categories of food products

6.1.1. Green logo

Products displaying our green logo are 'Fully certified' Organic All process, from growing to preparing to packing have been done according to National and International Standards, verified by accredited agencies.

6.1.2. Blue logo

Products displaying our Blue logo are 'In conversion'. This means that the farmer is using purely organic techniques, and has registered and is complying with set standards, farmland must be managed organically for about 3 years before it can be fully certified.

6.1.3. Yellow logo

Products displaying our Blue logo are 'Natural'. This category contains products produced by small farmers who use purely organic technique, but who have decided to not yet register for certification also includes some processed food, which do not contain synthetic preservatives, colours, flavours or additives.

Rainfed Organic Cultivation of Linseed Crop Variety T397 in Farmer's Field (Ranchi district, Jharkhand)

Raw Organic Product (Linseed)

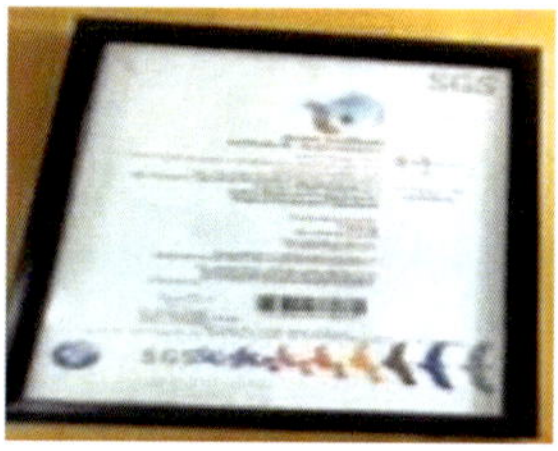

SGS- National Programme for Organic Production (NPOP) in FabIndia, Patliputra

Fig. 1. Organic Linseed Production Procedures

Every part of the linseed plant is utilized commercially, either directly or after processing. Seed contains 33 to 47% of oil. On a very small scale, the seed is directly used for edible purposes. About 20% of the total oil produced is used at farmers' level, and the rest 80% oil goes to industries in various forms, such as boiled oil, borated oil, eposidized oil, aluminated oil, urethane oil, isomerized oil etc. The oil is rich (> 66%) in linolenic acid, and is a perfect drying oil. Hence it is utilized in the manufacture of paints, oil cloth, varnish, pad-ink, printed ink, linoleum etc. The oil-cake is a good feed for milch cattle and poultries and hence priced 50% higher than rapeseed-mustard cake. It is good in taste and contains 36% protein, 85% of which is digestible. It is also used as organic manure. It contains about 5% N, 1.4% P_2O_5 and 1.8% K_2O. Linseed is globally cultivated for its fibres and is called flax. Fibres are used for the manufacture of linen. The stem yields fibre of good quality having high strength and durability. The weather resistant fibre is lustrous and blends well with wool, silk, cotton etc. Strong twines, canvas suiting's, shirtings and various indispensable products for defence purposes are manufactured from it. Linseed is nutritionally rich and have many health benefits for organic production. It can be grown in marginal lands under different agro- climatic conditions of India with assured harvest. It is short duration, drought resistant, easy for cultivation with limited external inputs, fewer pests and diseases and long storability of grains without store pests' damage are unique qualities and hence can be a step forward for doubling the income of farmers in India. Singh *et al.* (2018) opined that due to increasing demand of linseed seed in cities, organic linseed farming can be considered as a step towards doubling farmers' income. The cultivation practices of organically grown linseed seeds are described below:

6.2. National scenario

Bihar, Chhattisgarh, Jharkhand, Karnataka, Madhya Pradesh, Maharashtra, Nagaland, Odisha, Uttar Pradesh and West Bengal are the major linseed growing states

6.3. Climatic requirements

Linseed is a cool season crop. The temperature during the vegetative development of the crop should be moderate or cool. Temperature above 32 °C accompanied with drought during the flowering stage reduces the seed yield, oil content in seed and also the quality of the oil. Moderate temperatures (21°-26°C) are ideal. At the flowering, frost is very harmful to the crop. The crop is well suited to tracts of low rainfall and is generally raised where the average annual rainfall ranges from 45 to 75 cm.

6.4. Soils and its preparation

Linseed can be profitably raised in places where the other crops may fail. Hence, it is often grown on marginal and sub-marginal rainfed soils as pure crop, mixed crop, intercrop and paira or *utera* crop.

6.5. Seed rate, seed treatment, time of sowing and spacing

A seed rate of 25- 30 kg ha^{-1} is usually followed and sown by broadcast or by drilling in rows at row spacing of 25 to 30 cm. The seed requirement is more in bold-seeded varieties and in *utera* cropping system. Treat the seeds with *Trichoderma* @ 4 g kg^{-1} seed before sowing. This ensures a good stand by protecting seedlings against seed borne diseases. An inter-row spacing of 20-30 cm and intra-row spacing of 7-10 cm are ideal. The time of sowing varies from early October to mid-November in different states. Rainfed crop requires early sowing. Early sowing also helps the crop to escape from powdery mildew, rust and pod fly menaces.

6.6. Nutrient management

Under rainfed situations best crop is raised with use of organic source i.e., application of 2.5 tonnes of vermicompost per hectare at the final field preparation. But in irrigated situations, total amount of vermicompost is divided in two equal parts i.e., @ 1.25 tonne as basal and the remaining 1.25 tonne is applied with the first irrigation 35 days after sowing.

6.7. Water management

Linseed is a crop of rainfed areas (> 90%). However, it responds well to irrigation. Branching, flowering and capsule formation stages are critical for irrigation. Two irrigations are sufficient to obtain good yields. First irrigation should be applied 30 to 40 days after sowing and the second just before flowering. However, 3 irrigations (35, 55 and 75 days after sowing) have proved effective.

6.8. Weed control

This crop is usually dwarf statured, and therefore suffers severe competition by weeds. Initial 3-6 weeks after sowing is critical period of crop-weed competition. The uncontrolled weeds can reduce yields by 25 to 40%. The losses are more in rainfed and utera cropping systems primarily due to competition for moisture followed by nutrients. The important weeds of linseed include *Anagallis arvensis, Vicia hirsuta, Fumaria parviflora, Melilotus spp., Chenopodium album, Phalaris minor* etc. The crop is parasitized by *Cuscuta* sp. leading to heavy losses of yield. In line sown linseed crop, power weeder (covering 22 cm width of land between rows) when used twice (3 and 6 weeks of sowing) has been found most eco-friendly, efficient and economical

6.9. Yield potential

The crop yield varies from place to place depending on the climate, soil, technology and variety. A well-managed crop may yield 1.5 to 2.0 tonne of seeds ha^{-1}. In linseed oil to seeds crushed is 33% while cake to seeds crushed is 67%.

7. Research findings of organic cultivation in oilseed as component crop in cropping system mode

ICAR-Indian Institute of Farming Systems Research during X Five-year plan initiated a 'Network Project on Organic Farming (NPOF)' scheme, to study the location specific agronomic practices of organic farming and developed package of practices for arable crops by involving 11 SAU's and 8 ICAR Institutes that included important oilseeds as component crops in cropping systems perspective (Panwar *et al.*,2022).

Experiments involved, systematic evaluation of organic farming practices including screening of varieties of different crops under organic production system in a scientific way by combining tradition, innovation and science. Organic management practices were compared with integrated crop management practices comprising of integrated nutrient management, integrated weed management and integrated pest management. In all, package of practices covering organic production of cropping systems with oilseed crop as component crops have been developed suitable for different oilseed growing states (Panwar *et al.* 2022). Ecological benefits of organic farming in terms of soil microbes and natural enemy population along quality and sustainability parameters are also highlighted for selected cropping systems and salient findings are described in Table 3.

Table 3. Organic Farming packages for cropping systems with oilseed crop as component crops in different agro-eco regions (***Source:*** Panwar *et. al.*, 2022)

S . No	State/ Oilseed based cropping system	Recommended Practices			Salient Findings
		Crop/ Varieties	**Nutrient supply**	**Pest management**	
1.	**Karnataka** Soybean (*kharif*) – Wheat (*rabi*)	**Soybean** (JS-335, JS-9305, DSB 21)	Seed treatment with *Rhizobium and PSB* @ 1 kg ha^{-1} each + Enriched compost @ 3.33 t ha^{-1} + Vermicompost @ 2.66 t ha^{-1} + green leaf manure 5.33 t ha^{-1} + Neem oil cake @ 250 kg ha^{-1} asbasal application + Foliar spray of Cow urine @ 10% @100 l ha^{-1} + *Panchagavya* @ 3% (30 l ha^{-1})at 30 and 45 days after sowing (DAS).	Botanicals spray 5% @ 25 l ha^{-1} + NSKE(Neem seed kernel extract) 5% at 50 l ha^{-1} + *Nomuraea rileyi* (1 g l^{-1}) @ 1 kg ha^{-1}	Soybean-wheat system is dominant in 4 districts occupying 77,297 ha. Soybean recorded an average yield of 2130 kg ha^{-1}under organic and 1923 kg ha^{-1} under integrated crop management (ICM). Wheat recorded mean yield of 1131 kg ha^{-1}under organic and 1116 kg ha^{-1}under ICM. Mean B:C ratio is higher under organic (2.87) than ICM (2.46). Soil Organic Carbon improved by 16.4% in 8 years under organic management.
		Wheat (DWR-2006, Bijaga Yellow)	Seed treatment with *Azospirillum* and PSB@ 0. kg ha^{-1} each + Enriched compost @ 2.00 t ha^{-1} + Vermicompost @ 1.70 t ha^{-1} +green leaf manure 3.30 t ha^{-1} + Neem oil cake @ 250 kg ha-1 as basal application + Foliar spray of Cow urine @ 10% @ 100 lha^{-1} +*Panchagavya* 3% @ 30 lha^{-1} at 30 DAS and at boot leaf stage.	Seed treatment with *Trichoderma harzianum* @ 0.5kg ha-1 in soybean and @ 0.25 kg ha-1 inwheat to control soil borne diseases. Botanicals10% @ 500 lha^{-1} + NSKE (Neem seed kernel extract) 5% at 250 lha^{-1}+ *Verticilliumlecanii* (1 g l^{-1}) @ 0.5 kg ha-1, Foliar Spray of *Pseudomonas* fluorescans @ 5% + Neem oil @0.5% for rust and leaf spot management.	

2.	**Karnataka** Groundnut (*kharif*) – Sorghum (*rabi*)	**Groundnut** (GPBD - 4, JL – 24)	Seed treatment with *Rhizobium and PSB* @ 1kg ha^{-1} each+Enriched compost @ 3.0 t ha^{-1} + vermicompost @ 2.4 t ha^{-1} + green leaf manure 5.0 t ha^{-1} + Neem oil cake@ 250 kg ha^{-1} as basal application + Foliar spray of Cowurine @ 10% @ 100 l ha^{-1} + *Panchagavya* 3% @ 30 l ha^{-1} at 45 and 60 days after sowing + Soil application of Gypsum @ 500 kg ha^{-1} at 35 days after sowing.	Seed treatment with *Trichoderma harzianum* @ 0.75kg ha^{-1} in groundnut and @ 0.03 kg ha^{-1} in sorghum to control soil borne diseases. Neem seed kernelextract @ 5% or custard apple leaf extract @ 5% @ 100 l ha^{-1} + *Nomuraea rileyi* conidia (1 g l^{-1}) @ 2 kg ha^{-1}	Groundnut-sorghum system is dominant in 12 districts occupying 2.8 lakh ha. Groundnut recorded an average yield of 3064 kg ha^{-1} under organic and 2900 kg ha^{-1} under integrated crop management. Sorghum recorded mean yield of 1226 kg ha^{-1} under organic and 1176 kg ha^{-1} under ICM. Mean B:C ratio is higher under organic (3.80) than ICM (3.56). Soil Organic Carbon improved by 34 % in 7 years under organic management.
		Sorghum (M35-1, DSV-4)	Seed treatment with *Azospirillum* and PSB @ 0.5kg ha^{-1}each+ Enriched compost @ 2.0 t ha^{-1} + vermicompost@ 1.7 t ha^{-1} + green leaf manure 3.33 t ha^{-1} + Neem oil cake @ 250 kg ha^{-1} as basal application + Foliar spray of Cow urine @ 10% @ 100 l ha^{-1} + *Panchagavya* 3%@ 30 l ha^{-1} 30 and 45 days after sowing.	Botanicals 10% @ 50 l ha^{-1} + Neem Seed Kernel Extract 5% at 500 l ha^{-1} +foliar application of *Verticillium lecanii* (2 g l^{-1}) @ 1.0 kg ha^{-1}	

3.	**Madhya Pradesh** Soybean (*kharif*) – Mustard (*rabi*)	**Soybean** (RVS 2002-4, JS-20-41)	Cow dung manure (0.95% N) @ 5 t ha^{-1}	Neem oil (10000 ppm) @ 1 l ha^{-1} along with soapsolution	Soybean-mustard system is dominant in 9 districts occupying 1.9 lakh ha. Soybean recorded an average yield of 817 kg ha^{-1} under organic and 691 kg ha^{-1} under integrated crop management (ICM). Mustard recorded mean yield of 1250 kg ha^{-1} under organic and 1186 kg ha^{-1} under ICM. Mean B:C ratio is higher under organic (2.44) than ICM (2.00). Soil Organic Carbon improved by 71 % in 7 years under organic management.
		Mustard Pusa bold)	Cow dung manure (0.95% N) @ 1.5 t ha^{-1} + Vermicompost (1.41% N) @ 1.7 t ha^{-1} + Poultry manure (2.36% N) @ 1 t ha^{-1}	Neem oil (10000 ppm) @ 1 l ha^{-1} along with soapsolution	
4	**Madhya Pradesh** Soybean (*kharif*) – Chickpea (*rabi*)	**Soybean** (RVS 2002-4,JS-20-41)	Cow dung manure (0.95% N) @ 5 t ha^{-1}	Neem oil (10000 ppm) @ 1 l ha^{-1} along with soap solution	Soybean recorded an average yield of 791 kg ha^{-1} under organic and 726 kg ha^{-1} under integrated crop management (ICM) Chickpea recorded mean yield of 1532 kg ha^{-1} under organic and 1451 kg ha^{-1} under ICM. Mean B:C ratio is higher under organic (2.45) than ICM (2.00). Soil Organic Carbon improved by 75 % in 7 years under organic management
		Chickpea (JG-130, RVG203)	Cow dung manure (0.95% N) @ 1.7 t ha^{-1} + Vermicompost (1.41% N @ 1.3 t ha^{-1} + Poultrymanure (2.36%) @ 0.5 t ha^{-1}	Neem oil (10000 ppm) @ 1 l ha^{-1}along with soap solution	

5.	**Madhya Pradesh** Soybean (*kharif*) – Linseed (*rabi*)	**Soybean** (RVS 2002-4,JS-20-41)	Cow dung manure (0.95% N) @ 5 t ha^{-1}	Neem oil (10000 ppm) @ 1 l ha-1 along with soap solution	Soybean-linseed system is dominant in 4 districts occupying 23,037 ha. Soybean recorded an average yield of 814 kg ha^{-1} under organic and 675 kg ha^{-1} under integrated crop management (ICM). Linseed recorded mean yield of 1464 kg ha^{-1} under organic and 1333 kg ha-[1] under ICM. Mean B:C ratio is higher under organic (2.44) than ICM (2.01). Soil Organic Carbon improved by 19.2 % in 6 years under organic management.
		Linseed (JL-9)	Cow dung manure (0.95% N) @ 3.4 t ha^{-1} + Vermicompost (1.41% N @ 1.7 t ha^{-1} + Poultry manure (2.36% N) @ 1 t ha^{-1}	Neem oil (10000 ppm) @ 1 l ha^{-1} along with soap solution	
6.	**Maharashtra** Rice (*kharif*) – Groundnut (*rabi*)	**Rice** (Karjat-3, 4 and 7 Palghar-1, Sahyadri-3, 4 and 5)	Basal application: FYM @ 5 t ha^{-1} + Neem cake @0.50 t ha^{-1} (before puddling) + *Glyricidia* green leaves @ 4.5 t ha^{-1}+ Rice straw @ 4.2 t ha^{-1} (soil incorporation before transplanting);Top dressing + Cow urine and Vermiwash each @ 50 l ha^{-1} at 30 and 60 DAT.	Use of pheromone traps @20 nos. ha^{-1}; Release of*Trichogramma* @ 50,000 ha^{-1} for 4 times for stemborer management	Rice-groundnut system is dominant in 4 districts occupying 18,821 ha. Rice recorded an average yield of 3974 kg ha^{-1} under organic and 4156 kg ha^{-1} under integrated crop management (ICM). Groundnut recorded mean yield of 2486 kg ha^{-1} under organic and 2491 kg ha^{-1} under ICM. Mean B:C ratio is higher under organic (1.79) than ICM (1.56). Soil Organic Carbon improved by 18.6 % in 4 years under organic management.
		Groundnut (SB-XI, Konkan Gaurav, Konkan Trombay Tapora,TG 26)	Basal application: FYM @ 1.66 t ha^{-1} + Vermicompost @ 0.55 t ha^{-1} + Neem cake @ 0.16t ha^{-1}; Top dressing: Cow urine and Vermiwash each @ 50 l ha^{-1} at 30 and 60 DAS.	Application of nimbicide@ 3ml l^{-1} of water for aphids management	

7.	**Coimbatore** Chilies (*kharif*) – Sunflower (*rabi*) – Green manure(summer)	**Chillies** (PKM1 & K1)	FYM @ 7.05 t ha^{-1} + vermicompost @ 3.09 t ha^{-1} + *Azospirillum* @ 2 kg ha^{-1} + *Pseudomonas* @ 2.5 kg ha^{-1}, *Trichoderma* @ 2.5 kg ha^{-1}, seed treatment with *Trichoderma viride* @ 4g kg^{-1} or *Pseudomonasfluorescens* @ 10g kg^{-1}	Pheromone traps @ 10 nos.ha^{-1} for managementof fruit borer *Helicoverpa armigera or Spodoptera litura*, spray Neem oil@ 3% or Neem seed kernelextract @ 5%	Chillies-sunflower-greenmanure system is dominant in 2 districts occupying 3,700 ha. Chilli recorded mean yield of 4645 kg ha^{-1} under organic and 4791 kg ha^{-1} under ICM. Sunflower recorded mean yield of 1399 kg ha-[1] under organic and 1665 kg ha^{-1} under ICM. Green manure recorded an average biomass yield of 6399 kg ha^{-1} under organic and 6292 kg ha^{-1} under integrated crop management (ICM). Mean B:C ratio is found to be 1.72 under organic and 2.11 under ICM. Soil Organic Carbon improved by 26.02 % in 7 years under organic management
		Sunflower (TNAU sunflower hybrid-COSFV -5)	FYM @ 5.30 t ha^{-1} + vermicompost@ 3.37 t ha^{-1} + *Azospirillum* @ 2.0 kg ha^{-1} + *Phosphobacteria* @ 2.0kg ha^{-1}	Marigold as trap crop at 50 plants $acre^{-1}$, use of pheromone traps @ 4 traps $acre^{-1}$ to trap the male moths, Release of predators @ 1 larva $head^{-1}$ like *coccinellids, Chrysoperla carnea, release parasitoides* @ 20,000 acre $^{-1}$ like *Trichogramma sp.,(Bracon* spp., *Campoletis* sp.), spraying of 5% Neemoil or 5% Neem seed kernal extract before egg laying	

8.	**Uttar Pradesh** Maize (sweet corn) (*kharif*)– Mustard (*rabi*) – Sesbania (summer)	**Maize (Sweetcorn)** (Madhuri,VL sweetcorn 1)	FYM @ 10.0 t ha^{-1} + *Azotobacter*@ 10.0kg ha^{-1}+*PSB* @ 10.0 kg ha^{-1}+ *Trichoderma*@ 5.0 kg ha^{-1} + *Neem cake*@ 2.0 q ha^{-1} +Vermicompost @ 4.0 t ha^{-1} + *Panchagavya*@ 15.0 l ha^{-1} at 45 and 60 DAT	*Trichogramma @ 1.0 lakh (nos.) eggs ha-1 in* 5-6 release at 10 days' interval to manage the stemborer Ginger-Chilli-Garlic extract (10 kg Garlic +5 kg Ginger + 5 kg Green chilli in 70 l water) @ 60.0 l ha^{-1} to manage caterpillars/leaf folders	Maize– Mustard – sesbania system is dominant in 8 districts occupying 2 lakh ha. Maize recorded an average yield of 7894 kg ha^{-1} under organic and 9482 kg ha^{-1} under integrated crop management (ICM). Mustard recorded mean yield of 2290 kg ha^{-1} under organic and 2281 kg ha^{-1} under ICM. Mean B:C ratio is found to be 2.52 under organic and 2.83 under ICM. Soil Organic Carbon improved by 12.90 % in 5 years under organic management
		Mustard (Pusa bold, Pusa Vishal)	FYM @ 5.0 t ha^{-1} + Vermicompost @ 5.0t ha^{-1} + Seed treatment with Bioagent-3(mixture of *Pseudomonas* + *Trichoderma*)@ 10.0 g kg^{-1} seed + Cow urine (10.0 l)fortified with Neem green leaves (1.0 kg) @ 3-4 at 15 days interval after 20-25DAS @ 50.0 l ha^{-1}	Spray of *P. florescence* and *Trichoderma* at the time of field preparation, before and afterflowering @ 5.0 g l^{-1} water ha^{-1} to manage root rot, white rust, downey mildew. Cow urine(10.0 l) fortified with Neem green leaves (1 kg)or Cow urine (10.0 l) fortified with Neem oil (1.0 l) @ 50.0 l ha^{-1}	

9.	**West Bengal** Rice (*kharif*) – French beans (*rabi*) – Sesame (summer)	**Rice** (Satabdi / IET4786)	Vermicompost @ 4.5 t ha^{-1} + Foliar applicationof *Panchagavya* (3.0%) @ 125.0 l ha^{-1} in each spray	Panchaparni @ 300 l ha^{-1} at 15 days interval to manage Rice leaf folder.	Rice – French beans – Sesame system is grown in 14 districts occupying 2 lakh ha. Rice recorded an average yield of 5867 kg ha^{-1} under organic and 6025 kg ha^{-1} under integrated crop management (ICM). French beans recorded mean yield of 5537 kg ha^{-1} under organic and 5517 kg ha^{-1} under ICM. Mean B:C ratio is higher under organic (1.39) than ICM (1.08). Soil Organic Carbon improved by 37.3 % in 4 years under organic management
		Frenchbean (Falguni)	Vermicompost @ 6.75 t ha^{-1} + Foliar applicationof *Panchagavya* (3.0%) @ 187.5 l ha^{-1} in each spray	Neem oil (10000 ppm) @ 600 l ha^{-1} in 15 days interval to manage Bean Common Mosaic	
		Sesame (Tilottama)	Vermicompost @ 3.0 t ha^{-1} + Foliar applicationof *Panchagavya* (3.0%) @ 125 l ha^{-1}in each spray	No major insect-disease; use resistant varieties.	

8. Conclusion

Organic farming is a system approach utilizing the natural cycles and biological interactions for crop production and protection. Organic food/products including oilseed crops are rapidly gaining popularity all over the world as a result of growing concerns and awareness among consumers regarding nutrition and food safety and health issues. Most of the annual oilseed crops are grown in rainfed areas where most of these crops are ideally suitable for organic farming in view of the existing low use of chemical inputs and less drastic yield declines when farms are converted from chemical to organic management. There is good international demand for organic sesame seed for medicinal and confectionary uses in Gujarat, Rajasthan and Telangana; residue/toxin free table varieties of groundnut for confectionary in Gujarat; demand for organically produced de-oiled cake (DOC) of soybean for livestock feed in Madhya Pradesh is on the rise. The field research conducted on organic practices are limited and there is an urgent need to evaluate organic farming practices in oilseed crops systematically to identify the most remunerative systems/products

9. Way forward

There is a need to carry out a country wide survey/inventorisation of areas in arid, semiarid and dry sub humid regions about the level of chemical input use, productivity in selected oilseed crops and related commodities which have potential to fetch price premiums in international markets. Identification of contiguous blocks of areas with little or no chemical input use and where productivity can be enhanced by using permitted inputs to enable group certification to farmers. There is need to identify suitable oilseed crops/products on regional basis for organic production that has national/international market demands. There is a need to develop protocols for organic production of important commodities through farmer's participatory network research. These protocols should be based on the entire cropping system approach and not on individual seasonal crops. There is a need to leveraging technology for innovations, production systems, quality assurance, traceability, market access and delivery solution. Linking farmers to value addition and direct marketing is the need of the hour. Favouring and supporting the premium price of organic produce and providing accessibility to the marketplace would encourage the farmers to opt for organic farming.

10. References

Annual Report. 2022. ICAR-Indian Institute of Oilseeds Research, Hyderabad.

Aher, S.B., Brij Lal, L., Singh, A. B., Swami Kaleshananda., Ramana, S., Ramesh, K., Thakur, J. K., Rajput, P. S. and Yashona D.S. 2019.Effect of organic sources of nutrients on performance of soybean (Glycine max). Indian Journal of Agricultural Sciences 89(11): 1787-1791.

AICRP-RM. 2006. Annual Progress Report of All India Coordinated Research Project on Rapeseed-Mustard, pp. A1-16, 2006.

Gangwar, B. and Deepmala, J. 2023. Package of practices for organic mustard in Bundelkhand Indian Farming 73(07): 14-17.

Gopinath, K.A., Venkateswarlu, B., Venkateswarlu, S., Yadav, S.K., Balloli, S.S., Srinivasa Rao, Ch., Prasad, Y.G. and Maheswari, M. 2011. Organic Sesame Production, Technical Bulletin, Central Research Institute for Dryland Agriculture, Santoshnagar, Hyderabad, Andhra Pradesh, India, p 34.

Kuepper G. 2003. Organic soybean production. Appropriate Technology Transfer for Rural Areas, National Center for Appropriate Technology.

Liebhardt, W.C., Andrews, R.W., Culik, M.N., Harwood, R.R., Janke, R.R., Radke, J.K. and Rieger-Schwartz, S.L. 1989. Crop production during conversion from conventional to low- input methods. Agronomy Journal 81: 150-159.

Panwar, A.S., Ravisankar, N., Shamim, M., Bhaskar, S., Bhatt, B. P., Raghuveer Singh, A.K., Prusty, M.A. A., Raghavendra, K.J. and Poonam Kashyap. 2022. Organic Farming Packages for Major Cropping Systems Indian Council of Agricultural Research, New Delhi

Ramesh, P., Panwar, N. R., Singh, A. B., Ramana, S., Yadav, S. K., Shrivastava, R. and Rao, A. S. 2010. Status of organic farming in India. Current Science 98(9): 1190-1194.

Rao, K., Tejeswara, Rao, A., Upendra, Naidu, G. J. and Reddy, D. S. 2013. Organic farming studies in sunflower. International Journal of Plant Science 8(1): 173-178.

Reider, C., Herdman, W., Drinkwater, L.E. and Janke, R. 2000. Yields and nutrient budgets under composts, raw dairy manure and mineral fertilizer. Compost Science & Utilization 8: 328-339.

Singh, S.K., Manibhushan and Ajay Kumar. 2018. Organic linseed (Tisi) farming: a step towards doubling farmers' income. Indian Farming 68(01): 55-58.

Somasundaram, E. 2003. Evaluation of organic sources of nutrients and Panchagavya spray on the growth and sustainable productivity of maize – sunflower- greengram cropping system. Ph.D. Thesis, Tamil Nadu Agricultural University, Coimbatore, Tamil Nadu.

Venkateswarlu, B. 2008. Organic farming in rainfed agriculture: Prospects and limitations. In: Venkateswarlu B., Balloli SS and Ramakrishna YS (eds) Organic Farming in Rainfed Agriculture: Opportunities and Constraints. Central Research Institute for Dryland Agriculture, Hyderabad. pp. 7-11.

11

Organic Farming in Groundnut

A.V. Ramanjaneyulu[1], N. Nalini[1], K. Ramesh[2], P. Duraimurugan[2], T.L. Neelima[1] and M. Ramesh Naik[3]

[1]*Professor Jayashankar Telangana Agricultural University, Rajendranagar, Hyderabad-500 030, Telangana, India*

[2]*ICAR-Indian Institute of Oilseeds Research, Rajendranagar, Hyderabad-500 030 Telangana, India*

[3]*ICAR-National Academy of Agricultural Research and Management, Rajendranagar Hyderabad-500 030, Telangana, India*

Abstract

Organic farming in groundnut cultivation is gaining global recognition as an environmentally sustainable and socially responsible approach. Groundnut is a versatile crop with high nutritional value and economic importance. This chapter explores the principles and practices involved in organic groundnut farming. It focussed on soil management, organic fertilization, pest, disease and weed management without using synthetic chemicals. This approach minimizes impact on the environment and reduces the presence of pesticide residues in the final produce. Moreover, organic practices help maintain soil health, conserve biodiversity and reduce environmental pollution. Thus, farmers can produce high quality groundnuts and fetch higher price and profits.

Keywords: *Groundnut, organic farming, pesticide residues, quality*

1. Introduction

The groundnut (*Arachis hypogaea* L.) is known by different names such as monkey nut, goober, earth nut etc., and belongs to leguminosae family. It produces flowers above ground but the fruits/pods underground (Sanders, 2003). It can be grown successfully in tropical and sub-tropical regions of the world. Globally, groundnut covers 32.7 M ha with a production of 53.9 MT with a productivity of 1648 kg ha^{-1} (FAOSTAT, 2021). Among the developing countries, Egypt has the highest productivity of groundnut with 2.70 t ha^{-1} (Nautiyal, 2001). Globally, India ranks first in groundnut area (5.42 m-ha) and is the second largest producer in the world with 10.1 MT with a productivity of 1863 ha^{-1} in 2021-22 (agricoop.nic.in). India ranks second next to China (34%) with 19% of world groundnut production in 2021. Gujarat is the largest producer contributing 36% of the total production of groundnut followed by Rajasthan (17%) and Tamil Nadu (7.5%). Groundnut accounts for 18% area and 26.8% production of the total oilseeds in India (CARP, 2022).

Numerous number of products and by-products *viz.*, chikkis, groundnut butter, flour, oil etc., are prepared from groundnut. Its seeds are called kernels and are used for oil extraction. This oil is in high demand across the globe for edible purpose. The remaining portion after extraction of oil known as groundnut oil cake (7.3:1.5:1.3 N: P_2O_5:K_2O) can be used as organic source of supplying nutrients to various crops. Chikkis are sweet confectioneries prepared using kernels and jaggery or sugar syrup and oil. Groundnut butter is a popular groundnut spread with an oil content of 45-55%. It may contain salt, sugar or hydrogenated vegetable oils used in a variety of cuisines for cooking, frying and flavouring. Groundnut flour is made by crushing roasted or raw groundnuts. It is used in baking and cooking as well as a protein-rich substitute for standard flours. Groundnut flour contains 40 and 50% oil too. The pod shells can be used as firewood or as mulch for conserving soil moisture in dry land agriculture. The pod shells can be also used for vermicomposting purpose. This vermicompost inturn can be applied for organic crop production. The leftover haulms after harvesting pods can be used as a promising source of either fodder for cattle or mulch for soil moisture conservation.

2. Need for organic farming package in groundnut

The groundnut crop is rich nutritionally. The kernels present inside pods are the edible portion in groundnut. These kernels are rich in protein (22-30%) and edible oil (44 to 56%) (Savage and keenan, 1994). It is also a good source of carbohydrates (10 %), minerals (3%), niacin (17 mg/100g) and vitamin B thus high food value. Its kernels can be eaten directly *i.e.*, for table purpose (Rajagopal *et al.*, 2000). Organically produced groundnuts are abundant source of vitamins (E and B3) and minerals, Mn, P, Mg. They also act as small protein banks and are cost effective which are easily affordable by the common man. Its protein is >2.5 times more than that of milk. Furthermore, groundnut is a leguminous oil crop requiring less N (20-30 kg ha^{-1}) which can be supplied easily through organic sources. However, this crop is grown across varied edapho-climatic conditions. Hence, the availability of phosphorus, secondary nutrients like Ca and S and micro-nutrients like zinc, boron and molybdenum is a problem in some of the impoverished and problematic soils. It leads to poor growth and biomass production and less nodulation thus low yield of groundnut. Application of organic manures, vermicompost, neem products, poultry manure, growing green manures etc., as a part of organic farming help to supply these nutrients besides improving soil fertility and health (Palaniappan and Siddeswaran, 1994). Furthermore, the demand for organic groundnut kernels and its by-products (edible oil and oil cake) is increasing exponentially across the globe. Further, organic farming in the recent years is gaining impetus due to better price for the farm produce. Thus, there is an ample scope for organic groundnut. But, need of the hour is to develop and upscale organic production technology and associated benefits for health, among various stakeholders like farmers, oil millers, policy makers, consumers etc.

3. Package of practices for organic production of groundnut

3.1. Selection of soil

In tropical and subtropical conditions of India, groundnut is extensively grown in light textured red sandy, red loamy, alluvial and coastal alluvial soils as well as on mixed black and red and medium black soils. Both bunch (more than 50%) and runner (about 80%) type of groundnuts are grown on sandy, sandy loam and mixed red and black soils in Southern India. Groundnut grown in medium black soils, deep alluvial loams and on sandy and gravelly soils with low organic matter content will give poor yields.

Groundnut grows best in well-drained soils with a sandy or very loose surface later. If the moist soil is rubbed between the index finger and the thumb, it should not form a ribbon but should fall apart easily. Such soils are best for groundnut growth and development. Seed germination and seedling emergence are favoured in loose, friable and sandy to loam soils. The pegs can penetrate the soil easily and clean pods can be harvested with minimum losses from such soils. Heavy and fine textured black (clayey) soils cause serious difficulty in groundnut harvesting with pod loss. Further, soil also adheres to groundnut pods reducing their marketable quality. Adequate drainage is required for groundnut production. Fields with shallow top soil, poorly drained areas and those subject to excessive erosion should be avoided. The ill drained acidic, alkaline and saline soils are not suitable for groundnut production. Otherwise, it adversely affects root growth, no. of pods per plant, plant growth and development, nitrogen fixing bacteria and nutrient uptake.

3.2. Land preparation

Summer ploughing (need based only) is advantageous to kill perennial weeds like *Cynodon* or *Cyperus* besides hibernating insects and disease organisms by exposing them to the high summer temperatures. Initial ploughing should be carried out at optimum moisture range to get good tilth and should avoid when moisture is in excess. Number and depth of ploughing depends on weed intensity. Further, care has to be taken avoid loss of organic matter by repeated ploughings.

For a rainy season crop, the field has to be ploughed once or twice to obtain a good tilth with onset of rains in May-June. Harrowing of soil should invariably be done to reduce the clod size and bring the field to fine tilth. In case of *rabi* season, the stubbles of preceding crop have to be incorporated (maize, jowar, mungbean using rotary mulcher) or removed from the site (rice) to facilitate timely sowing.

3.3. Host Plant Resistance (HPR)

Groundnut is generally affected by multiple pests and diseases leading to quantitative and qualitative yield loss. As synthetic chemicals can't be used in organic production of groundnut, it is difficult manage pests and diseases. Hence, host plant resistance plays a major role to manage the same and reduce the indiscriminate use pesticides thus reduce the cost of production. Host plant resistance (HPR) is sound, economical, practical, specific to target pests or group of pests, socially acceptable and eco-friendly method of pest control. Cultivation of varieties with tolerance/resistance to endemic pest or disease is very important, hence, an effort was made for compilation of such varieties as detailed in Table 1.

3.4. Seed treatment, sowing methods and time, seed rate and spacing

Groundnut kernels have to be treated with bio-fungicides followed by biofertilizers. First, they should be treated with *Trichoderma viride* @ 8-10 g kg^{-1} before sowing. It helps to overcome soil and seed borne diseases. Groundnut is a leguminous crop. So, *Bradirhizobium* bacteria residing in its root nodules will fix atmospheric N_2. Its kernels have to be treated with *Rhizobium* strain and PSB species like *Pseudomonas striata* or *Bacillus polymyxa* (200-250 g each for every 10 kg kernels and jaggery solution or rice glue) followed by shade drying before sowing. This process helps to meet N and P needs of the crop partially. Groundnut is sown during the warm season when the soil temperature is favourable for germination and growth. In India, for example, groundnut is typically sown from May to July, depending on the specific state and agro-climatic zone. For early sowing of *kharif* crops, FYM/ compost and green leaf manures have to be applied 10-15 days before sowing and vermicompost at the time of sowing. Sowing has to be done at a spacing of 30cmx10 cm in *kharif* (June-August) and 22.5 cmx 10cm in *rabi* (Oct-Nov) and summer (2nd fortnight of January) season.

Table 1. Groundnut varieties with host plant resistance

S. No	Variety	Season	Pod yield (kg ha^{-1})	Area suitability	Kernel yield (kg ha^{-1})	Shelling (%)	Oil (%)	Maturity duration (d)	Testa colour	100-kernel mass (g)	Special characters
Spanish Bunch											
1	Kadiri 6 (K 1240)	*Rabi*-summer	3100-3800	Andhra Pradesh & Telangana	2170-2660	70.0	49.0	100-105	Tan	39.0	Tolerant to leaf spots
2	ICGV 91114	*Kharif*	2000	Andhra Pradesh & Telangana	1500	75.0	48.0	90-95	Light tan	41.0	Early maturity, tolerant to rust and LLS
3	Kadiri 9	*Kharif*	2750	Andhra Pradesh & Telangana	1870	68.0	50.0	105-110	Rose	31.0-35.0	Tolerant to early and late season drought

4	Kadiri Hari-thandhra (K 1319)	*Rabi-summer*	3720	Andhra Pradesh, Telangana Karnataka, Maharashtra	2530	68.0	48.0	122	Rose	41.0	Fresh seed dormancy upto 20 days; multiple diseases (rust, ELS, LLS, stem rot, PBND) and insect pests (thrips, *Spodoptera litura,* jassid, *Heli-coverpa)* resistance
5	Dharani (TCGS 1043)	*Kharif* and Summer	1887 *(kharif);* 2666 (*rabi-*summer)	Andhra Pradesh, Telangana, Kerala, Tamilnadu, Orissa, West Gujarath, Rajasthan etc.,	1434 *(kharif);* 2026 (*rabi-*summer)	76.0	48.0-50.0	105-110	Light rose	38-43	Drought tolerant; tolerant to stem and dry root rot

6	Jagtial Palli 1 (JGC 2141)	*Kharif* and summer	3330-3500	Andhra Pradesh & Telangana	2131-2240	64.0	51.5	110	Rose	29.0	Tolerant to early season drought; resistant to leaf spots and rust diseases, tolerant to thrips and jassids
7	K 1812 (Kadiri Lepakshi)	*Kharif*	3514	Andhra Pradesh, Telangana, Karnataka, Tamilnadu pods of Maharashtra and Odisha	2460	70.0	51.0	107-113	Rose	40.0	Tolerant to drought, multiple disease And pests resistant
8	TAG 24	*Kharif* and Summer	2000-2500	Maharashtra Andhra Pradesh Telangana Gujarat Odisha Tamilnadu Rajesthan	2400	72.0	51.0	95-100	Light pink	35.0-38.0	Short duration

9	TCGS 1694	*Kharif* and *rabi-*	3300-3500 *(kharif)* 4000-4500 *(rabi)*	Andhra Pradesh	2800-3600	73.0-75.0	50	105-110	Pink	42.0	Short stature, LLS and rust tolerant, drought tolerant
Virginia Bunch											
10	Kadiri 7	*Kharif*	1640	Andhra Pradesh	1115	68.0	47.0	120-125	Rose	65-75	Large seeded, tolerant to sucking pest and leaf spots
11	Kadiri 8	*Kharif*	1520	Andhra Pradesh	1049	69.0	47.0	120-125	Rose	65-75	Large seeded; tolerant to sucking pests and leaf spots

12	Girnar 4 (ICGV 15083)	*Kharif*	3218	Rajasthan, Gujarat, Tamil Nadu and Andhra Pradesh	2156	67.0	53	110-112	Rose	43.0	78.5% oleic acid and 4.8%linoleic acid; tolerent to LLS, rust, stem rot and Groundnut Bud Necrosis diseases, leafhopper, leaf miner, thrips and *Spodoptera litura*

13	Girnar 5 (ICGV 15090)	*Kharif*	3124	Rajasthan, Gujarat, Karnataka, Tamil Nadu and Andhra Pradesh	2093	67.0	53	110-115	Rose	41.0	78.4% oleic acid and 4.6% linoleic acid; tolerant to Late Leaf spot, Rust, Stem rot and Collar rot, Leaf hopper, Leaf miner, thrips and *Spodoptera litura*
14	Gujarat Groundnut 41 (Padma)	*Kharif*	2722	Gujarat	2014	74.0	51.4	120	Rose	56.0	High shelling and oil percent

LLS: Late leaf spot**; ELS:** early leaf spot**; PBND:** Groundnut Bud Necrosis Disease

(*Source*: Pal *et al.*, 2021)

A seed rate of 125 kg kernel ha^{-1} for small kernel types and 150 kg ha^{-1} for bold kernles types can be adopted. Seed cum ferti drill can be used for sowing irrespective of season. Ensure depth of sowing within 5 cm of the soil under sufficient moisture conditions and deeper placement at 8-10 cm in light soils with insufficient surface moisture. Broad bed and furrow method (BBF) of sowing in groundnut can also be adopted under irrigated conditions using BBF maker which can facilitate sowing too. Depending on the bed width, three to four rows of groundnut can be accommodated. A furrow width of 30-45 cm has to be left between beds to facilitate irrigation and drainage. The bed height should be 15-20 cm.

3.5. Nutrient management

Healthy soil is the foundation for environmentally sound groundnut production system. Soil fertility is very important for production of good quality groundnuts. Organic matter of about 2% can help improve the water and nutrient supplying capacity of the soil, especially the micro-nutrients like Zn, Cu, Fe, Mn, B and Mo to meet the plant needs.

***Organic Manures*:** Organic manures (FYM, compost, vermicompost, oil cakes *etc.,*) have profound effect on improving soil physical, chemical and biological properties and enhancing productivity of groundnut crop. In groundnut, FYM @ 10-15 t ha^{-1} increased pod and haulm yields and also is reported to improve quality parameters like shelling percentage and 100 kernel weight. Vermicompost (1.5 t ha^{-1}) was as effective as FYM (5 t ha^{-1}) in increasing groundnut yield.

***Role of biofertilizers*:** In groundnut, *Rhizobia* sp. has a major role for enhancing biological nitrogen fixation (BNF), besides phosphate solubilizing microbes (PSM) for augmenting phosphorus nutrition and plant growth promoting rhizobacteria (PGPR) for growth promotion and biocontrol. Two groundnut rhizobia namely, IGR 6 and IGR 40 were found to enhance pod yield by 11-18%. *Pseudomonas striata* produced 11% higher pod yield besides application of PSM in rainfed groundnut (Dey and Pal, 2014).

Three PGPR namely; PGPR 1, PGPR 2 and PGPR 4, belonging to fluorescent pseudomonads (especially *Pseudomonas fluorescens* and *Pseudomonas putida*), obtained from groundnut rhizosphere have been identified as effective biofertilizer for groundnut under rainfed condition by AICRP Groundnut. The groundnut-microbe generally establish an intimate, dynamic relationship that allows them to coexist to boost groundnut health.

Patra and Sinha (2012) recommended combined application of 2.5 t ha^{-1} each of phosphocompost (PC), poultry manure (PM), neem cake (NC) and vermicompost (VC) or combination of any two sources @ 2.5 t ha^{-1} each, over their solitary application of 5.0 t ha^{-1} each, for significantly higher pod and haulm yield of groundnut besides two and three seeded pods and stomatal conductance under organic management under Cooch Behar conditions of West Bengal. They also emphasized on phosphocompost as critical for meeting P needs of this leguminous oilseed crop. Furthermore, gypsum @ 500 kg ha^{-1} obtained from mineral source only has to be applied at 35-40 days after sowing, to the plant rows followed by earthing up will help meet the crop Ca and S requirements which inturn helps in better pod and kernel development (NCONF, 2013). Adequate supply of Ca and S in the soil is very essential for the production of well-formed groundnut pods with sound and mature kernels and oil content, respectively.

Fertigation with chemical fertilizers should be done away. However, jeevamrith (500 lit ha^{-1}) can be scheduled at 15 days interval.

3.6. Weed management in organic groundnut

Organic groundnut production is risky and laborious due to difficulties in managing weeds using methods acceptable for certified organic production. Hence, successful weed management in certified organic groundnut production will depend on an integrated weed management (IWM) approach but excluding chemicals. Implementation during critical period of crop-weed competition i.e. 30-45 DAS is very important for timely and effective weed control.

Cultural weed control includes practices that promote vigorous early-season groundnut growth and lessen chances for weed escapes.

Groundnut plants closely spaced had fewer number of weeds and less weed dry weight. Those spaced at 25 cm x 25 cm or 30 cm x 10 cm had weed density 12 times less than 100 cm x 50 cm (Ansa and Joseph, 2016). Groundnut seedlings are slow to achieve canopy closure in this row spacing. There has been a continual effort to develop production systems using multiple narrow-spaced rows to quicken canopy closure and suppress weed emergence (Buchanan and Hauser, 1980; Johnson *et al.,* 2005).

Mechanical weed control is based on repeated cultivation using a tine weeder and sweep cultivator to control weeds before they emerge. Tractor drawn groundnut intercultivation implements are available for successful weed control. After 40-45 DAS, intercultivation equipment should not be run in groundnut which otherwise damages pegging leading to poor yields. Hand weeding/hoeing, a form of mechanical weed control has to be done supplement other weed control efforts.

In organic groundnut production, limited selection of herbicide oils which are derived from natural products are suitable for controlling of weeds. These are called as essential oils. These include clove oil, d-limonene, lemongrass oil and cinnamon oil herbicides derived from essential oils are foliar applied. These are contact herbicides that disrupt cellular membranes (Tworkoski, 2002). These are non-selective and post emergence only (Dayan *et al.,* 2009). Fortunately, ground nut is somewhat tolerant and quickly recover from direct applications of essential oil herbicides with no yield effects (Johnson and Mullinix, 2008a).

Organic acids: Considerable weed control research has been conducted in several crops using vinegar with high concentrations (20%) of acetic acid. However, there has been no reported research with vinegar for weed control in organic groundnut. Earlier trials in Tifton, GA on organic groundnut evaluated a commercial mixture of citric acid and acetic acid (listed as an inert carrier) (Johnson and Mullinix, 2008a), but, this mixture did not adequately control any weed species.

Thermal weed control: Propane flaming does not incinerate weeds. A propane flamer was configured by USDA-ARS personnel in Tifton, gibberellic acid (GA) to direct heat from an array of eight propane burners directly to young groundnut soon after crop emergence. It will kill weeds due to cell rapid expansion that ruptures cell walls and membranes. Seedlings of dicot weeds were effectively controlled by this method was effective only on dicots weeds but not perennial nutsedges and large dicot weeds. (Boyd and Brennan 2006; Johnson and Mullinix, 2008b).

Corn gluten meal: In 1985, corn gluten was accidentally discovered to have herbicidal properties by inhibiting root formation in susceptible plants (Mc Dande and Christians, 2000). It is strictly a pre-emergence herbicide. In greenhouse studies, corn gluten-controlled pigweed (*Amaranthus* spp.), crabgrass (*Digitaria* spp.), plantain (*Plantago* spp.), lambsquarters, and curly dock (*Rumex crispus* L.). However, in field trials conducted in Tifton, GA during the 2006 and 2007 seasons, corn gluten did not adequately control annual grasses in organic groundnut (Johnson *et al.*, 2013).

3.7. Water management

Groundnut requires an average of 400 to 450 mm of water depth, which varies depending on soil type, agro climate and genotype. However, during *rabi*/ summer, the groundnut's water needs can range between 450 and 600 mm. A good groundnut crop requires 8 to 9 irrigations at 10-day intervals beginning on 25 DAS. After the crop is established, irrigation needs to be stopped for roughly 25 days to induce stress, which aids in blooming synchronization. Irrigation is vital during the flowering, pegging and pod formation phases, when adequate soil moisture is required. The first irrigation should occur immediately after the crop is planted. When 50% of the plants begin to flower, the second irrigation is applied. Irrigation should be administered at 8-10 days interval throughout pegging and pod filling stages. Last irrigation can be given at 90 days after sowing. Sprinkler use can reduce irrigation water use by about 24–30% and it has to be scheduled at 1.0 Epan (https://rarstpt.org) (Ramanjaneyulu *et al.*, 2022).

Patil *et al.* (2009) proved micro-sprinkler irrigation or drip inline system as superior to broad bed and furrow method in achieving higher pod yield by 53-73% in summer groundnut. Aydinsakir *et al.* (2016) recommended drip irrigation equivalent to 100% Epan when water is not a constraint and 75% Epan under water scarce conditions for better yields in groundnut. Groundnut drippers (4 lph) positioned at 40 cm (sandy loam) or 60 cm (clayey soils) are appropriate for inline or tape irrigation. Each bed requires only one drip line (1.2 m wide) for proper irrigation. Ramanjaneyulu *et al.*, (2022) recommended operating drip irrigation @ 0.8 Epan in alfisols of Telangana for enhancing growth and reproductive attributes and dry pod yield by 8-10% besides higher water use efficiency (6.31 kg ha^{-1} mm^{-1}) and higher energy efficiency in *rabi* groundnut.

In the case of sprinkler irrigation, the application rate per irrigation should be equivalent to the infiltration rate of the soil of each location and approximately this would be 15-20 mm at a time for various soil types. Sprinkler irrigation results in an increase of yield by 100% and also a reduction of water use by 55% (http://jains.in).

3.8. Harvesting, threshing and storage

Inside shell colour turning black is the best indication of maturity in groundnut. Traditionally groundnut is harvested by pulling or digging from the soil using manual labour or bullock-drawn blade harrows or tractor-drawn diggers or country plough. Then they are heaped in the field exposing the pods to the sun for subsequent drying. It is better to do harvesting in sunny days with less humidity to avoid mold development. The pods may contain 25 to 50% moisture when freshly harvested.

Wet pod thresher can be employed for immediate threshing. However, with dry pod thresher two or three days of drying in the field is required to separate the pods from the vines. Further, the moisture level has to be lowered to 8 to 10% by the drying process for safe storage. This is very important to avoid aflatoxin contamination. Fully dried pods have to be stored in mesh bags in a well-ventilated room till it is used further for different purposes. Groundnuts have high oil content and will eventually grow rancid. Hence, they have to be stored in a sealed container in the refrigerator for several months or in the freezer for several years to extend their shelf life. Care has to be taken to avoid contamination of organic groundnuts with that of inorganic groundnuts throughout the production, storage and consumption process.

3.9. Quality parameters

Several characteristics are evaluated while determining the quality of groundnuts as detailed below.

3.9.1. Moisture content

The moisture content should not exceed 8-10% for safe storage.

3.9.2. Size and uniformity

Groundnuts are evaluated based on pod size and is an indicator of quality. Larger groundnuts are preferred due to superior taste and texture.

3.9.3. Shell integrity

The shell's integrity is a significant quality criterion. Undamaged shells with no fractures or pest/disease infestation indicate high quality.

3.9.4. Kernel colour

Well-roasted or processed groundnuts have a nice golden brown appearance, indicating adequate heat treatment and flavour development.

3.9.5. Oil content

Groundnut oil concentration normally varies from 40% to 50%, with variations depending on type and growth conditions. Higher oil content indicates higher quality and flavour.

3.9.6. Aflatoxin contamination

Aflatoxin is a poisonous substance produced by certain moulds (*Aspergillus flavus* and *Aspergillus parasiticus*) that can infect groundnuts. Its pollution causes health problems. It should be within acceptable limits i.e., 10 (μg/kg or ppb) in India either for consumption or export purpose and fetch higher market price.

3.9.7. Texture and taste

Groundnuts with a crisp with a deep, nutty flavour, fresh and sweet indicate good quality.

The quality criteria of groundnuts may differ depending on whether they are intended for direct consumption, oil extraction, or processing into goods such as groundnut butter or confectionary items. Quality requirements for groundnuts in different locations and markets are determined by regulatory authorities, industry standards and market demands.

The CODEX standard applies to groundnuts intended for processing for direct human consumption. As per the standards, the maximum level of moisture content should be 10% in Groundnuts (in-pod) and 9.0% in Groundnut kernels. Lower moisture limits should be required for certain destinations in relation to the climate, duration of transport and storage. The maximum percent of mouldy, rancid or decayed kernels should not me more than 0.2% m m^{-1}. Rancid kernels are defined as those which have undergone oxidation of lipids (should not exceed 5 meq active oxygen kg^{-1}) or the production of free fatty acids (should not exceed 1.0%) resulting in the production of disagreeable flavours. Impurities of animal origin (including dead insects) should not exceed more than 0.1% m m^{-1}. Other organic and inorganic extraneous matter should not exceed 0.5% m m^{-1} in both groundnut pods and kernels.

3.10. Plant protection

The groundnut crop is invaded by a large number of insect-pests and diseases across all the phenophases of it's growth. Among the insect pests, defoliators such as the groundnut leaf miner (*Aproaerema modicella*), tobacco caterpillar (*Spodoptera litura*), red hairy caterpillars (*Amsacta albistriga* and *A. moorei*), Bihar hairy caterpillar (*Spilosoma obliqua*) and gram pod borer (*Helicoverpa armigera*) are the most important yield barriers in India (Patel *et al.*, 2015; Basappa and Duraimurugan, 2018). Aphids, leafhoppers (*Empoasca kerri*) and thrips are the major sucking pests. Apart from causing direct damage to the crop, aphids (*Aphis craccivora*) and thrips (*Scirtothrips dorsalis*, *Frankliniella schultzei*, *Thrips palmi* and *Caliothrips indicus*) are the vectors of important viral damage *viz*., Groundnut Stripe Virus (PStV) and Groundnut Bud Necrosis Disease (PBND), respectively (Sundar *et al.*, 2020). Among soil insect pests, white grub (*Holotrichia consanguinea* and *H. serrata*) and termites (*Odontotermes obesus* and *Microtermes obesi*) are important. Early leaf spot (*Cercospora arachidicola*), late leaf spot (*Phaeoisariopsis personata*) and rust (*Puccinia arachidis*) are the economically significant foliar diseases in groundnut. Of different soil borne diseases affecting its production, collar rot (*Aspergillus niger*), stem rot (*Sclerotium rolfsii*) and bacterial wilt (*Ralstonia solanacearum*) are the devastating ones. The following are some of the non-chemical methods of pest and disease management in groundnut (Singh *et al.*, 2002; Vimala Devi and Duraimurugan, 2013; Sujatha *et al.*, 2017; Basappa and Duraimurugan, 2018; Vimala Devi *et al.*, 2020 and 2021; Patel *et al.*, 2022)

- Avoid monocropping of groundnut and adopt crop rotation with maize, wheat, sorghum, pearl millet or minor millets
- Summer ploughing to incorporate plant debris deep into the soil and to expose the dormant structures of fungi to direct sunlight
- Soil amendment with castor cake @ 500 kg ha^{-1} (preferably 15 days before sowing) for soil borne diseases like stem rot and collar rot
- Soil application of vermicompost @ 1 t ha^{-1} + *Matarhizium. anisopliae* 1.15 WP (1×10^8 $cfug^{-1}$) @ 2 kg ha^{-1} was found most effective for the management of white grub in groundnut.
- Soil application of *Trichoderma viride*/*T. harzianum* @ 25 to 62.5 kg ha^{-1}, preferably in conjunction with organic amendments such as cake or FYM
- Seed treatment with *Trichoderma viride*/*T. harzianum*/*Pseudomonas fluorescens* against seed and soil-borne pathogens @ 10 g kg^{-1} seed
- Intercropping of groundnut with pigeonpea/pearlmillet/sorghum (3:1 or 5:1) depending upon the locally recommended intercrops against foliar fungal diseases (early and late leaf spots, rust) and insect pests (leaf miner and thrips)
- Grow trap crops such as cowpea/soybean (leaf miner and leafhoppers)/ castor (*Spodoptera* and leafhoppers) for controlling sucking/defoliating insect pests
- Vegetative trapping of migratory caterpillars of red hairy caterpillars by keeping twigs of Jatropha or Calotropis and neem branches for trapping adult-white grubs
- Set bonfires on community basis on the field bunds from 7-10 pm starting from the next day of the rain. Setup the light traps to attract and kill the lepidopteran moths during June-August
- Erection of bird perches @ 50 ha^{-1}
- Pheromone traps @ 10 ha^{-1} for *Spodoptera* and *Helicoverpa* and 25 ha^{-1} for leaf miner
- For white grubs, aggregation pheromone dispenser be placed (2-3 $tree^{-1}$) on neem or host trees in evening (continuously for three days) after beetle emergence as mass trapping method (on community basis)
- Removal of volunteer groundnut plants, self-sown groundnut plants, reservoir hosts and destruction of infected crop debris is important in reducing the primary source of infection
- Collection and destruction of egg masses, early instar larvae (mesh like leaves with larvae) of *Spodoptera*, *Spilosoma*; late instar larvae of *Helicoverpa* and defoliators.

- Release of *Trichogramma chilonis* @ 50000 ha^{-1}, two times at 7-10 days interval followed by release of *Bracon hebetor* @ 5000 ha^{-1}, two times at 7-10 days against leafminer and defoliators
- Release *Telenomus remus* @ 50000 ha^{-1} in 4 times (7-10 days interval) against defoliators (*S. litura*) based on pheromone trap catching
- Spray of aqueous neem leaf extract (2-5%) for the management of leaf spots and rust diseases
- Spray of neem seed kernel extract (5%), or crude neem oil (2%) against defoliators, sucking pests and foliar pathogens
- Apply nuclear polyhedrosis virus for the management of *Spodoptera litura* (Sl NPV*), Helicoverpa armigera* (HaNPV), *Amsacta albistriga* (AaNPV) @ 250 $LEha^{-1}$
- Spray *Metarhizium (Nomuraea) rileyi* @ 2 x 10^{11} conidia lit^{-1}against *Spodoptera litura*
- Spray of Bt (*Bacillus thuringiensis*) @ 300-500 gha^{-1} against *S. litura, H. armigera* and red hairy caterpillars
- Store pods in polythene lined gunny bags to reduce infestation of storage pests
- Botanical extracts like garlic, ginger and castor oil found eco-friendly, in-expensive and easily available to the farmers with no harm to any non-target organisms. Hence, their application using a jute rope as a barrier around the crop fields can be recommended for containing the wild boar damage (Srinivas Rao, *et al.*, 2019).
- Vasudeva Rao (2015) has suggested raising safflower or castor in four rows around groundnut crop, arranging barbed wire fence or circular razor fence or chain link fence are effective to reduce wild boar menace in groundnut. Further, he also advocated establishment of thorny, xerophytes like *Euphorbia caducifolia*, *E. meriifolia, Opuntia elatior*, O. *dillenii, Ziziphus oenopolia, Z. mauritiana and Agave americana, Caesalpinia cristata* on the bunds around the crop. He also reported that few farmers are practicing tying of old sarees of different colors around the crop to repel the wild boars. He suggested that the birds like crows can be controlled to some extent by using bio-acoustics, hanging dead crows, pitcher effigy etc.

3.11. Export of organic groundnut

Most of organic groundnut is exported to United Arab Emirates, United States and Singapore. The top three exporters of organic groundnut are India followed by United States and South Korea (https://www.volza.com).

India has exported 5.04 lakh tons of groundnut worth of Rs. 3386 crores during the year 2017-18. The varieties include Bold, Java and Red Natal. Apart from raw edible groundnuts, India also supply blanched groundnuts, roasted salted groundnuts, dry roasted groundnuts, groundnut butter and a variety of groundnut based products (IOPEPC).

Maximum Levels of aflatoxins, other safety and quality requirements and quarantine concerns are yet to harmonize amongst countries for groundnut export. It is, therefore, essential to monitor these requirements for exports of groundnuts and groundnut products. The Government of India, Ministry of Commerce & Industry, Department of Commerce vide Notification No. 28 (RE-2012)/2009-2014 dated 3rd January, 2013 issued under the Section 5 of the Foreign Trade (Development & Regulation) Act, 1992 as published in the Gazette of India conferred powers to APEDA permitting export of groundnuts (groundnuts) subject to registration with APEDA along with controlled aflatoxin level certificate issued by APEDA recognized laboratories. Subsequently, Ministry of Commerce & Industries advised APEDA vide letter No. 11/1/2013-EP (Agri.IV) dated 4.3.2015 to issue necessary orders for export of groundnuts (groundnuts).

Groundnuts for edible use require considerable processing and sorting to ensure high quality removal of stones and other foreign matter; removal of shell, removal of kernels that are shrivelled, off-colour, or otherwise unsatisfactory; and grading for size and uniformity. In this process a large measure of hand sorting is needed to get the required quality. Therefore, edible groundnuts are traditionally called 'Hand Picked and Selected Groundnuts' or 'HPS Groundnuts' (HPSG) in trade circles. In developed countries, machines are also used for sorting. HPSGs are graded into sizes which confirm to counts per ounce. For instance, 55-60 count means 55 to 60 kernels per ounce. Based on this count, kernels are classified into four broad groups i.e. small (60-80 counts), medium (40-60 counts), large (30-40 counts) and very large kernels (20-30 counts). Usually, counts have a range of 10 for small kernels, a range of 5 for medium and large kernels, and a range of 2 for very large kernels. Similarly, a range of 2 is preferred for nuts-in-shell. HPSGs are normally indicated by country of origin, varietal group and count. For instance, Indian Bold 55/60; US Virginias 28/30; South Africa Natals 60/70 etc.

3.12 Recommended package of practices for organic groundnut in India

Chavan *et al.* (2014) recommended following agronomic package for organic groundnut production in rice-groundnut system in Maharashtra

- Tikka leaf spot and rust tolerant varieties *viz.,* SB XI (95 kg ha^{-1}), Konkan Guarav (110 kg ha^{-1}) and Konkan Trombay (125 kg ha^{-1})
- Seed treatment with 5 g *Trichoderma*, 25 g each of *Rhizobium* and PSB per kg seeds
- Spacing: 30 cm x15 cm
- Basal application of organic manures like FYM 1.5 t ha^{-1}, neem cake 0.16 t ha^{-1} and vermicompost 0.56 t ha^{-1}
- Spraying cow urine or vermi wash @ 50 lit ha^{-1} at 30 and 60 DAS

The Institute of Organic Farming, UAS, Dharwad has recommended the following for achieving 30-35 q ha^{-1} of pod yield and 35-40 q ha^{-1} of haulm yield. (https://agricoop.nic.in).

a) The varieties such as GPBD-4 (zone 8, June-July, Resistant to leaf spot and rust diseases. 125 kg ha^{-1}), JL 24 (all zones, June-July, big pods, 150 kg ha^{-1} kernels), TAG-24 and DH-86 (all zones, Dec 15 - Jan 15, resistant to bud necrosis, 150 kg ha^{-1} kernels) have to be grown

b) Kernels are to be treated with *Rhizobium* 1.0 kg, PSB 1.0 kg and *Trichoderma harzianum* 0.75 kg ha^{-1}

c) Enriched compost 3.0 t ha^{-1}, vermicompost 2.4 t ha^{-1}, green leaf manure 5.0 t ha^{-1} and neem cake 0.25 t ha^{-1} have to be applied. Gypsum @ 500 kg ha^{-1} has to be applied at 35-40 DAS

d) Timely weed control through intercultivation at 25 and 40 DAS and hand weeding at 30 DAS is critical

e) Cow urine @ 10 % and Panchagavya spray @ 3% as a source of nutrient and growth promoter at 45 and 60 DAS have to be sprayed.

f) Intercropping with *Setaria* or bajra (7:1) and castor as trap crop for *S. litura* management is highly advocated. Pheromone traps @ 5 ha^{-1} for monitoring of *S. litura* have to be arranged. The egg masses of *S. litura* from the leaves have to be collected and destroyed. Neem seed kernel extract @ 5% or custard apple leaf extract @ 5% + Metarhizium *Nomuraea rileyi* 1011 conidia ha^{-1} @ 1 glit can be sprayed at 45 and 60 DAS against management of defoliators.

Ramesh Naik *et al.* (2023) reviewed the literature and reported the following packages.

In case of groundnut in Gujarat, TG 37 A variety with vermicompost @ 5.0 t ha^{-1} and castor cake 0.12 t ha^{-1}, arranging yellow sticky traps 20 ha^{-1} to control sucking pests, soil application of *Metarhizium anisoplae* @ 1.0 kg with 500 kg FYM ha^{-1} to reduce white grub meance and foliar spray of cow urine @

10 lit ha^{-1} to reduce chlorosis are recommended. Similarly, in groundnut of cassava-groundnut system in Kerala, Co-7 variety with FYM @ 3.0 t ha^{-1}, Phosphobacteria and K solubilizer @ 3.0 kg ha^{-1} each, 40 kg P_2O_5 ha^{-1} through rock phosphate+wood ash 0.75 t ha^{-1}+Panchagavya 3% @ 15 lit ha^{-1} twice at 30 and 60 DAS+vermiwash 10% @ 50 lit ha^{-1} twice at 30 and 60 DAS are advocated.

3.13. Constraints in organic production and processing of ground nut

Lack of low cost and location specific production and protection technology and lack of availability and accessibility to suitable machines for organic production of groundnut are the important constraints. No premium price of organic produce in the market is discouraging the farmers to take up organic groundnut. The information regarding demand and supply of organic groundnuts is meagre, hence, planning for it's production is difficult. The certification procedures and multiple certifications for various countries are the cumbersome procedures.

4. Conclusion

Organic farming of groundnuts has various advantages including boosting soil health and fertility by eliminating the use of synthetic fertilizers and pesticides thus reduces agriculture's environmental effects. Organic inputs, crop rotation and integrated pest management measures can help reduce pest and disease outbreaks, improve soil structure and increase the nutritional content of groundnut crops. Organic farming in groundnut has possible problems and limits, such as the requirement for extensive knowledge and skill, potential output variations and limited access to organic markets. These hurdles, however, can be solved with sufficient training, support and market incentives, leading to increased adoption and success of organic groundnut growth.

5. Way forward

Development of eco-friendly production and protection strategies for organic groundnut production is the need of the hour. Custom hiring centres must be in place to make organic farmers to access the required machinery for efficient weed control and moisture conservation. The governments have to announce separate MSP for organic products. The importance of organic groundnuts in human and animal nutrition has to be explained to the consumers in order to create demand. The certification procedures must be made simple and farmer friendly. The facilities for processing of organic groundnut by-products have to be strengthened.

6. References

Ansa and Joseph, E.O. 2016. Effects of groundnut spacing on yield and weed control in the rainforest agro-ecological zone of Nigeria. Direct Research Journal of Agriculture and Food Science (12): 374-377.

Aydinsakir, K., Dinc, N., Buyuktas, D. and Bastug, R. 2016. Assessment of different irrigation levels on groundnut crop yield and quality components under Mediterranean conditions. Journal of Irrigation and Drainage Engineering 142: 04016034.

Basappa, H. and Duraimurugan, P. 2018. Management of pests of oilseed crops. In: Pests and their management, Omkar (Ed.), Springer Nature, Singapore, pp 223-240.

Boyd, N.S. and Brennan, E.B. 2006. Burning nettle, common purslane, and rye response to a clove oil herbicide. Weed Technology 20: 646-650.

Buchanan, G.A. and Hauser. E.W. 1980. Influence of row spacing on competitiveness and yield of groundnuts. Weed Science 28: 401-409.

Chavan, L.S., Jondhle, D.G., Mhatre, S.J., More, S.S. and Kshirsagar, S.K. 2014. Package of practices for organic crop production. In: Scientific Package of Practices (PoPs) for organic production of crops in cropping systems. ICAR-Network Project Organic Farming ICAR-Indian Institute of Farming Systems Research, Modipuram, Meerut - 250 110 (UP), pp 91-97.

CARP, 2022. Groundnut outlook report. https://angrau.ac.in.

Dayan, F.E., Cantrell, C.L. and Duke, S.O. 2009. Natural products in crop production. Bioorganic Medicinal Chemistry 17: 4022-4034.

Dey, R. and Pal, K.K. 2014. Biofertilizers: for enhancing groundnut productivity. ICAR – Directorate of Groundnut Research, Ivnagar Road, PB No. 5, Junagadh, Gujarat, India. p.30.

FAOSTAT. 2021. https://www.fao.org.

https://agricoop.nic.in.

http://jains.in.

https://rarstpt.org

https://www.volza.com.

Johnson III, W.C and Mullinix, Jr. B.G. 2005. Texas panicum (Panicum texanum) interference in groundnut (Arachis hypogaea) and implications for treatment decisions. Groundnut Science 32: 68-72.

Johnson, W. C., III and Mullinix Jr. B.G. 2008a. Groundnut response to naturally-derived herbicides used in organic crop production. Peanut Science 35: 73-75.

Johnson, W.C., III and Mullinix Jr. B.G. 2008b. Potential weed management systems for organic peanut production. Peanut Science 35: 67-72.

Johnson, W. C. III, Boudreau, M.A. and Davis, J.W. 2013. Combinations of corn gluten meal, clove oil, and sweep cultivation are ineffective for weed control inorganic groundnut production. Weed Technology 27: 417-421.

McDande, M.C. and Christians, N.E. 2000. Corn gluten meal – a natural pre emergence herbicide: effect of vegetable seedling survival and weed cover. American Journal of cover. Americanriculture 15: 189-191.

NCONF, 2013. Organic production technologies developed in crops. Network Project on Organic Farming Dharwad Centre. Institute of Organic Farming Directorate of Research University of Agricultural Sciences.

Nautiyal, P. 2001. Groundnut post-harvest operations. Post - Harvest Compendium. pp 166.

Pal, K.K., Ranvir Singh and Singh, A.L. 2021. Handbook of groundnut varieties of India released during 2001-2021. AICRP on groundnut, ICAR - Directorate of Groundnut Research, P.B.5, Junagadh – 362 001, Gujarat, India, pp.124.

Palaniappan, S.P. and Siddeswaran, K. 1994. Integrated nutrients management in rice based cropping systems. Proceedings of the XIII National Symposium on integrated input management for efficient crop production, February 22-25, 1994, TNAU, Coimbatore, India, 41-53.

Patel, B.V., Basappa, H., Duraimurugan, P., Ghante, V.N., Santhalakshmi, P. M., Prasad, R.D. and Gururaj Sunkad. 2015. Protection technologies for major oilseed crops. Compendium of Lead Papers, National Seminar on Technologies for enhancing oilseeds production through NMOOP, 18-19 January 2015, Hyderabad, India, pp 39-60.

Patel, P.S., Deb, S., Rabari, P.H. and Joshi, M.J. 2022. Field evaluation of the entomopathogenic fungi enriched with organic amendments against Holotrichia consanguinea Blanchard (Coleoptera: Scarabaeidae) infesting groundnut crop. Egyptian Journal of Biological Pest Control 32: 7.

Patil, S.T., Mane, M.S. and Ayare, B.L. 2009. Effect of irrigation methods on yield and water requirement of summer groundnut. Journal of Soils and Crops 19: 147-152.

Patra, P.S. and Sinha, A.C. 2012. Studies on organic cultivation of groundnut (Arachis hypogaea) in Cooch Behar. Indian Journal of Agronomy 57(4): 386-389.

Rajagopal, K., Chandran, K., Mishra, J.B., Bhalodia, P.K. and Mathur, R.S. 2000. Evaluation of bold seeded groundnut accessions for confectionary attributes. International Arachis Newsletter 20: 18-20.

Ramanjaneyulu, A.V., Ramulu, V., Ramana, M.V. and Mamatha, K. 2022. Crop performance, water use, economics and energy indices in rabi groundnut (Arachis hypogaea L.) under micro irrigation methods. Journal of Environmental Biology 43: 326-337.

Ramesh Naik, M., Umesh, H., Gopinath, K.A., Ramanjaneyulu, A.V., Ravisankar, N., Srinivas, T., Ansari, M.A., Singh, V.K., Venkateshwarlu, G. and Srinivasa Rao, Ch. 2023. Organic farming in India: Production constraints and policy recommendations. ICAR-NAARM Policy Paper2023/1 pp 30.

Sanders, T.H. 2003. Encyclopedia of food sciences and nutrition (Second Edition). Academic Press, pp. 4420-4427.

Savage, G.P. and Keenan, J.I. 1994. The composition and nutritive value of groundnut kernels. (Ed. J. Smartt), the groundnut crop: A scientific basis for improvement, Chapman and hall, London, United Kingdom, 173-213.

Singh, H., Basappa, H., Chander Rao, S., Prasad, R.D. 2002. IPM in oilseed crops. Directorate of Oilseeds Research, Hyderabad, India, pp 25.

Srinivasa Rao, N., Praveen Kumar, K. and Sakthivel, P. 2019. Efficacy of botanical repellents against wild boar in field crops. Green Farming (3): 368-371.

Sujatha, M., Chander Rao, S., Vishnuvardhan Reddy, A. 2017. Significant achievements of 50 years of AICRP on Oilseeds. ICAR-Indian Institute of Oilseeds Research, Hyderabad, India, pp 46.

Sundar, B., Navya, M., Duraimurugan, P. and. Surya Prakash Reddy, M. 2020. Sucking pests of oilseed crops. In: Sucking pests of crops, Omkar (Ed.), Springer Nature Singapore Pvt. Ltd., Singapore, pp 1-39.

Tworkoski, T. 2002. Herbicide effects of essential oils. Weed Science 50: 425-431.

Vimala Devi, P.S. and Duraimurugan, P. 2013. Exploitation of Nomuraea rileyi and Beauveria bassiana for the management of lepidopteran pests. Kavaka 41: 43-67.

Vimala Devi, P.S, Duraimurugan. P., Chandrika, K.S.V.P. and Vineela, V. 2021. Development of a water dispersible granule (WDG) formulation of Bacillus thuringiensis for the management of Spodoptera litura (Fab.). Biocontrol Science and Technology 31 (8): 850-864.

Vimala Devi, P.S., Duraimurugan, P., Poorna Chandrika, K.S.V., Vineela, V. and Hari, P.P. 2020. Novel formulations of Bacillus thuringiensis var. kurstaki: An eco-friendly approach for management of lepidopteran pests. World Journal of Microbiology and Biotechnology 36(5): 78.

Vasudeva Rao. V. 2015. Management strategies for birds and wild animals in organic crop production. International Journal of Economic Plants 2(4): 186-191.

12

Role of Agroforestry in Organic Farming

A.V. Ramanjaneyulu[1*], B. Joseph[1], Asha Ram[2], T. Chaitanya[1], R.V.T. Balazzii Naaiik[1] and S.N. Sudhakar Babu[3]

[1]*Professor Jayashankar Telangana Agricultural University, Rajendranagar Hyderabad-500 030, Telangana, India*

[2]*ICAR-Central Agroforestry Research Institute, Jhansi-284 001, Uttar Pradesh, India*

[3]*ICAR-Indian Institute of Oilseeds Research, Rajendranagar, Hyderabad-500 030 Telangana, India*

Abstract

The unsustainable, faulty, environmentally detrimental modern agricultural practices and their plausible deleterious effects on one side and challenges like global climate change, ever shrinking land and water resources and loss of biodiversity on the other hand are the main problems that Agriculture is facing. Hence, there is a need to adopt a set of ecofriendly and climate smart practices to achieve sustainable development goals (SDGs). Agroforestry and organic farming (farming without chemicals) are the two important concepts projected to mitigate the effects of climate change, enhance soil health, encourage crop diversification, promoting biodiversity and achieving climate resilience. We made an attempt to synthesize the available information on the role of agroforestry in promotion of organic farming.

Keywords: *Agroforestry, organic farming, carbon sequestration, GHG emissions*

1. Introduction

Agroforestry is a land use system that integrates trees, crops and animals in a way that is scientifically sound, ecologically desirable, practically feasible and socially acceptable to the farmers (Nair, 1979). It is a collective name for land use systems and technologies, where woody perennials (trees, shrubs, palms, bamboos, etc.) are deliberately used on the same piece of land management units as agriculture crops and/or animals in some form of spatial arrangement or temporal sequence. It includes both ecological and economical interactions among the different components. Nearly 43% of world's agricultural land has at least 10% tree cover, which suggests that agroforestry is widespread and infact, it is present in all continents. Agroforestry accounts for nearly 45.4 m-ha globally and 23.25 m-ha of land in India (Rizvi *et al.*, 2019; Anonymous, 2020).

Agroforestry provides 6Fs i.e. food, fruit, fodder, fuel, fertilizer, fibre besides timber, medicine, gums etc., to meet households needs. It is also important from the view point of bio-fuel, bio-energy, bio-pesticides, bio-remediation and industrial needs. It has strong potential for carbon sequestration, climate change adaptation and mitigation besides combating desertification and rehabilitation. Further, protecting and stabilizing the ecosystems, producing a high level of output of economic goods and improving income and raw materials to rural population are added advantages. It is well known for conserving natural resources, supporting rhizosphere microorganisms and the biodiversity. It is identified as the most vital option for enhancing income, climate resilience and ecosystem services (Handa *et al.*, 2020). Agroforestry systems with N fixing leguminous trees enhance soil fertility and crop productivity and tree fodders supply good quality fodder even during lean season too. They also control the micro-climate through shade, addition of biomass, erosion and run-off control, stabilising slopy lands etc., Organic farming envisages use of non-synthetic inputs and eco-friendly practices in producing crops. Thus, Agroforestry and organic farming can be mutually inclusive and complementary, so, the concept of organic agroforestry has come into existence. Though Agroforestry systems and practices help in promoting of organic farming and make the whole ecosystem as organic, but, the information on mechanism, extent, opportunities and challenges involved is not reviewed thoroughly.

2. Constraints in promotion of agroforestry in organic farming

There are many eventual benefits of organic agroforestry like improved soil quality, less energy and pollution, better employment opportunities, ecosystem services, less exposure to synthetic hazardous inputs, improvement in biodiversity, high productivity and profitability and sustainability (Reganold and Wachter, 2016). However, the following hindrances or challenges have been identified for its promotion.

- Trees offer competition for space, light, moisture and nutrients for intercrops and reduce food or fodder crop yields
- Trees act as hosts to insect pests, birds, animals etc., which in turn feed on food crops. This may further augment the menace of biotic stresses in organic farming which excludes use of synthetic chemicals.
- Lack of quality planting material (QPM) and marketing for agroforestry trees products is the major constraint for it's promotion

- It is laborious to prune/lop the branches and incorporate the same in the field
- Harvesting of tree products interfere with the crop maintenance
- Limited mechanization and severe labour scarcity, poor yield and theft, poor access to extension service (Ibrahim *et al.,* 2019)
- Decomposition of tree parts i.e., leaves, small twigs, loppings is slow and may interfere with crop management.
- Carbon monetization: Though many people speak on this aspect, but less is done on the ground. Lack of knowledge and awareness among farmers and other stake holders, 25-30% share of revenue on C-monetization has to be paid to the middle men and imperfect rules and regulations on extending carbon credits to the beneficiary farmers are the main issues to be addressed.
- The research and generation of quantitative data on benefits of nexus between Agroforestry and organic farming is limited (Rosati *et al.,* 2020).
- Inadequate policy and uncertainties in tree felling, transit and marketing regulations and differential rules in different states

3. Role of agroforestry in organic farming

3.1. Green manures and green leaf manure

With the onset of monsoon, green manure crops can be grown in the alleyways under Agri-horti, Hoti-pasture, Silvi-pasture, Agri-silviculture and Agri-horti-silviculture systems, and incorporated *insitu at* 50% flowering stage (45-60 DAS). Under limited irrigated conditions, it can be done even during off season as a cover crop to avoid loss of organic carbon and moisture. Ramanjaneyulu *et al.,* (2021) reviewed, compiled and reported dhaincha, sunhemp, pilli pesara, cowpea, cluster bean, mung bean, urd bean, sweet clover, berseem clover, crimson clover, milk vetch, hairy vetch etc., are suitable for green manuring before actual crop season starts. The N fixation ranges from 30-150 kg N ha^{-1} and save 25-50% N requirements of crops grown under organic farming. They further indicated that *Glyricidia maculate, Pongamia glabra, Leucaena leucocephala, Azadirachta indica, Delonix alata/regia, Cassia auriculata, Peltoforum ferrugenum, Moringa oleflorea, Agave sisalana, Thesphesia populanea, Cassia siamea, Vitex negundo, Bauhina vahii, Lentil, Pisum sativum, Chrysopogon, Musa sapietum, Quercus leucotrichophora/ semicarpifolia, Daphiniphyllum himalayaense, Bridelia retusa, Nyctanthes arbortristis, Artocarpus spp., Adina cardifolia, Terminalia paniculata,*

Pterocarpus marsupium, Syzgium cumini, Terminalia bellerica, Cleistanthes collinus can be promoted as green leaf manures (GLMs) depending on the soil and climate. Farmers have to grow these GLMs on their field boundaries/bunds only so that their prunings/loppings can be used for incorporating into the soil. The pruning can be done 2-3 times per year. On the contrary, transporting leaves, twigs, branches etc., from elsewhere grown trees is costly, energy and labour intensive. GLMs help in partial nutrient requirement of crops.

The biomass can be used as mulch for controlling soil temperature, moisture and suppressing weed growth and enhancing organic carbon. GMs and GLMs improve organic matter and nutrient availability (N, P, K, Ca, Mg, Zn, Fe, Mn etc.) by regulating soil pH through release of organic acids besides decomposition making the soil amenable for organic production (Katyal, 1977; Takkar and Nayyar, 1986; Swarup, 1987; Hundal *et al.,* 1987; Khind *et al.,* 1987, Nagarajah *et al.,* 1989).

3.2. Role of Agroforestry in nutrient recycling under organic farming

Nutrient recycling (NRC) refers to continuous transfer of nutrients from soil to plant and back to soil through the weathering of minerals, activities of soil biota and other transformations occurring within different components of the ecosystem. NRC and litter production are high under Agroforestry due to presence of woody perennials which are characterized by more extensive and deeper root systems than herbaceous pants.

The main principle of soil management in agroforestry is to make the best use of its resource-conserving and resource sharing potentials. Therefore, it is extremely useful to have a proper nutrient budget for the whole system based on nutrient requirements of individual components and nutrient dynamics within the system.

3.2.1. Comparison of pattern of nutrient cycling in forest, agriculture and agroforestry

Among three ecosystems *viz.*, Agriculture (Ass), Forestry (FSs) and Agroforestry (AFSs), the difference lies in the mobility of nutrients from one component to the other within the system (Fig. 1). In case of Agriculture, the depth of root system is limited thus extraction of moisture and nutrients from deeper layers is also limited. Further, the loss of soil organic carbon (SOC) especially during lean period is very high. In addition, burning of crop residue leading to less or negligible addition of biomass to the soil, hence, the magnitude of nutrient mining especially in tropical regions is high resulting in net negative balances if nutrients are not replenished. Thus, common ASs are open type. So, depletion of soil fertility is the fundamental biophysical reason for decreasing factor productivity (Sanchez *et al.,* 1995).

Though FSs and AFSs are similar with regard to addition of biomass, nutrient cycling, extraction of moisture from deeper layers due to very deep root system and maintenance of micro-climate are better, the AFSs have dual advantage of addition of prunings, leaf litter, roots and biomass from tree as well as intercrops. It preserves SOC and moisture besides controlling soil erosion and alkalinity better than FSs (Fig. 2). AFSs are closed systems with in which more NRC occurs. Thus, among three systems, AFSs with self-sustainability and high nutrient capital, are best for maintaining nutrient recycling, positive nutrient balance and soil health to make it suitable for organic farming.

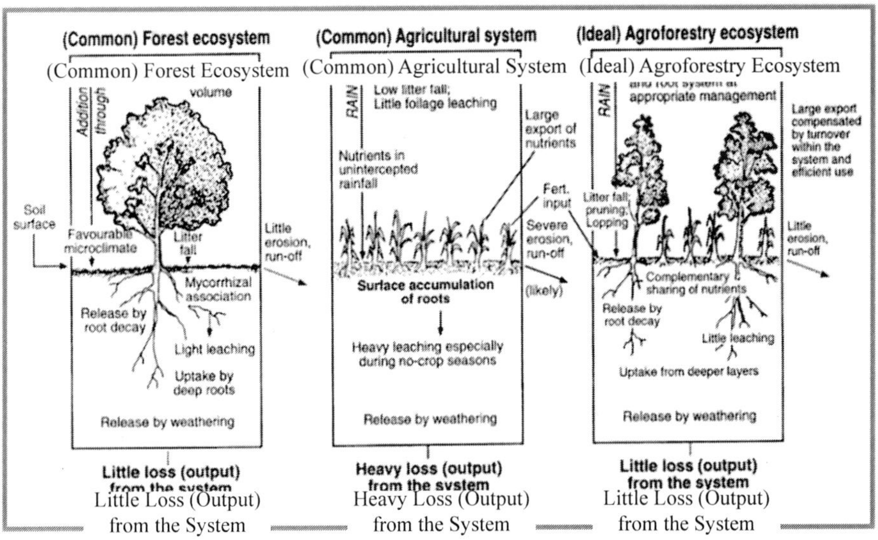

Fig. 1. Comparison of pattern of nutrient cycling in forestry, agriculture and agroforestry systems

(*Source:* Nair, 1984 and 1993)

Juo and Lal (1977) compared the effect of *Leucaena leucocephala* fallow and a bush fallow in an Alfisol in western Nigeria. In the former system, the biomass of *L. leucocephala* was added annually to the soil as mulch due to which the cation exchange capacity (CEC) and the levels of exchangeable calcium and potassium were found significantly higher than that of bush fallow.

Agamuthu and Broughton (1985) reported efficient nutrient cycling in oil palm plantations with leguminous cover crops (*Centrosema pubescens* and *Pueraria phaseoloides*) than in plantations without cover crops. Further, this system also avoided loss of nitrate nitrogen through leaching besides addition of about 150 kg N $ha^{-1}yr^{-1}$. Young (1989) reported addition of 100-200 kg N $ha^{-1}yr^{-1}$ in various agroforestry systems of humid and moist sub-humid climates, if all the prunings are left on the field. It was found to be equal to that of N removed in cereal/legume intercropping system. Similarly, the N return from litter and prunings was 100-300 kg N $ha^{-1}yr^{-1}$ in coffee and cacao plantations with shade trees.

3.2.2. Mechanism of nutrient recycling

NRC is influenced by climate, the land use capability class, type of tree species and field crop and their combinations, type of AFSs (silvi-pasture, agri-silvi-pasture, horti-pasture and agri-horti-pasture), arrangement directions, biomass production, rate of litter decomposition, microbiome etc., Leguminosae trees and crops are efficient in nutrient recycling with greater input of nutrients and less loss. Nitrogen transfer from nitrogen fixing trees (NFTs) occur both above and below ground through decomposition of litter and root residues and via soil organic matter. Mulches of non-nitrogen fixing trees can also supply nitrogen. Thus, it helps to enhance organic matter status which inturn will lead to improved nutrient recycling, soil structure, water holding capacity and soil productivity. Inclusion of leguminous cover crops (*Centrosema pubescens* and *Puueraria phaseloides*) in oilpalm plantations helped in nutrient recycling compared to no cover crop. Several leguminosae trees such as *Leucaena leucocephala*, *Acacia nilotica*, *Dalbergia sissoo*, *Gliricidia sp.*, *Sesbania sp.* etc., and some non-legume species such as *Casuarina equisetifoia*, *Alnus sp. etc.,* are important to fix about 50-500 kg N ha^{-1} (Agamuthu and Broughton, 1985). Agroforestry systems help to improve soil microbial activity due to greater amounts of litter thus promotes recycling of nutrients. Trees reduce nutrient losses by arresting soil erosion and intercepting the nutrients by means of their deep and extensive root systems, which otherwise leach out of the crop root zone. Besides their role in above-ground carbon sequestration, AFSs also have a great potential to increase carbon stocks in the soil.

The soil surface is the entry point for inputs (nutrients, rainfall, stem flow and stream flow) into the soil profile. While the surface layer is considered as the zone of intensive root activity, the sub-soil is the extensive root activity zone. The lower end of the extensive root layer is the boundary of the ecosystem between hydrosphere and lithosphere.

Nutrients taken up by the trees are either stored or used for the production of non-storage organs. Organic manure application acts as an input of nutrients and the accompanying ions will go into the solid phase of the uppermost layer of the soil. They pass into the solution phase of the soil layer and then spread simultaneously depending the mobility of the ions in different layers of the soil into the plant stand via uptake. Based on the water content, flow rate of percolating water and the soil properties, bio-elements transported beyond boundary layer are lost from the ecosystem and appear as output from the system (Fig. 3). Thus, a suitable combination of nitrogen fixing fodder/fruit trees with field crops can play a prominent role in enhancing productivity, soil nutrient status and microbial population and nutrient cycling to maintain the ecosystem. This kind of mechanism in agroforestry helps to convert entire ecosystem into organic.

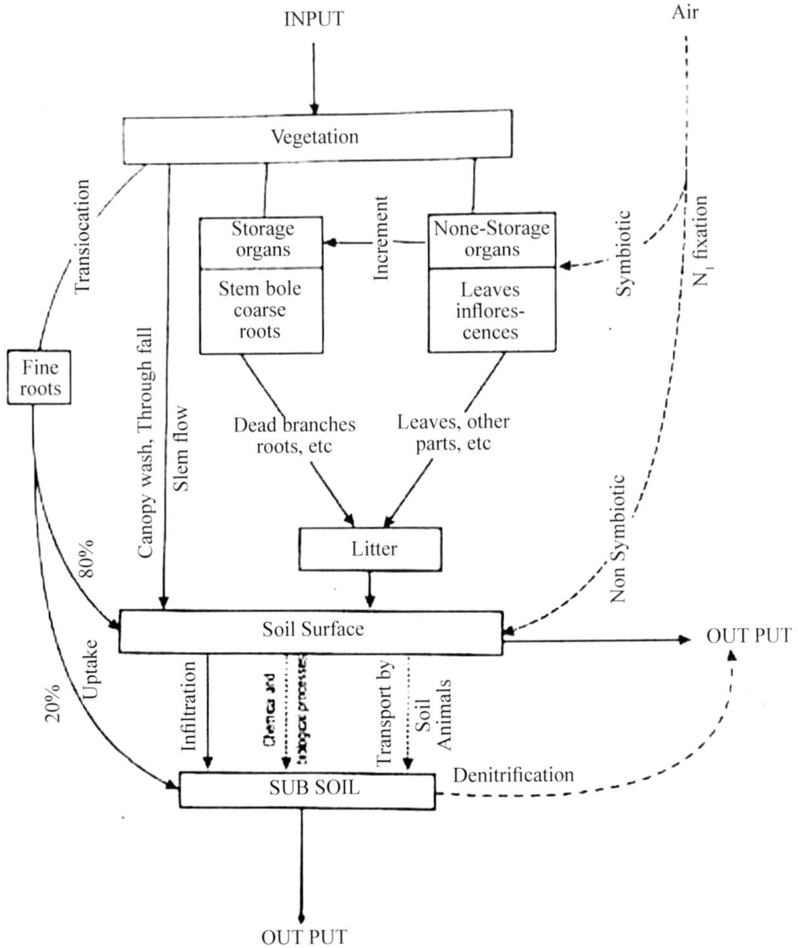

Fig. 2. A simplified model of nutrient cycling in tree based ecosystem (Dotted lies indicate the biological process taking place with the involvements of other organisms)

(*Source:* Nair, 1984)

3.3. Agroforestry for erosion control

Approximately, 75 BT of soil is eroded annually from arable lands globally leading to financial loss of US $400 billion per year (GSP, 2017). In India, about 120.7 m-ha' land suffers from various forms of degradation (NAAS, 2010). Among these, the soil and water erosion are one of the biggest threats to sustainability of agriculture as it affects productivity, profitability, ecological balance and environmental security. Agroforestry is one of the best options to counteract such problems. Trees when planted across the slope can help reduce soil erosion along streams and gullies (Fig. 3 and Fig.4). Their roots strongly hold the soil and reduce the impact of storm water. The tree species like *Paraserianthes falcataria*, *Gmelina arborea*, *Leucaena leucocephala*, *Sesbania grandiflora* and *Moringa oleifera are* suitable for stabilising stream banks and gullies (Nair, 1989). Further, runoff and soil loss were significantly less when small watersheds with agriculture were replaced either by trees and grasses (silvipasture) or with mechanical measures (Singh *et al.*, 1990). Alley cropping or hedgerow cultivation with tree species like *Leucaena leucocephala*, *Gliricidia sepium*, *Indigofera tysmani, Fleminigia spp. and Desmodium rensonii and* two grass species viz., *Vetiveria zizanoides* and *Thysanolaena maxima were found* to control runoff and erosion in the hilly areas (Khisa *et al.*, 2002).

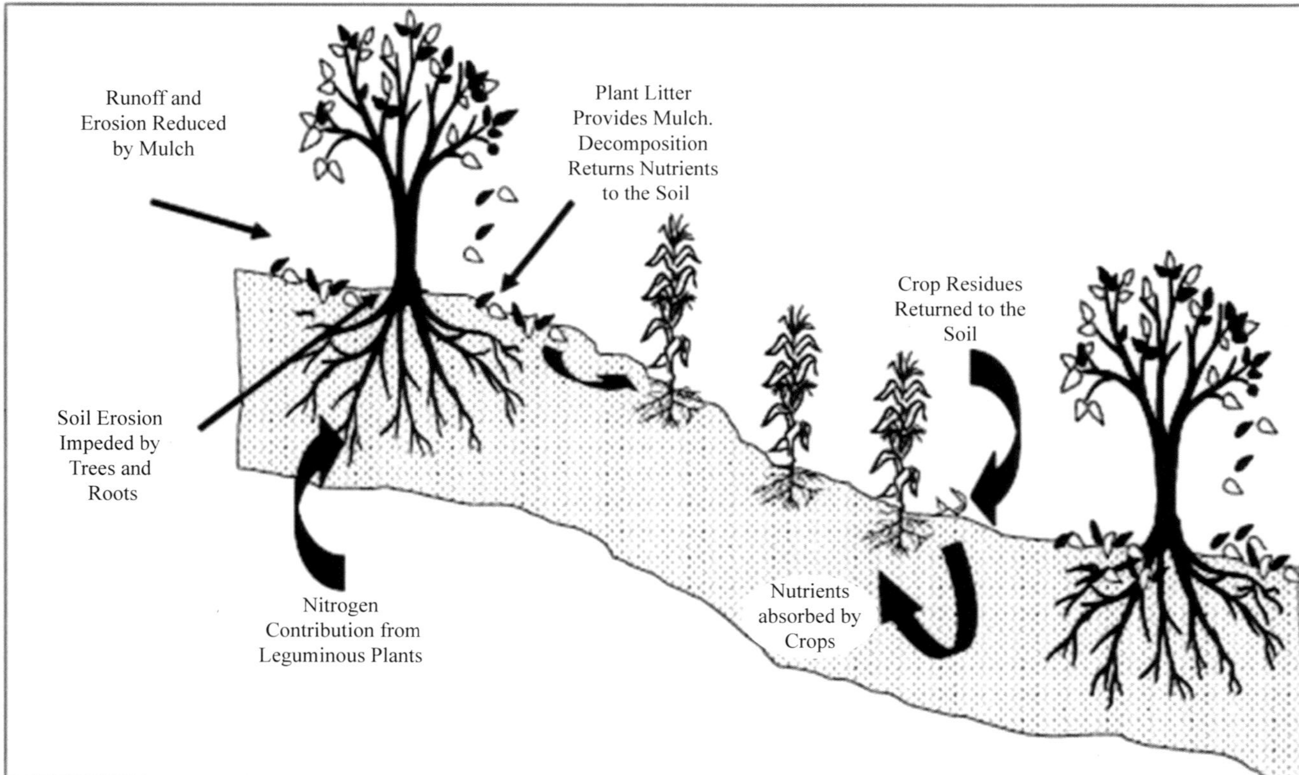

Fig. 3. Erosion control and soil fertility improvement due to Agroforestry in hill ecosystem (*Source*: Jinger *et al.*, 2023)

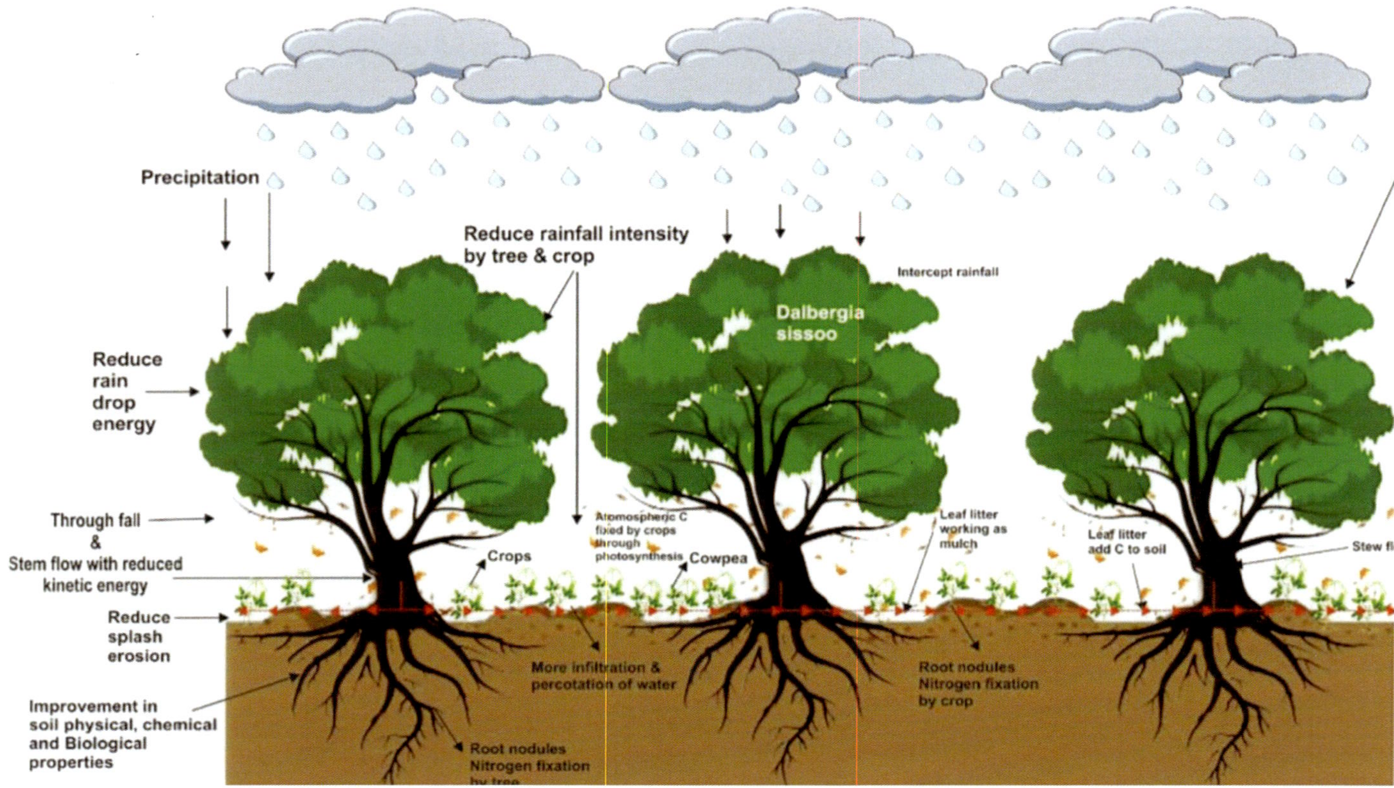

Fig. 4. Agroforestry system and its impact on carbon sequestration, soil erosion and soil fertility

3.4. Role of Agroforestry systems in building organic matter and microbial activity

Soil organic matter (SOM) is very important indicator of better soil fertility. It plays a greater role in enhancing soil health and productivity (Minasny *et al.,* 2017). The excretion of root exudates into the soil throughout the year by perennial tree species, presence of litter on the soil surface beneath different canopy layers and high biomass production are used as an energy source by macro and microorganisms leading to build up of organic carbon and biota in the soil. This further helps in supplying nutrients to the crops during subsequent seasons. A minimum of 20% tree crown cover is advocated for organic matter build up. A continuous soil cover of living plants and trees facilitates the capture and infiltration of rainwater and protects the soil and differential root depth of different plants permit an effective uptake of nutrients. Furthermore, the nutrients recycled by deep-rooting plants are replenished by tree species and soil macrofauna and microfauna. In addition, agroforestry techniques have the potential to increase soil porosity, reduce runoff, and increase soil cover, which can improve water infiltration and retention in the soil profile thereby reducing moisture stress in low rainfall years. Thus, AFSs enhance soil physical, chemical and biological properties and also recycles soil organic matter and nutrients as required for promoting organic farming.

3.5. Role of Agroforestry in improving biodiversity

Declining biodiversity (BD) is affecting food security, agricultural sustainability, and environmental quality (Ranjith *et al.,* 2019). Though Agroforestry systems use only 5% of the farm land area, but, contributes to more than 50% of the biodiversity, improving wildlife habitat and harbouring birds and beneficial insects which feed on crop pests. Agroforestry has potential in reducing species loss in agricultural landscapes and also help in endangered species conservation (Torralba *et al.,* 2016; Udawatta *et al.,* 2019). The agroforestry helps in biodiversity conservation by providing supplementary and secondary habitat for species and reducing rate of conversion of natural habitat. The earthworm density was found higher in AFSs especially in the proximity to tree vis-à-vis traditional cropping systems in southern Ontario, Canada (Price and Gordon, 1999). Further, wooded pasture fields were rich in earthworm population than that of open pasture lands (Moreno *et al.,* 2016). High SOC, less disturbance and heterogenous soil properties near trees are attributed to higher earthworms near the tree. Further, AFSs also help in improving faunal diversity. Tree based AFSs were found rich in arbuscular mycorrhizal fungi (AMF) colonization than in a tree less system in Canada (Bainard *et al.,* 2011). The enhanced biodiversity in Agroforestry systems can be ascribed to food, shelter, habitat, protection, refuge, favorable microclimate, improved soil-plant-water relationships, and other resources provided by multi-species vegetation of AF. However, strategic planning of the agroforestry systems based on soil-site-climate suitability and consideration of social and local needs is required to reduce negative effects on all stakeholders.

3.6. Agroforestry for ecosystem services, carbon sequestration and reducing green house emissions

Tree farming and agroforestry provide wide range of ecosystem services (Nair *et al.*, 2009; Murthy, 2013; Dhyani *et al.*, 2016) as detailed below.

3.6.1. Provisioning ecosystem services

These services are direct benefits or refer to the products obtained from ecosystems, including genetic resources, fuel wood, food, energy, fibre and fresh water. Agroforestry tree species provide non-timber forest products such as fruits, medicinal products, mushrooms, *honey (Akinnifesi et al, 2006;* Makonda and Gillah 2007). *Sclerocarya birrea* bark extract is used to treat malaria, dysentery, diarrhea, and rheumatism (Hall *et al.*, 2002). *Grevillea* robusta trees planted on contours on an average farm size of 1.64 hectares could provide the entire household's fuel wood needs annually (Nyadzi *et al.*, 2003; Kimaro *et al.*, 2007). Throughout the world, fodder trees play an important role as feed sources for livestock. *Gliricidia, Ailanthes, Leucaena* and *Albizia spp.* are considered as important fodder tree species in India. A field study at ICAR-CAFRI on *Albizia procera+ C. fulvus* yielded 10 tons of biomass per ha.

3.6.2. Regulating Services

These include regulating climate, water cycle, soil erosion and controlling some diseases. In recent years, models on the regulatory services provided by agroforestry systems have been developed for hydrological processes (water quality, erosion regulation), carbon sequestration and for the impacts of extreme meteorological events (flood and coastal regulation). It also plays an important role in carbon sequestration.

In India, carbon sequestration potential of agroforestry systems is estimated between 0.25 – 19.14 MgC ha^{-1} yr^{-1} for tree component and between 0.01 to 0.60 MgC ha^{-1} yr^{-1} for crop component. The contribution of agroforestry in soil carbon sequestration was also estimated which varied between 0.003 to 3.98 Mg C ha^{-1} yr^{-1} (Dhyani *et al.*, 2016). Pandey (2007) estimated the carbon stock in agroforestry systems between 12 to 228 Mg C ha^{-1} in various region of world with a mean value of 95 Mg ha^{-1}. Carbon sequestration potential of various agroforestry systems are given in Table 1. Trees and shrubs in agroforestry systems can contribute to provide better microclimate as shade and windbreak to field crops. Reduction of air and soil temperature and conserving moisture during heat waves help the associated crop to survive (Sileshi *et al.,* 2007). Agroforestry system in general and silvopastoral system in particular plays an important role in soil and water conservation.

3.6.3. Supporting services

Several tree species increase nutrient cycling by extracting subsoil inorganic nutrients from deep soils and cycling them to crops through decomposing biomass (Gindaba *et al.,* 2005). The most effective method for accumulating nitrogen in the soil is biological nitrogen fixation. According to Ram *et al.,* (2017), agroforestry add 0.250 Tg N in Indian soils out of which 0.195 Tg N is added through N-fixing trees and rest (0.055 Tg) through litter fall. Trees contribute 9% of total N addition in AFS in India. As trees' biomass is applied to the soil repeatedly, soil organic matter increases, providing a good environment for soil microbes and plants. Further, it also increases soil fertility. Thus, agroforestry plays an important role in preserving biological diversity. By providing a more productive, sustainable alternative to traditional agricultural systems, agroforestry systems preserve germplasm of sensitive species, reduce the rates of natural habitat conversion, erosion control and water recharge, thereby preventing habitat degradation and loss. Chilufya and Tengnas (1996) found that agroforestry could improve beekeeping by introducing improved beehives in orchards, woodlots or fodder banks that produce pollen for nectar.

Table 1. Carbon sequestration potential (CSP) of trees in India

Location	Agroforestry System	Tree species	No. of tree per hectare	Age (year)	CSP (Mg C $ha^{-1}yr^{-1}$)
Uttarakhand	Agrisilviculture	*D. hamiltonii*	1000	7	15.91
Himachal Pradesh	Agrihorticulture	Fruit trees	69	-	12.15
Khammam, Andhra Pradesh	Agrisilviculture	*L. leucocephala*	4444	4	14.42
			10000	4	15.51
Uttarakhand	Agrisilviculture	*P. deltoids*	500	8	12.02
SBS Nagar, Punjab	Agrisilviculture	*P. deltoids*	740	7	9.40
Dehradun, Uttarakhand	Silviculture	*E. tereticornis*	2500	3.5	4.40
			2777*	2.5	5.90
Kurukkhetra, Haryana	Silvipasture	*A. nilotica*	1250	7	2.81
		D. sissoo	1250	7	5.37
		P. juliflora	1250	7	6.50
Chandigarh	Agrisilviculture	*L. leucocephala*	10666	6	10.48
Tripura	Silviculture	*T. grandis*	444	20	3.32
		G. arborea	452	20	3.95
Tarai Region Uttarakhand	Silviculture	*T. grandis*	570	10	3.74
			500	20	2.25
			494	30	2.87
Jhansi, Uttar Pradesh	Agrisilviculture	*A. procera*	312	7	3.70
Jhansi, Uttar Pradesh	Agrisilviculture	*A. pendula*	1666	5.3	0.43

Jhansi, Uttar Pradesh	Silviculture	*A. procera*	312	10	1.79
		A. amara	312	10	1.00
		A. pendula	312	10	0.95
		D. sissoo	312	10	2.55
		D. cinerea	312	10	1.05
		E. officinalis	312	10	1.55
		H. binata	312	10	0.58
		M. azaderach	312	10	0.49
Hyderabad, Andhra Pradesh	Silviculture	*L. leucocephala*	2500	9	10.32
		E. camaldulensis	2500	9	8.01
		D. sissoo	2500	9	11.47
		A. lebbeck	625	9	0.62
		A. albida	1111	9	0.82
		A. tortilis	1111	9	0.39
		A. auriculiformis	2500	9	8.64
Hyderabad, Andhra Pradesh	Agrisilviculture	*L. leucocephala*	11111	4	2.77
			6666	4	1.90
Raipur Chhattisgarh	Agrisilviculture	*G. arborea*	592	5	3.23
Coimbatore Tamilnadu	Agrisilviculture	*C. equisetifolia*	833	4	1.57
Kerala	Home garden	Mixed tree spp.	667	71	1.60

(*Source*: Dhyani *et al.*, 2020)

3.6.4. Cultural services

A cultural service is a non-material benefit people obtain from ecosystems through spiritual enrichment, cognitive development, reflection, recreation and aesthetic experience, including knowledge systems, social relationships and aesthetic values. Since ancient times, many communities maintaining the sacred grooves which provide various cultural services to the community.

3.7. Tree fodders

The tree species which yield fodder are called as 'tree fodders'. These are perennial and best alternative to annual grasses. They can supply fodder throughout the year and their role will be more promising especially during lean season when no fodder will be available from annual grasses. They supply fodder in the form of seeds, fruits, nuts, acorns, flowers or foliage, grains etc., The idea of feeding the livestock with tree fodders was promoted by J. Russel Smith in his classic book 'Tree crops' (Smith, 1950).

Farmers across the world who maintain livestock like cattle, buffalo, goat and sheep for obtaining various animal products and income, try to supply good quality feed. Tree fodders which are produced naturally without addition of synthetic chemicals have a greater role to play in organic production of livestock. The leaves of tree fodders contain water and dry matter. The dry matter contains carbohydrates, protein and other substances like minerals needed for growth and development of livestock and maintenance stability of the animal. Further, tree fodders provide shade and shelter thus protects them from abiotic stresses like hot, cold or windy weather. Some of the examples for tree fodders are given in Table 2. Tree fodders should possess the following characters for including them in production of organic fodder (Abel *et al.*, 1997).

- Deep and extensive root system
- Resistant to frequent pruning (lopping, pollarding and coppicing) and browsing
- Less nutrient demanding
- Capacity to maintain large pool of nutrients within living tissues such as bark, stems and roots and then recycling internally from season to season (Youssefi *et al.*, 1999).
- Leaves should have high protein content
- Faster leaf growth with higher production
- The leaves should not contain toxins and (or) should be with in permissible limits

- Palatable and digestible
- Should be resistant to drought, pests and diseases

Table 2. Details of important fodder trees, duration of flowering and harvesting

S. No.	Local name	Botanical name	Period of flowering	Period of harvesting leaves
1	Babul	*Acacia nilotica*	June-September	April-June
2	Siris	*Albizzia lebbeck*	April-June	January-February
3	Rela	*Cassia fistula*	April-June	March-April
4	Sarugudu	*Casuarina equisetifolia*	Feb-April	June-July
5	Sissoo	*Delbergia sissoo*	March-April	Dec-Jan
6	Rose wood	*Dalbergia latifolia*	---	March-April
7	Peddamanu	*Ailanthus excels*	Feb- March	May - June
8	Jamun	*Syzgium cumini*	March-May	June-August
9	Chintha	*Tamarindus indica*	April-June	Feb-April
10	Usiri	*Emblica officinalis*	March-May	Nov-March

(*Source:* Joseph *et al.*, 2016)

Based on the quantity of leaves required per day depending on the number of animals, no. of feeding days in a year, leaf yield per tree, the total no. of fodder yielding trees have to be planted and maintained. In case of problematic, poor and impoverished soils, planting has to be done with more tree density duly following reclamation measures. Furthermore, few tree fodders should be left without pruning to enable them to produce seeds, which can be used for future fodders cultivation in their farm organically.

3.7.1. Role of fodder trees in organic farming

Trees are the most important components of Agroforestry. They play a pivotal role in supplying naturally produced fodders rich in proteins, minerals and fibre as detailed below (Table 3).

3.7.1.2. Source of protein

Protein is required for growth and milk production in the daily feeding ration. But often, the livestock is often fed with low-quality roughage or grasses, such as napier grass, which contain less amounts of protein. Therefore, it is recommended to supplement the basic diet of dairy cows and cattle with concentrates (dairy meal) which contains enough protein (about 160 grams in 1.0 kg). Economically weak farmers can't afford to supply protein rich fodders. Hence, tree fodders grown organically in the farm are rich in proteins can be easily harvested and fed to the cattle. Among many, *Sesbania sesban*

(33.40%), *Melia azadirach* (29.25%) and *Leucaena leucocephala* (23.33%) can be preferred due to high protein content.

3.7.1.3. Dry season supplement

The trees and shrubs remain green for longer period due to deep root system and can provide the required energy and protein when other sources such as grasses are dry in off season.

3.7.1.4. Other uses

Tree species enhance soil fertility by providing mulch, fixing nitrogen fertilizer trees).

Table 3. Crude protein, ether extract, crude fibre, nitrogen free extract, P & Ca in trees (%)

S. No	Tree *sps.*	Crude protein	Either extract	Crude fibre	Nitrogen free extract	Phosphorus	Calcium
1	*Acacia nilotica*	16.90	5.50	23.90	-	-	-
2	*Ailanthus exelsa*	19.87	3.53	12.82	51.81	-	2.11
3	*Azadirachta indica*	17.04	2.74	29.41	46.27	-	-
4	*Dalbergia sissoo*	13.09	3.35	20.67	53.81	-	2.24
5	*Ficus benghalensis*	10.90	3.90	36.90	-	-	-
6	*Glricidia sepum*	22.33	4.20	19.70	-	2.30	11.90
7	*Leuceana leucocephala*	23.33	1.92	12.83	-	-	-
8	*Melia azadirachta*	29.25	-	15.47	-	-	-
9	*Pongamia pinnata*	19.36	-	30.85	-	-	-
10	*Prosophis cineraria*	12.32	1.60	17.71			
11	*Sesbania sesban*	33.40	-	-	-	0.34	-

(*Source*: Anthony Joseph Raj and Lal, 2014)

3.7.2. Role of N fixing Agroforestry tree fodders in organic farming

Trees which have the capacity to produce fodder besides N fixation are best for organic farming as they can act as natural N suppliers to the ecosystem. Nitrogen fixing fodder trees (NFTs) grown in Agroforestry systems are capable of fixing approximately 50-100 kg N ha^{-1} yr^{-1} (Table 4). These are the most important component of leaf litter which forms humus after decomposition, releases nutrients and improves soil properties (Roshetko and Gutteridge, 1996).

Table 4. Nutritious nitrogen fixing Agroforestry fodder trees suitable for integration into organic farming

S. No.	Local Name	Botanial name	Family	N fixation ($kg\ N_2\ ha^{-1}\ yr^{-1}$)
1	Erythrina	*Erythrina sppecies*	Papilionacea	57-66
2	Inga	*Inga edulis*	Mimosoideae	35-40
3	Subabul	*Leuceaena leucocephala*	Mimosoideae	100-500
4	Babul	*Acacia mangium*	Fabaceae	50-100
5	White acacia	*Faidherbia albida*	Fabaceae	20-50
6	Casuarina	*Casuarina equisetifolia*	Casuarinaceae	60-110
7	Glyricidia	*Glyricidia sepium*	Fabaceae	13 -50
8	Siris	*Albezzia lebbeck*	Fabaceae	260
9	Mesquite	*Prosopis species*	Fabaceac	25-26
10	Silveroak	*Grevellea robusta*	Proteaceae	200-300
11	Agati	*Sesbania rostrata*	Fabaceae	83-109

(*Source*: Suzette Searle, 2005; Joseph *et al.*, 2016)

The different fodder tree based agroforestry systems for promotion of quality fodder production are detailed hereunder (Rai and Ajit Sharma, 2007).

3.7.2.1. Boundary plantation

In this system, fodder trees are planted all along the field boundaries (Fig. 5) to get quality green leaf fodder during lean periods without jeopardizing the yield of grain crops. They can also act as live fence, wind breaks, shelter belts or fire breaks. In addition to protection and stabilization of site they also produce green manure as organic fertilizer, fodder for farm and animals. Eg. *Acacia nilotica*

0	0	0	0	0	0	0
0	X	X	X	X	X	0
0	X	X	X	X	X	0
0	X	X	X	X	X	0
0	X	X	X	X	X	0
0	0	0	0	0	0	0

0 = Trees; X = Crops

Fig. 5. Boundary plantation with fodder trees

3.7.2.2. Alley cropping

Alley cropping or hedgerow intercropping is the planting of agronomic, horticultural or forage crops in the alleys formed due to planting of fodder trees. The rows act as contour hedge rows which help in stabilization and conservation of soil if planted across the slope in high rainfall (humid and sub-humid) areas especially. Further, it is effective in reducing runoff, increasing infiltration and reducing soil and nutrient loss. Additionaly, it also provides leaf fodder and fuel wood through lopping frequently. Besides, it also provides organic fertilizer upon coppicing, if nitrogen fixing trees are planted. For instance, alley cropping groundnut crop in pollarded subabul (Fig. 6).

Fig. 6. Alley cropping Groundnut crop in Pollarded Subabul

3.7.2.3. Agri-silvi-horticultural system

In this system, between two rows of fruit trees, one or two fast growing forest trees (quality rich fodder yield trees) and agricultural crops are introduced to utilize the wider spaces more efficiently without affecting the growth and yield of fruit trees. eg. Groundnut in Guava plantation (Fig. 7)

Fig. 7. Groundnut in guava plantation

3.7.2.4. Agri-silvi-pastoral system

Under this system, specific fodder trees are integrated with grain crops, pasture and animals. This system provides opportunity in arable high land, humid tropics for mixed tree crop livestock husbandry. The crop, trees and grasses are grown on the same unit of land simultaneously which provide food and fuel wood to human being and feed and fodder to the livestock. Sometimes, arable crops are grown as under-storey crops with the trees and grasses like *Panicum* and Elephant grass are being grown along the rows of trees which serves as vegetative barrier to control the erosion. In European countries, panicum grass is being allowed to grow under the coniferous trees like Larches and even under *Alnus sp.* where the cattle will graze. For example: Cowpea +Tamarind + Seethaphal (Fig. 8).

Fig. 8. Cowpea +Tamarind + Seethaphal

3.7.2.5. Horti-pastoral system

It is a two tier model for fruit-cum-fodder based Agroforestry system. In this system, fruit tree species are integrated with livestock as they are usually. These are planted with wide spacing, which provides good light penetration especially in young crops. The shade level is regulated by lopping of trees branches, which are fed to livestock. The major objective of this system is to yield fruits. However, tree leaves are obtained as by-product apart from fruits to feed the animals. This system works best in medium soils. Slow growing and late bearing fruit tree species like tamarind, jamun, aonla, bael, cashew and wood apple are highly suitable. In a low input management system with continuous grazing. For example: Aonla+*stylosanthes hamata*, Custard apple+*Cenchrus ciliaris*, *Amla* + *Stylosanthes hamata* (Fig. 9).

Fig. 9. *Amla + Stylosanthes hamata*

3.7.2.6. Silvi-pastoral system

It is an ideal combination of grasses, legumes and trees for optimizing land productivity, conserving plants, soils and nutrients and producing forage, fuelwood and timber on sustainable basis. Thus, this involves replantation, substitution or intervention in existing vegetation by desirable species. Actually silvipastoral systems are land use systems in which fodder trees are combined with livestock and pasture for forage and fuelwood on the same unit of land. Suitable for areas with severe fodder scarcity. It is advantageous in semi-arid tropic (SAT) where 60% of the cropped area is occupied by shallow soils and these soils support only *kharif* crops and crop failures are quite common. For example: *Acacia nilotica+Stylosanthes hamata,* (Fig. 10).

Fig. 10. *Acacia nilotica + Stylosanthes hamata*

The details of important fodder tree species along with suitable region, fodder availability period and preference to livestock are listed in Table 5.

Table 5. Fodder tree species, region, fodder availability period and preference to livestock

Tree species	Region	Fodder availability Period		Preferences to livestock
		Leaf Fodder	Pod Fodder	
Faidherbia albida	SAR	May-Feb	Mar-June	Goat Sheep, Camel
Aacia nilotica	AR,SAR,IGP, CZ,COZ	May-Feb	Apr-June	Cattle, Goat, Sheep, Camel
Ailanthus excels	AR,SAR	May-Mar	-	Goat, Sheep
Albizia lebbeck	AR,SAR,IGP, CZ,NER	Apr-Nov	Aug-Nov	Goat, Sheep, Cattle
Azadirachta indica	AR,SAR,IGP, CZ,NER	Throughout the year	Jul-Aug	Goat, Sheep, Cattle
Dalbergia sissoo	SAR,IGP	Feb-Dec	Apr-Sep	Goat, Sheep, Cattle
Ficus religiosa	SAR,CZ,IGP	July-Nov	-	Cattle, Sheep
Gliricidia sepium	SAR	Mar-Nov	-	Goat and Sheep
Gmelina arborea	SAR,CZ,IGP	July-Nov	-	Cattle, Sheep, Goat
Hardwickia binata	SAR,CZ	Apr-Jun	Feb-Mar	Cattle, Sheep, Goat

Leucaena leucocephala	SAR,IGP,CZ	Throughout the year	Dec-May	Cattle, Goat, Sheep
Melia azedarach	IGP,SAR,STZ	July-Nov	-	Goat, Sheep, Cattle
Morus alba	IGP,CZ,STZ	Throughout the year	-	Cattle, Goat, Sheep
Prosopis cineraria	AR,SAR	Throughout the year	May-June	Goat, Sheep, Camel, Cattle
Sesbania grandiflora	IGP,CZ,NER	July-Nov	-	Cattle, Sheep, Goat
Sesbania sesban	SAR,IGP,CZ,NER	Throughout the year	Dec-Mar	Cattle, Goat, Sheep

(AR=Arid region, SAR=Semi arid region, IGP=Indogangetic plains, CZ=Central zone, COZ=Coastal zone, NER=North Eastern region, STZ=Sub-temperate zone)

(*Source*: Anthony and Lal, 2014)

The parts of trees most preferably leaves, but, pods, seeds and small twigs may also be fed to the livestock. These fodders can be browsed or 'cut-and-carried' to stall-fed (zero-grazed) animals depending on the need, facilities available, financial conditions and time required. In case of abundant fodder trees availability at the farm, the leaves can be fed to the animal throughout the year in limited quantity. But in case of limited number of trees, it is recommended to feed a dairy cow, dairy goat and sheep the month before and the month after giving birth. Oxen should be fed one or two months before and during the period of land preparation. It is always beneficial to feed cattle, goats and sheep with organically produced tree fodders. However, research efforts are needed to overcome the problem of toxins, tannins, undigestible protein present in fodders to be fed to various types of animals (Roothaert, 2000). Further, excess feeding should be avoided duly following following dosages (Table 6).

Table 6. Fodder tree species and the dosage to be given to animals

S.No.	Name of the tree species	Quantity to fed (kg day^{-1})
1	*Calliandra calothyrsus* (in humid and sub-humid areas with a minimum annual rainfall of 700 mm; can be grown in acid soils)	3-4 kgs for heifers; 8-9 kgs dairy cows; 50% of their total intake of food for sheep and goats
2	*Gliricidia sepium* (in less rainfall areas i.e. 600 mm; can be grown in acid soils and tolerate waterlogging)	10-30% of ration weight with other grass or roughages

3	*Leucaena leucocephala* (in dry and wet areas with annual rainfall ranging from 650-1500 mm)	30-50% of the total feed day^{-1} for Sheep and Goats; 10-30% of the total amount of feed consumed daily for cattle (contains Mimosine. So over feeding with leaves, young twigs and pods leads to hair loss, low weight gains, low birth weight and infertility in animals)
4	*Sesabania sesban* (in areas with low to high annual rainfall of 500-2000 mm; tolerate waterlogging)	30 % of the total feed taken by the cattle, goats and sheep per day. Excess feeding of this fodder is harmful as it contains "Saponin" which is a poison when consumed and it leads to bloat.

(*Source*:www.agromisa.org)

4. Conclusion

Agroforestry provides numerous benefits for improving the livelihoods of farmers. Besides promoting crop diversification, it also ensures carbon sequestration, nutrient recycling, soil health, productivity and profitability. Inclusion of N fixing fodder trees, tree residue incorporation, addition of green leaf manures etc., will ensure soil quality, animal health, biodiversity and sustainability. Agroforestry practices without chemicals will help convert entire ecosystem organic in with few years. Organic agroforestry has high scope in the areas where agroforestry systems are already in vogue. The farmers have to take advantage of provisions of National Agroforestry Policy (2014), Sub mission on Agroforestry (2016), National Bamboo Mission (2018), Watershed development programmes, Pradhan Mantri Krishi Sinchayee Yojana (PMKSY), Haritha Haram programme, Paramparagat Krishi Vikas Yojanan (PKVY) etc., to adopt and promote organic agroforestry.

5. Way forward

Organic farmers rarely adopt agroforestry practices. There is a need to identify the researchable issues and extension gaps in organic agroforestry without foregoing the basic principles of organic farming. Early maturing tree species with resistance to biotic and abiotic stresses have to be identified and promoted. Further, the inter row distance has to be adjusted such that intercultivation equipment, harvesters and threshers can be run easily. Low cost mechanized pruning and lopping methods are needed. Quantitative data on GHGs emission, carbon sequestration and monetization of organic agroforestry is needed for it's further promotion. Monetization or incentivization of beneficiaries may further help to upscale it. Capacity building to improve the awareness of all stakeholders is required. Furthermore, bolstering marketing framework is the promising strategy to foster organic agroforestry on the lines of U.K., Hungary, Poland, Ukraine and Italy.

6. References

Abel, N., Baxter, J., Campbell, A., Cleugh, H., Fargher, J., Lambeck, R., Prinsley, R., Prosser, M., Reid, R., Revell, G., Schmidt, C., Stirzaker, R. and Thorbum, P. 1997. Design principles for farm forestry: A guide to assist farmers to decide where to plant trees, RIRDC/L. WPRDC, Joint venture agroforestry program.

Agamuthu, P.J. and Broughton, W.J. 1985. Nutrient cycling within the developing oil palm-legume ecosystem, Agriculture, Ecosystems and Environment 13: 111-123.

Akinnifesi, F.K., Makumba, W., Kwesiga, F.R. 2006. Sustainable maize production using gliricidia/maize intercropping in southern Malawi. Experimental Agriculture 42:441–457.

Anonymous, 2020. Global Forest Resources Assessment 2020, Food and Agriculture Organization of the United Nations, Rome, Italy.

Anthony, J.R. and Lal, S.B. 2014. Agroforestry Theory and practices; Scientific Publishers, Jodhpur, India, pp 603-605.

Bainard, L.D., Klironomos, J.N. and Gordon, A.M. 2011. Arbuscular mycorrhizal fungi in tree-based intercropping systems: A review of their abundance and diversity. Pedobiologia 54: 57-61.

Chilufya, H. and Tengnas, B. 1996. Agroforestry extension manual for northern Zambia. Reg. Soil Conserv. Unit (RSU), Tech. Handbook. Ser. 11:124.

Dhyani, S.K., Ram, A. and Dev, I. 2016. Potential of agroforestry systems in carbon sequestration in India. Indian Journal of Agricultural Sciences 86: 1-10.

Dhyani, S.K., Ram, A. and Newaj, R. 2020. Agroforestry for Carbon Sequestration in Tropical India. In: Ghosh P, Mahanta S, Mandal D, et al. (eds) Carbon Management in Tropical and Sub-Tropical Terrestrial Systems Singapore Pte Ltd, Singapore, Springer Nature pp 313–331.

Gindaba, J., Rozanov, A. and Negash, L. 2005. Trees on farms and their contribution to soil fertility parameters in Badessa, eastern Ethiopia. Biology and Fertility of Soils 42: 66–71.

GSP. 2017. Global Soil Partnership. Endorses guidelines on sustainable soil management.
http:// www. fao. org.

Hall, J., O'Brien, E. and Sinclair, F. 2002. Sclerocarya birrea: a monograph. Univ Wales, School of Agricultural Sciences and Technology 143-157.

Handa, A.K., Chavan, S.B., Veeresh Kumar, Vishnu, R., Ramanan S.S., Tewari, R.K., Arunachalam, A., Bhaskar, S., Chaudhari, S.K. and Mohapatra, T. 2020. Agroforestry for income enhancement, climate resilience and ecosystem services. Indian Council of Agricultural Research, New Delhi.

Hundal, H. S., Biswas, C. R. and Vig, A. C. 1987. The utilization by rice of P from different 32P labelled green manures. Biological Wastes 22: 97-105.

Ibrahim, A.O., Adedeji, A.S. and Meduna, P.N. 2019. Constraints facing agroforestry practices among farmers in New Bussa, Nigeria. Journal of Research in Forestry, Wildlife & Environment 11(3): 133-141.

Jinger, D., Kaushal, R., Kumar, R., Paramesh, V., Verma, A., Shukla, M., Chavan, S.B., Kakade, V., Dobhal, S., Uthappa, A.R., Roy, T., Singhal, V., Madegowda, M., Kumar, D., Khatri, P., Dinesh, D., Singh, G., Singh, A.K., Nath, A.J., Joshi, N., Joshi, E. and Kumawat, S. 2023. Degraded land rehabilitation through agroforestry in India: Achievements, current understanding, and future prospectives. Frontiers in Ecology and Evolution 11:1088796.

Joseph, B., Hemalatha, S., Madhavi, L.A. and Laxminarayana, P. 2016. Practical manual on Principles and Practices of Social Forestry; BS Publication, Hyderabad, India, pp 55-56.

Juo, A.S.R. and Lal, R. 1977. The effect of fallow and continuous cultivation on the chemical and physical properties of an Alfisol in western Nigeria, Plant and Soil 47: 567-584.

Katyal, J. C. 1977. Influence of organic matter on the chemical and electro-chemical properties of some flooded soils. Soil Biology and Biochemistry 9: 259- 266.

Kimaro, A.A., Timmer, V.R., Mugasha, A.G., Shabani., Kimaro. and Deborah. 2007. Nutrient use efficiency and biomass production of tree species for rotational woodlot systems in semi-arid Morogoro, Tanzania. Agroforestry Systems 71:175–184.

Khind, C. S., Jugsujinda, A., Lindau, C. W. and Patrick, W. H. J. 1987. Effect of sesbania straw in a flooded soil on soil pH, redox potential and water soluble nutrients. International Rice Research Institute Newsletter, 12(3): 42-43.

Khisa, S.K and Hossain, A.T.M.E. 2002. Upland settlement programme with small holding rubber based agroforestry farming systems in the CHT. In: Khan NA, Alam MK, Khisa SK, Millat-e-Mustafa M (eds) Farming practices and sustainable development in the Chittagong Hill Tracts. CHTDB and VEFP-IC, pp 123–134.

Minasny, B., Malone, B.P., McBratney, A.B., Angers, D.A., Arrouays, D., Chambers, A., Chaplot, V., Chen, Z.S, Cheng, K., Das, B.S., Field, D.J., Gimona, A., Hedley, C.B., Hong, S.Y, Mandal, B., Marchant, B.P., Martin, M., McConkey, BG., Mulder, V.L., O'Rourke S, Richer-de-Forges AC, Odeh I, Padarian J, Paustian K, Pan G, Poggio L, Savin, I., Stolbovoy, V., Stockmann, U., Sulaeman, Y., Tsui, C-C., Vågen, T-G., van Wesemael. B. and Winowiecki, L. 2017. Soil carbon 4 per mille. Geoderma 292:59–86

Murthy, I.K. 2013. Carbon Sequestration Potential of Agroforestry Systems in India. Journal of Earth Science and Climatic Change 04:1000131.

Makonda, F.B.S, Gillah, P.R. 2007. Balancing Wood and Non-Wood Products in Mi-ombo Woodlands. MITMIOMBO-- Management of Indigenous Tree Species for Ecosystem Restoration and Wood Production in Semi-arid Miombo Woodlands in Eastern Africa. Morogoro, Tanzania 6–12.

Nagarajah, S., Neue, H. U. and Alberto, M. C. R. 1989. Effect of Sesbania, Azolla and rice straw incorporation on the kinetics of NH_4, K, Fe, Mn, Zn and P in some flooded rice soils. Plant and Soil 116: 37-48.

NAAS-National Academy of Agricultural Sciences. 2010. Degraded and wastelands of India -Status of spatial distribution. ICAR: New Delhi.

Nair, P.K.R.1979.Intensive multiple cropping with coconuts in India. Verlaug Pau Parey, Berlinand Hamburg, Germany, p 147.

Nair, P.K.R. 1984. Soil productivity aspects of agroforestry, International Council for Research in Agroforestry, Nairobi, Kenya. Experimental Agriculture 21(3): 299-299.

Nair, P.K.R. 1989.Agroforestry Systems in the Humid Tropics. Dordecht, Netherlands: Kluwer Academic Publishers, p 249.

Nair, P.K.R., Kumar, B.M. and Nair, V.D. 2009. Agroforestry as a strategy for carbon sequestration. Journal of Plant Nutrition and Soil Science 172: 10-23.

Nyadzi, G.I, Otsyina, R.M, Banzi, F.M., Bakengesa, S.S., Gama, B.M., Mbwambo, L. and Asenga, D. 2003. Rotational woodlot technology in northwestern Tanzania: Tree species and crop performance. Agroforestry Systems 59: 253-263.

Moreno, G., Gonzalez-Bornay, G., Pulido, F., Lopez-Diaz, M.L., Bertomeu, M., Jua'rez, E. and Diaz, M. 2016. Exploring the causes of high biodiversity of Iberian dehesas: The importance of wood pastures and marginal habitats. Agroforestry Systems 90: 87-105.

Pandey, D.N. 2007. Multifunctional agroforestry systems in India. Current Science 92:455-463.

Rai, P. and Ajit, Sharma, A.K. 2007. Tree leaves their production and nutritive value for ruminants: In review, Animal Nutrition and Feed Technology 7(2): 135-159.

Ram A, Dev I, Uthappa AR, Dhiraj, K., N. Kumar., Chaturvedi, O.P., Dotaniya, M. L. and Meena, B.P. 2017. Reactive Nitrogen in Agroforestry Systems of India. Elsevier Inc. pp 207-218.

Ramanjaneyulu, A.V., Sainath, N., Swetha, D., Reddy, R.U. and Jagadeeshwar, R. 2021. Green Manure Crops: A Review. Biological Forum 13(2): 445-455.

Ranjith, P.U., Lalith, M., Rankoth. and Shibu, J. 2019. Agroforestry and Biodiversity. Sustainability 1-11.

Reganold, J.P. and Wachter, J.M. 2016. Organic agriculture in the twenty-first century. Nature Plants 2(2): 1-8.

Rizvi, R.H., Newaj, R., Handa, A.K., Sridhar, K.B. and Kumar, A. 2019. Agroforestry mapping in India through geospatial technology: Present status and way forward. National Research Centre for Agroforestry, Jhansi.

Rosati, A., Borek, R. and Canali, S. 2020. Agroforestry and organic agriculture. Agroforestry Systems 4: 1-17.

Roshetko, J.M. and Gutteridge, R.C. 1996. Nitrogen fixing trees for fodder production-A Field Manual, Morrilton, Arkansas USA: Forest, Farm, and Community Tree Network (FACT Net), Winrock International- Institute for Agricultural Development.

Roothaert, R.L. 2000. The potential of indigenous and naturalized fodder trees and shrubs for intensive use in central Kenya, Doctoral thesis, Wageningen University, The Netherlands.

Sanchez, P. 1995. Science in agroforestry, Agroforestry systems 30: 5-55.

Sileshi, G., Schroth, G., Rao, M.R. and Girma, H. 2007. Weeds, diseases, insect pests, and tri-trophic interactions in tropical agroforestry. Ecol Basis Agroforestry pp 73-94.

Singh, G. S., Shastri, G., & Bhardwaj, S. P. 1990. Watershed response to conservation measures under different agro-climatic regions of India. Indian Journal Soil Conservation 18(3): 16-22.

Smith Suzette, Searle. 2005. Nitrogen fixation in Acacias: an untapped resource for sustainable plantations, Farm forestry and land reclamation. pp 68.

Swarup, A. 1987. Effect of pre-submergence and green manuring (Sesbania aculeata) on nutrition and yield of wetland rice (Oryza sativa) on a sodic soil. Biology and Fertility of Soils: 5: 203-208.

Takkar, P. N. and Nayyar, V. K. 1986. FAI seminar on growth and modernization of fertilizer industry, New Delhi, pp 1-16.

Price, G.W. and Gordon, A.M. 1999. Spatial and temporal distribution of earthworms in a temperate intercropping system in southern Ontario, Canada. Agroforestry Systems 44: 141-149.

Torralba, M., Fagerholm, N., Burgess, P. J., Moreno, G., and Plieninger, T. 2016. Do European agroforestry systems enhance biodiversity and ecosystem services? A meta-analysis. Agric. Ecosystems and Environment 230: 150-161.

Udawatta, P. R., Rankoth, L. and Jose, S. 2019. Agroforestry and biodiversity. Sustainability 11:2879.

Young, A. 1989. Agroforestry for soil conservation. C A B International International Council for Research in Agroforestry. Printed in the UK by BPCC Wheatons Ltd, Exeter, pp 160.

Youssefi, F., Brown, P.H, and Weinbaum, S.A. 1999. Regulation of nitrogen uptake at the whole plant level: a study in almond trees, Hort Technology 9: 598-600.

www.agromisa.org.

13

Organic Livestock Production and Certification in India

P. Baswa Reddy

ICAR-National Meat Research Institute, Hyderabad-500 092, Telangana, India

Abstract

Organic agriculture in India is growing steadily as more and more farmers are adopting organic practices to meet the increasing demand for organic products in the country. The number of food categories sold as organic is growing rapidly in India. Starting with organic tea and spices, it's grown to organic flour, breakfast cereals, fruits, vegetables and many more. Of late, the consumer demand for organically produced livestock products like milk, eggs and meat is also on the rise.

Keywords: *Livestock production, milk production, organic meat, production standards, certification*

1. Introduction

In India, though organic crop production and certification system is reasonably well established in India, the organic livestock and poultry sector is in its budding stage. Organizations such as APEDA, FSSAI and BIS along with research establishments like ICAR- National Meat Research Institute are now putting serious efforts to establish the protocols and promote organic livestock production and certification in India. As a result, in the last few years, livestock farmers are gradually getting encouraged to take up organic production and around 3000 farmers mostly from Uttar Pradesh, Madhya Pradesh and Himachal Pradesh have taken up organic dairy, sheep and poultry farming. Export of organic milk and ghee from India to different countries has just begun and is expected to grow very fast in volumes and revenues in the years to come.

2. Organic Meat and Milk Production

Organic livestock production and marketing of organic livestock products like milk, meat and eggs have witnessed an upward trend in growth rate worldwide in the last decade. In 2020 the organic dairy market crossed the USD 20 billion and as per the estimates of market experts, it is expected to reach USD 32 billion by 2026 with CAGR of more than 6%. In India, organic milk market is still in its infancy and it was worth Rs 450 crores in 2020 and the market analysts expect it to grow at more than 30% CAGR up to 2026.

The global organic meat products market is expected grow from USD 15 billion in 2020 to USD 20 billion in 2025 at a CAGR of around 7%. In India, the organic meat sector is yet to take off. A few organic sheep, poultry and pig units have recently been established and the certified organic meat from these units is likely enter the markets in the next one or two years. Since India already occupies number one position in buffalo meat export in the world, using these established channels for male buffalo calf rearing on organic lines and exporting such organic meat has the potential to keep India in a prominent position in world organic meat map.

As conventional livestock production practices are frequently criticized in the context of food safety concerns and animal disease outbreaks, there is growing consumer demand to move away from the widespread use of antibiotics and feed additives and to strengthen animal welfare standards. With increasing concern for animal welfare and environmental preservation consumers are now looking for alternative livestock production systems, which follow natural process of animal production with utmost regards for food safety and food security. This has led to the concept of organic meat, egg and milk consumption by the public. Organic animal production system is emerging as an effective alternative to address all these issues. Owing to this, in India, there are bright prospects for tapping this opportunity to develop it into a business model in Animal Husbandry sector. However, among many other issues, the profitability of organic livestock and poultry farming *vis a vis* conventional farming, is the crucial one from the point of view of the Indian farmers, particularly the small and marginal.

3. Organic Livestock and Poultry Production Standards

The broad guidelines for organic livestock production under National Programme for Organic Production (NPOP) of India are as under:

3.1. Organic management plan

Before starting organic livestock farming, producer has to make a detailed organic management plan and present it to the certifying agency for approval.

3.2. Breed/strain selection

Local breeds/strains which can easily be acclimatized to local climatic conditions shall be preferred over the exotic breeds

3.3. Source/Origin

Initial stock should be sourced from organically certified units. If the animals are introduced from non-organic units, they should undergo mandatory conversion period as specified for each category of livestock/ poultry.

3.4. Housing and management

Livestock and poultry shall be maintained under natural conditions as far as possible. This shall include utilizing natural breeding methods, housing and management conditions to minimize stress, health management system to prevent diseases. Minimum shaded as well as open area as specified for each category of livestock / poultry shall be followed.

3.5. Conversion period

Animals brought from non-organic sources shall undergo conversion period as specified for each category of livestock and poultry. During the conversion period the animals should be maintained under organic conditions. After completion of conversion period, the animals can be sold as organic.

3.6. Feeding

Livestock and poultry shall be provided with organically produced feed during the entire rearing period. The agriculture land committed for cultivation of feed / fodder crops intended to be used as feed for livestock and poultry shall be organic. The overall feeding practices shall satisfy the daily nutrient requirements of the concerned animals.

Feed additives and supplements used for feeding shall be from natural sources and as permitted under organic livestock guidelines.

3.7. Healthcare

The organic livestock & poultry, in general, should follow the basic principles of preventive health and productivity management wherein the focus would be on preventing diseases, detecting underlying fertility and production problems and its correction primarily on correcting management, nutrition and sanitation. Antibiotics and other allopathic drugs are not permitted. For purpose of treatment and prevention of diseases and under-performances, herbal/ phyto-therapeutic (excluding antibiotics), homeopathic or ayurvedic products shall be preferred to allopathic veterinary drugs or antibiotics, provided that their therapeutic effect is effective for the species of animal and the condition for which the treatment is intended. Antibiotic and other allopathic drugs can be used when other modes of cure are not effective to save the life of the animal and in such cases withdrawal periods as prescribed for each drug should be strictly followed. Hormonal treatment and use of growth promotors shall not be permitted. All vaccinations required by law of the land shall be permitted.

3.8. Breeding and management

The preference for reproduction shall be through natural methods, although artificial insemination may be used. Embryo transfer techniques and the use of hormonal reproductive treatment shall not be used unless prescribed for therapeutic purpose directed towards correcting the physiological problem. The breeding techniques employing genetic engineering shall not be used.

3.9. Manure and urine management

Manure and urine excreta collection and management practices in the organic livestock & poultry farm are a critical component. The collection, handling and disposal of the dung and urine from shed, paddock, open run or grazing areas shall be implemented in a manner that minimizes soil and water degradation, does not significantly contribute to contamination of water, optimizes recycling of nutrients and does not include burning or any practice inconsistent with organic practices.

3.10. Transport

During transport, the producer shall prevent stress, injury, hunger, thirst, malnutrition, fear, distress, physical & thermal discomfort, pain, disease during the transport and shall observe all the conditions set in law of the land for animal transportation. Animals must be fit for the intended journey. All necessary arrangement shall be made in advance to minimize length of the journey and meet the animal's need during the journey. Means of transport as well as the loading and unloading facilities must be designed, constructed, maintained and operated so as to avoid injury and suffering and ensure the safety of the animals.

3.11. Slaughter

The slaughter of livestock and poultry shall be undertaken in a manner, which minimizes stress and suffering, and shall be in accordance with the territorial/ national rules framed for the purpose. Minimum age of slaughter for each species of livestock should be followed as prescribed for the respective species.

4. Organic livestock certification

Production of crops or livestock like sheep and goats organically without validation or auditing has limited reach to the consumers. The consumers of organic products cannot always verify the production practices followed by the organic producers and hence they need authenticity to rely upon. Certification of organic products and processes serves this purpose and improves the consumer confidence and there by widens the market for the organic produce.

In India, there are different avenues available for certification of organic produce. Each system of certification has its own purpose, merits and constraints. Farmers interested in organic farming need to assess the pros and cons in each system of certification and choose the one which fulfils their requirement.

The different avenues of organic certification in India are

1. NPOP certification
2. Participatory Guarantee System (PGS)

The NPOP (National Programme for Organic Production) certification system of India involves a third party inspection to ascertain the compliance of organic production system. The certification is supervised by the APEDA at national level. APEDA authorizes the eligible certification agencies through a process of accreditation and these certification bodies can take up the certification at farm/ producer/ processor/ grower group level. Certification process is carried out as per the laid down standards and guidelines of organic livestock and poultry production. The produce certified under NPOP is entitled to carry the 'India Organic' logo along with 'Jaivik Bharat' logo as specified by FSSAI and can be sold within India and also can be exported to other countries as organic product after complying with the requirements of the importer.

National Programme for
Organic Production

Participatory Guarantee System (PGS) is a quality assurance initiative that is locally relevant, emphasize the participation of stakeholders, including producers and consumers and operate outside the frame of third party certification. This certification is based on active participation of stakeholders and is built on a foundation of trust, social networks and knowledge exchange. It is for farmers or communities that can organize and perform as a group within the village or in close-by villages with continuous territory. PGS ensures traceability only up to end till it is in the custody of the group. Once the product leaves the custody of PGS group there is no control of PGS on its integrity. The produce certified under PGS can use the logo of 'PGS- Green' (for farm/ produce under conversion) or 'PGS- Organic' (for farm/ produce completely concerted to Organic) along with Jaivik Bharat logo as specified by FSSAI and can be sold within India as organic product.

PGS-INDIA GREEN	PGS-INDIA ORGANIC
Logo used while marketing on the products produced from the field under coversion	Logo used while marketing on the products produced from field of completely converted into organic

Certified organic fodder production or certification of grazing land is in similar lines to the certification of crops which is fairy established in India. Organic certification of livestock products (value added products of meat, milk etc.,) is considered in the similar category as that of certification of processed foods and other value added food products which requires organic processing certification.

The process of becoming organically certified can be expensive, but it is an essential step for farmers and manufacturers wanting to meet the growing demand for certified organic food among the quality conscious consumers which fetches premium price for their produce.

4.1. Promotion of organic production by the government

The Government of India has introduced multiple schemes and incentives to encourage the adoption of organic farming by the farmers. The most prominent schemes of the government for promotion of organic farming in India are Paramparagat Krishi Vikas Yojana (PKVY), Mission Organic Value Chain Development in NE Region (MOVCDNER) National Project on Organic Farming(NPOF), Rashtriya Krishi Vikas Yojana (RKVY), Mission for Integrated Development of Horticulture (MIDH), ICAR-Network Project on Organic farming, Namami Gange Project etc.,

Paramparagat Krishi Vikas Yojana (PKVY) is an elaborated component of Soil Health Management (SHM) of major project National Mission of Sustainable Agriculture (NMSA). Under PKVY Organic farming is promoted through adoption of organic village by cluster approach and PGS certification. The

Scheme envisages Promotion of commercial organic production through certified organic farming. Groups of farmers are encouraged to take up organic farming under PKVY and they are motivated for natural resource mobilization for input production. There will be no liability on the farmers for expenditure on certification. Every farmer will be provided with some monitory support on per acre basis for three years. Organic farming will be promoted by using traditional resources and the organic products will be linked with the market.

5. Challenges

5.1. Certification of both fodder and livestock

Unlike organic crop certification where a producer is required to get his only the concerned crop certified, a farmer who wishes to take up organic small ruminant production needs to obtain certification for both fodder as well as the animals. Thus it requires double certification compliances in case of organic small ruminant production.

5.2. Awareness

Lack of awareness among the farmers about the guidelines of organic livestock & poultry farming and the processes available for their certification is the main hurdle hindering the development of this sector in the country.

5.3. Non availability of certified feed and fodder

Lack of availability of organically certified feed in the market is a major setback for the beginners till they establish their own certified feed and fodder production unit to meet the feed requirements of the livestock unit.

5.4. Lack of awareness on ethno-veterinary medicine

Since organic livestock production does not permit usage of synthetic products like antibiotics and hormones for animal health care, it entirely relies on traditional systems of medicine like ethno-veterinary medicine (EVM) for animal healthcare. However, not many veterinarians in the country have expertise in EVM procedures for animal healthcare and this system of health care for various disease conditions of animals needs to be standardized and popularized in the country to effectively take care of animals under organic production.

5.5. Certification

Very few organic certifying agencies in India have accreditation for livestock certification. In case of organic crop certification under PGS, in many states, the machinery of state agricultural departments is involved in organic certification but the same is lacking in case of livestock certification.

5.6. Labour

Currently, organic food production is assumed to be more labour intensive since under this system the farmers are advised to be more animal welfare friendly and adopt traditional health care systems which involves relatively higher labour when compared to allopathic health care.

5.7. Documentation

Documentation of all the processes and activities in organic production and marketing of the produce like under NPOP system is a cumbersome process for an ordinary farmer

5.8. Productivity

Initially, the productivity of organic livestock farms is likely to be lower than the traditional commercial farms till the organic livestock management, nutrition, health care and reproduction practices are fine tuned for higher productive yields.

5.9. Costs

Organic certification is expensive and organic feed for animals can cost twice that of normal feed. Organic farms tend to be smaller than conventional farms, which means fixed costs and overhead must be distributed across smaller produce volumes without government subsidies.

5.10. Market

Organized market for organic livestock products is yet to evolve in the country. Since there is no stringent enforcement mechanism, the certified organic produce has to face competition from non-certified and dubious produce sold in the market. However, as the markets for organic products get matured and certified organic livestock products gain consumer confidence, these products fetch premium price which can nullifying the higher costs involved in production and certification.

6. Conclusion

Despite the growing interest in organic livestock production, it still faces several challenges, including a lack of awareness among producers about the production standards and certification process, lack of proper infrastructure, high production costs, and limited marketing opportunities. Nonetheless, the government's continued support for organic farming and livestock production, combined with increasing consumer demand for organic products, is expected to drive the growth of the sector in the coming years.

7. Way Forward

Since organic livestock rearing and organic livestock products preparation are yet to take a leap in India, there are possibilities of entrepreneurial opportunities in all spheres of organic value chain right from production through certification, value addition, marketing till consumption. All the products and services in the organic food value chain are likely to fetch premium price in the market as they are targeted for niche consumers who are quality conscious and would be willing to pay premium price for quality products.

India already occupies number one position in buffalo meat export in the world. This can be potentially extended to organic meat production by rearing male buffalo calves on organic lines to export certified organic buffalo meat and put India in a prominent position in world organic meat map.

Certified organic feed and fodder is a crucial input for organic livestock rearing. Some farmers can produce certified organic feed and fodder and supply it to organic livestock producers. Those who have resources may produce and market both organic fodder and livestock as well. This sector also throws huge opportunities for veterinarians who specialize in traditional and ethno-veterinary practices of animal health care. Other opportunities in organic livestock value chain include certification agencies, inspection executives, Organic livestock production process advisors, organic product quality testing, organic products preparation like meat & milk products, wholesale and retail marketing, establishment of organic meat & milk outlets and export of organic meat, milk, eggs etc.

14

Organic and Natural Farming Research, Extension and Policy Needs

A.V. Ramanjaneyulu[1], M.V. Ramana[1], M. Ramesh Naik[2] K.A. Gopinath[3] and R.V.T. Balazzii Naaik[1]

[1]*Professor Jayashankar Telangana Agricultural University, Rajendranagar Hyderabad-500 030, Telangana, India*

[2]*ICAR-National Academy of Agricultural Research Management, Rajendranagar Hyderabad-500 030, Telangana, India*

[3]*ICAR-Central Research Institute for Dryland Agriculture, Rajendranagar Hyderabad-500 090, Telangana, India*

Abstract

Agricultural sector, with the technological advancements, is also contributing to global climate change, loss of biodiversity and degradation of resources. Further, there is a serious global concern about chemical residues, healthy food and healthy planet earth. Organic and natural farming are acclaimed to be the best alternative non-chemical production pathways to reverse the ill effects of modern agriculture. However, these systems are not contemplated to meet the food security and other requirements of ever-growing human population, particularly in developing countries. In this chapter, based on the available and documented protocols and production practices of these systems, the additional research, extension and policy needs to strengthen these production pathways to attain economic, ecological and social sustainability are discussed while considering the Sustainable Developmental Goals (SDGs) of United Nations.

Keywords: *Food security, natural farming, organic farming*

1. Introduction

The use of modern technologies involving high yielding cultivars, synthetic chemicals, irrigation and mechanization helped to achieve food self-sufficiency and security in India. However, it has resulted in yield plateauing and long-term sustainability issues. Further, soil degradation, environmental pollution, pesticide residues, loss of biodiversity, for future of microbiome and ecological imbalance are the negative consequences. Hence, eco-friendly methods of farming are the need of the hour. Among many, organic and natural farming are emerging fast as alternatives to overcome present day problems that agriculture is facing.

A vigorous campaigning is going on to accentuate the benefits of organic and natural farming *vis-à-vis* the modern farming among farmers and consumers in the country. Interestingly the area under both these methods is fast increasing. Moreover, a congenial socio-economic and cultural environment also prevails in India for adoption of organic and natural farming as environment friendly agricultural practices were adopted by our forefathers till the advent of the 'green revolution'. Furthermore, a large no. of small and marginal farmers especially in rainfed and hilly areas apply low levels of synthetics. In India, at present, urban people because of higher purchasing power, are consuming organic products even by paying premium price. With the increasing awareness and health consciousness the demands for organic and natural foods is increasing day by day. This throws a great challenge to the agricultural scientists to find ways and means to enhance productivity under non-chemical system. Hence, research has to be done to understand the repercussions of organic and natural farming methods on soil, environment, food and health systems and also in meeting the SDGs.

2. Differential Practices in Organic, Natural, Inorganic and Modern Farming

Before enlisting the research, extension and policy needs of organic farming, one has to understand differential practices under these concepts as detailed below in Table 1.

Table 1. Differential practices in organic, natural and modern farming

S.No.	Particulars	Organic farming (OF)	Natural farming (NF)	Inorganic (chemical) farming (IF or CF)	Modern farming (ICM)
I	*Planting material*				
1	Use of genetically engineered seeds, pollen, transgene plants or plant material	X	X	√	√
2	Seeds/planting material of improved varieties, hybrids etc.,	√	X (only indigenous varieties are accepted)	√	√

II	***Chemical inputs***				
1	**Chemical fertilizers**	**X**	**X**	√	√
2	**Growth regulators and synthetic dyes**	**X**	**X**	√	√
III	***Organic inputs***				
1	**FYM/compost/ vermicompost**	√	**X**	**X**	√
2	**Green manures/Green leaf manures**	√	**X**	**X**	√
3	**Biofertilizers**	√	**X**	**X**	√
4	***Trichoderma* for seed treatment/soil application**	√	**X**	**X**	√
5	**Rock phosphate as P source (from mined source only)**	√	**X**	√ ***in acid soil only**	√ ***in acid soil only**
6	**Gypsum**	√ **(from mined source)**	**X**	√	√
7	**Micronutrients***	√ **(spray in case of deficiency symptoms)**	**X**	√	√
IV	***Weed, pest and disease management***				
1	**Herbicides (recommended label claim chemicals in respective crops)**	**X**	**X**	√	√
2	**Hand weeding**	√	√	√	√
3	**Intercultivation (cattle drawn/mechanical)**	√	√	√	√
4	**Neem oil**	√	**X**	**X**	√
5	**Need based insecticides and fungicides**	**X**	**X**	√	√
6	**Physical and mechanical methods of controlling weeds, pests and diseases**	√	√	√	√
7	**Biological control of weeds, pests and diseases using parasites, predators, natural enemies etc.,**	√	√ **(only natural process and no manual release)**	**X**	√

V	*Others*				
1	**Deep ploughing once in three years**	√	X	√	√
2	**Mulches in rainfed crops for moisture conservation**	√	√	X	√
3	**Whapahasa for soil aeration**	X	√	X	X
4	**Bird perches**	√	X	X	√
5	**Pheromone traps**	√	X	X	√
6	**Seed treatment with recommended chemicals**	X	X	√	√
7	**Seed treatment with *beejamrit***	X	√	X	X
8	**Nutrition with *jeewamrit* (solid and liquid)**	X	√	X	X
9	**Foliar spray with 19-19-19/urea/DAP/13-0-45 etc., for drought mitigation**	X	X	√	√
9	***Neemastra***	X	√	X	X
10	***Dasaparni***	X	√	X	X
11	***Brahmasta***	X	√	X	X
12	**Fermented butter milk, dried ginger extract, cow urine, germinated seeds extract**	X	√	X	X
13	***Panchagavya***	√	X	X	X

Source: Trinadh (2014); Joshi (2019); Gopinath and Ramanjaneyulu (2019); Kumar *et al.* (2019); Grover *et al.* (2019); Mishra and Singh (2019); Neelima *et al.* (2019); Nataraj *et al.* (2019); Shivanna (2019); Naik *et al.* (2023).

X: Not included; √: Included;

IF: Inorganic farming; CF: Chemical farming; OF: Organic farming, NF: Natural farming; ICM: Integrated crop management

3. Organic Farming

3.1. Research needs in organic farming

3.1.1. Identification of crops and cultivars

The varieties and hybrids bred for high external input conventional farming are not optimal under organic agriculture. Hence, screening, selection and development of cultivars through organic seed breeding and low input ecosystem have to be carried out in order to identify and popularize pest and disease tolerant cultivars (Aurebach, 2020). Further, as a part of host plant resistance, identification and adoption of location specific tolerant/resistant cultivars for endemic pest/disease (For instance, rust in wheat, powdery mildew in mustard, blast, bacterial leaf blight, stem borer etc., in rice, fusarium wilt in pigeonpea and castor, boll worms in cotton, tikka leaf spot in groundnut, fall army worm in corn, etc.) are crucial for reducing cost of cultivation and enhance net farm income under organic farming.

3.1.2. Seed treatment

Seed treatment with synthetic chemicals are generally done to protect them from biotic stresses under conventional farming. However, to promote organic farming, compulsory seed treatment with these chemicals should be waived for producers and sellers (Aurebach, 2020). Further, alternate, acceptable and eco-friendly seed treatment methods should be developed.

3.1.3. Buffer zones

Ways and means of protecting organic farms from contamination from inorganic farms and genetically modified (GM) cropped farms need to be clearly spelt out. There is a need to create/maintain buffer zones to overcome this problem.

3.1.4. Integrated organic farming systems (IOFS)

Research on development of location specific integrated organic farming system (IOFS) models has to be conducted by including livestock and other suitable supplementary and complementary enterprises along with crops for small holder farms and large farms for better farm income and climate resilience. The farmers should generate all the required organic inputs, to the maximum extent possible on their own within the system through proper recycling of resources. The Integrated Farming Systems over a period of time can be easily converted into complete organic systems and better profits may be ensured to farmers.

3.1.5. Role of Agroforestry

Sustainable and profitable organic based Agroforestry models have to be developed to reduce erosion, conserve moisture, improve soil health, green leaf manure supply, enhance carbon sequestration and biodiversity, reduce greenhouse gas emission, supply food, fruit, fodder, fuel, fibre, fertilizer, medicine, wood, timber etc., Thus, Agroforestry is important for improving the livelihoods of rural communities and environmental security. Further, it is also important for furthering research on plant-based extracts for controlling weeds, pests and diseases. Furthermore, Agroforestry has the potential to improve drylands, barren lands, marginal and sub marginal lands, problematic soils etc., through nitrogen fixation and addition of huge amount of biomass and organic matter.

3.1.6. Alternative eco-friendly inputs

The available nutrient sources like FYM, compost, green (and green leaf) manures etc., are bulky in nature have very low to low in nutrient concentration. They may not meet the crop nutrient requirement even when they are applied on equivalent recommended nutrient dose basis due to low nutrient recovery. Besides, they have to be applied in large quantities to meet the nutritional requirement of crops and their availability and logistics is a problem. Further, as they are also expensive, there is a need to search for alternative inputs that are relatively high in nutrient concentration so that bulk of the input can be reduced. The research data need to be generated on such novel inputs and their suitability for organic farming for higher nutrient use efficiency. The possibility of utilizing sea weed, panchagavya. etc., for nourishment and plant-based extracts for pest and disease control may be explored.

3.1.7. Package of practices

Location specific eco-friendly best management practices for organic production of various crops (cereals, millets, pulses, oilseeds, fiber crops, narcotics, fruits, vegetables, medicinal and aromatic plants, etc.) and cropping systems have to be developed and popularised. The best bet practices should combine tradition (indigenous technical knowledge), science and innovation. Further, the packages should ensure more diversity in crops and biodiversity including livestock, agroforestry trees and other suitable enterprises.

3.1.8. Mechanization

In view of steady decline in labour availability, escalating labour cost and lack of efficient biological weed control methods, mechanization assumes importance in general and weed management in particular. Hence, energy efficient machinery depending on the farm size (small, medium, large) have to be designed and promoted for arresting the weed growth and managing the crop residue. Renewable and green energy sources may be explored for doing all the farm operations with machinery.

3.1.9. Livestock

Research on comparative study of indigenous and exotic breeds of livestock for their efficacy on various products is the need of the hour. Further, development of new breeds and better livestock feeding techniques with less greenhouse gas emission potential have to be given due importance.

3.1.10. Incentivization

Research on carbon credits and ecosystem services has to be strengthened in order to quantify them and also benefit the organic farmers. The rules, regulations and benefit sharing mechanism have to be clearly worked out to benefits farmers practicing organic farming.

3.2. Extension needs in organic farming

- The proven organic farming practices should be upscaled in rainfed areas and zones with low to negligible synthetic inputs use
- Imparting knowledge about pesticide residues in food grains and products due to excessive use of pesticides, is the need of the hour
- Though demand for organic products has been increasing in the country, many people are not coming forward to purchase organic products in view of higher market price. However, India is exporting approximately 85% of organic production to USA, European Union and Japan indicating high demand in the western countries. Hence, there is a need to improve the awareness among Indians about health benefits of consuming organic foods to create further demand
- The growers have to be educated regarding maintenance of buffer zones between organic and other farms to avoid flow of contaminants and pollutants into organic farms
- Capacity building to enhance the knowledge and capacities of the organic growers about various sources of nutrition, advanced composting techniques, novel organic inputs, benefits of incorporation rather than burning of crop residues, mechanization, biopesticides etc., Furthermore, farmers must be equipped and trained on meeting the crop nutritional requirements through local mass multiplication and utilization.
- Conducting exposure visits for all stakeholders to the hot spots of organic production
- Promote organic farming through Non-Government Organizations (NGOs), Farmers Producer Organisations (FPOs), Corporates etc., for production, processing and marketing of organic produce
- Inculcate knowledge on organic farming and its benefits among students
- Farmers should develop forward linkages for marketing purpose or they should create local market based on trust

3.3. Policy needs

- Bioinoculants and biopesticides did not become popular as expected, hence, marketing and distribution network have to be strengthened by the government as the retailers are not selling them owing to less demand and low profits unlike synthetic chemicals. Compulsory marketing and usage of these eco-friendly bio products with stringent regulatory mechanism will help reduce the production cost, enhance profitability and preserve ecosystem

- Formulating the government programmes to grow green manure crops and tree based green leaf manures on government lands on the lines of social forestry
- Government has to pass and implement strict rules on '*No to Crop residue burning*'. Further, the institutions and governments have to promote mechanization as an instrument for efficient crop residue *in situ* incorporation and (or) baling thus avoid burning and environmental pollution.
- The central government should introduce minimum support price (MSP) for promising tree species to encourage farmers to adopt agroforestry. Many farmers are not coming forward to take up Agroforestry as a regular enterprise due to lack of suitable market. Hence, market outlets should be established at least in Agroforestry hotspots. Furthermore, suitable modifications may be made in EXIM policies to promote Agroforestry. NGOs and corporate sectors may be invited to do research and promote this sector.
- The governments may create and encourage *Special Organic Zones* (SOZs) for the production and export of organic products (Naik *et al.*, 2023)
- Preferences should be given to organic growers while delivering incentives (for instance, subsidies)
- Formulate only a single and common certification for all exporting countries as multiple certifications for different countries would be cumbersome, difficult and costly
- Proper regulatory mechanisms must be in place for verification and validation of organic products as multifarious organic products are being sold in the open and online markets
- Suitable courses on organic farming may be introduced in educational institutions
- The organizations may encourage *start-up incubators* in organic farming research and promotion.

4. Natural Farming (NF)

4.1. Research needs under natural farming

4.1.1. Suitable cultivars

The natural farming model advocates the use of indigenous varieties only as they perform better even under adverse weather conditions too. But, these varieties are tall growing and thus prone to lodging. Further, they are susceptible to various biotic and abiotic stresses. This is the reason why, many improved varieties and hybrids have been developed for drought, pest and disease tolerance or resistance through various breeding and molecular techniques. Hence, a comparative performance analysis of indigenous varieties and improved cultivars is needed in various crops across different agro-climatic zones. Further, a separate breeding programme must be initiated to breed cultivars with the required architecture suitable for natural farming.

4.1.2. Indigenous animals

It is proposed to adopt only indigenous cow, but, no other cows, for practising natural farming. Scientific validation may be done on the comparative performance of indigenous cows and other cows on the quality of cow dung, urine, milk and other by-products, formulations/concoctions and their impact on crop productivity and soil health. Further, it is also claimed by the proponents of NF that one indigenous cow is sufficient for practicing natural farming in 30 acres (12 ha) land, which needs scientific affirmation. Furthermore, the percent nutrient requirement from components of NF and other nutritional sources has to be worked out.

4.1.3. Efficacy and impact of NF components

There is an urgent need to understand the seasonal and long-term effect of

4.1.3.1. Beejamrita

On germination, seedling vigour, plant growth and development and soil microbia.

4.1.3.2. Jeewamrita

On the extent of nutrient supplying capacity, micro and macro-organisms, nutrient fixation and mobility, availability to the plants and changes in the long-term soil fertility status, seed quality, greenhouse gas emissions etc.

4.1.3.3. Mulching

On soil physical, chemical and biological properties and crop productivity

4.1.3.4. Whapahasa

On soil physical, chemical and biological properties and crop productivity under diverse edapho-climatic conditions

4.1.3.5. Brahmastra, Neemastra, Dasaparni, Agniastra, Fermented buttermilk, Dried ginger extract, Gobanam, Cow urine, Germinated seeds extract etc.,

On the pest and disease dynamics, productivity levels, seed quality, pesticide residues in soil and plants etc.,

4.1.4. Dosification and longevity of formulations/concoctions

The dosification of formulations/concoctions as claimed by the proposers of natural farming is as detailed in the Table 2. But, these were suggested irrespective of the type of crop and stage of crop. Hence, there is a need to conduct research to standardize these doses for technical and cost effectiveness.

Table 2. Details of dosage and waiting periods for various formulations of NF

S.No.	Particulars	Dosage (lit or kg ha^{-1})	Purpose	Stage of application or periodicity	Waiting period after preparation
1	Beejamrit (lit)	2.5	Seed treatment	Before sowing	2
2	Liquid Jeewamrit* (soil drenching or foliar spray) (lit)	500.0	Nutrition	15 days	7
	Liquid Jeewamrit* (soil application) kg	500.0	Nutrition and insect control	15 days	7
3	Neemastra (lit)	250.0	To control sucking pest and small insect pests	Need based	7
4	Agniastra (lit)	12.5-15.0	To control leaf folder, stem borer, root grub	Need based	90
5	Brahmastra (lit)	12.5	Controlling larval pests, boll worms etc.,	Need based	180
6	Dasaparni (lit)	15.0-25.0	To control sucking pests in rice, powdery mildew in mango	Need based	180

7	Sour or fermented butter milk (lit)	25.0	Controlling fungal diseases	Need based	-
8	Grain mixture extract	200.0	For better grain quality	Before maturity	-
9	Gobanam	100.0	Controlling fungal diseases	Need based	Not specified
10	Dried ginger extract	200.0	Controlling fungal diseases	Need based	2
11	Cow urine (lit)	200.0	Better crop growth and repelling pests		Longer the storage period better is the quality

(*Source*: Trinadh, 2014)

Further, the waiting period seems to be specific for all formulations. However, scientific validation is required for understanding the fate of formulation with regard to beneficial and harmful microbiome after expiry of waiting period and subsequent impact on the crop growth and soil health.

4.1.5. Solutions to mitigate sudden flare up of pest and diseases

Unlike immediate solutions against sudden flare up of pest/diseases in modern farming, no such readymade formulations are available in natural farming. Hence, research need to be done on finding out cost effective control measures for such problems.

4.1.6. Productivity

The food security which is being enjoyed by all of us need to be supported by sustainable and viable measures for the benefit of forthcoming generations. Though, natural farming has been projected to enhance soil health and reduce the production costs, but, it is leading to decline in crop yield. Hence, there is a dire need to work upon

- Sustainability of yields in the long run
- Magnitude of yield reduction over the years
- Fate of overall food production at national level
- National food and nutritional security
- Possibility of conversion of Agricultural land to natural farming without impacting the food security

4.1.7. Economic returns

Reduced cost of production are being claimed by the proponents of NF. However, there is a need to work out scientifically by the agri-economists duly taking into account the cost of maintaining cow, fodder, various ingredients to be used for preparing natural farming formulations and concoctions. Finally, the economic returns to the farmers in the long run has to be worked out to understand the economic feasibility.

4.1.8. Quality of economic produce

The data need to be generated on the impact of natural farming practices vis-a-vis modern farming and organic farming, on the qualitative aspects of economic produce *viz.,* cooking quality, baking quality, sensory, protein, carbohydrates, fats, fatty acids, starch, oleoresin, curcumin, amylose, sugars, fiber, moisture, minerals, vitamins etc.

4.1.9. Soil health

The data sets need to be generated to understand the impact of natural farming practices on soil health, quality and productivity. The soil physical (bulk density, particle density, porosity, WHC, infiltration), chemical (pH, EC, organic carbon, available nutrient status) and biological properties (dehydrogenase, b-glucosidase, urease, cellulase, and acid/alkaline phosphatases) have to be analysed in natural farming adopted fields and compared with that of organic and modern farming. Further, studies are to be conducted on nutrient release patterns from liquid and solid *Jeewamrit*, thus extent of supply of various nutrients can be factored in. The other benefits like various ecosystem services also have to be unraveled.

4.1.10. Biodiversity

NF is projected to preserve and enhance biodiversity. The research may be oriented to understand the impact of NF practices on the biodiversity in different agro-ecosystems. SP-Zero budget natural farming (ZBNF) discourages use of exotic earthworms and favours *in situ* multiplication of indigenous earthworms for improving soil health. Research data on suitability of exotic earthworms in NF is not available. Hence, comparative studies on the efficacy of exotic and indigenous earthworms may be conducted to understand the impact on soil health and crop productivity.

4.1.11. Climate change

Natural farming has been projected as one of the best Agro ecological practice for mitigating the ill effects of climate change and achieving the United Nations

Sustainable Development Goals (UN-SDGs) besides harmonizing yield, sustainability and nutritious quality food. Hence, systematic research must be done on the extent of release of greenhouse gas emissions (CH_4, CO_2, N_2O), carbon sequestration potential in natural, organic and modern farming across various crops and cropping systems, farming systems in different edaphic-climatic conditions.

4.1.12. Ecosystem services and carbon credits

The research has to be carried out on quantification of different ecosystem services (provisioning, cultural, regulatory and supporting) and carbon monetization that can be realized under natural farming. Further, additional monetary benefits to the farmers may be assessed.

4.1.13. Integration of organic and natural farming

Natural farming formulations alone may not be sufficient to enhance the production and productivity. Hence, integrated organic models have to be developed duly combining the most effective organic and natural farming practices for long term sustainability, enhancing productivity, profitability and achieving food security. Further, the production systems should include formulations with zero pesticide residues.

4.1.14. Human health

Research on human health benefits due to consumption of food produced under NF, OF and conventional modern farming need to be conducted to generate required data. Comparative study of these systems in boosting immunity to humans in the event of COVID-19 and other health issues, will go a long way in Agriculture and health sector.

4.1.15. Scientific verification

There is a growing demand for organic food products in the market across the country. It led to proliferation of manufacturing and marketing companies. However, there is no authenticity that they are organic or natural in nature. Hence, there is a need for scientific verification of such products to ensure availability of organic or natural products to the consumers. The research institutions should develop simple procedures and tools for testing the same (Priyadarshini *et al.,* 2022).

4.1.16. Creation of data base

The successful practices of natural farming practicing farmers have to be collected, documented and validated scientifically. Further, identification and delineation of potential zones, niche area and crops for adoption of natural farming have to be carried out.

4.2. Extension needs under natural farming

4.2.1. No to deep ploughing

The proponents of natural farming are against the deep ploughing claiming that it leads to infiltration of rain water into deeper layers and the same is not available to crop plants. However, this claim is not correct, as it is scientifically proved that deep ploughing in summer once in three years helps in breaking the hard pans in the sub soil thus improves infiltration which in turn improve water holding capacity of the soil and delays the effect of dry spells. Besides, it also helps in exposing the propagules of difficult to control weeds, pests and diseases. Hence, controlled deep ploughing can be recommended even in natural farming, as deep tillage leads to loss of organic carbon due to tropical climate.

4.2.2. Increase the no. of indigenous cows

Special efforts must be made to increase the number of indigenous cows as they are very important for implementation of natural farming.

4.2.3. Kitchen gardens

Kitchen gardens should act as an entry point for natural farming as seen in the case of Andhra Pradesh. This will help farmers in building up the confidence in adopting natural farming in their fields and reach a larger section of the community.

4.2.4. Convergence

It is high time that convergence among the line departments (Department of Agriculture, Department of Horticulture, Department of Animal Husbandry etc.,), State Agricultural Management and Extension Training Institutes (SAMETIs), Krishi Vigyan Kendras (KVKs), and emerging Non-Governmental Organization (NGOs), Farmer Producer Companies (FPCs), and other innovative farmers and entrepreneurs is critical for strengthening the implementation of natural farming effectively. But, it has to be done only after development of profitable and sustainable model by the researchers.

4.3. Policy needs

4.3.1. Standards

A separate set of standards should be finalized to assure better market price for products produced from natural farming.

4.3.2. Branding and marketing

Developing an exclusive brand and market for natural farming is essential to differentiate the produce grown under natural farming from that of organic and chemical farming. This will enable the farmers to get the assured market and price for their produce. Further, certification of farms under natural farming is essential to differentiate them from the fields under chemical farming.

4.3.3. Benefits to the farmers

A policy may be made to give subsidy benefits to the natural farming farmers too on the lines of incentives given for farmers adopting synthetic fertilizers under modern farming.

5. Conclusions

The organic and natural farming are being promoted across the country with a view to get good quality food, reduce dependence on synthetic chemicals thus curtail the loss of foreign exchange. However, there is a dire need to study the long term impacts of natural farming in comparison with organic and modern farming on soil health, soil and crop productivity and economic benefits. Further, the soil physico-chemical and biological aspects including micro and macro biomes, crop nutritional profiles, seed quality aspects and greenhouse gas emissions need to be studied. The carbon, water, nutrient, energy foot prints and food, nutritional and environmental security issues need to be assessed. National level projections have to be made on food production and food security if organic and natural farming are upscaled and out scaled across the country. Otherwise, the per cent cropped area that can be diverted to OF and NF may be worked out without affecting the food security. The concept of "*Produce locally, Shop locally and Eat locally*" for organic and natural products should be promoted.

6. Way Forward

Despite variations in different organic farming methods, the primary goal of all is to provide healthy food without using chemicals. Hence, we feel that there is greater need to design a non-chemical production pathway encompassing all the best practices from all these non-chemical systems to attain better farm income, climate resilience and ecological sustainability. This further may help in achieving zero hunger and zero poverty goal.

7. References

Aurebach, R. 2020. Organic research contributes to sector development and good organic policy: The Danish, Swiss, American and African Case Studies. In: Aurebach, R. (eds) Organic Food Systems: Meeting the needs of South Africa. Nelson Mandela University, Georgia, South Africa. Centre for Agriculture and Bioscience International, pp 32-41.

Gopinath, K.A. and Ramanjaneyulu, A.V. 2019. Organic Farming. Divya Publishing House, A Division of Astral International Pvt. Ltd, New Delhi, pp 555.

Grover, M., Lata Naina and Sreedevi Sankar 2019. Microbial resources – An important component of organic farming. In: Gopinath, K.A. and Ramanjaneyulu, A.V. (eds) Organic Farming, Divya Publishing House, A Division of Astral International Pvt. Ltd, New Delhi, pp 113-136.

Joshi. 2019. Management of disease in organic production. In: Gopinath, K.A. and Ramanjaneyulu, A.V. (eds) Organic Farming, Divya Publishing House, A Division of Astral International Pvt. Ltd, New Delhi, pp 195-216.

Kumar, S.R., Spandana Bhatt, P., Ramanjaneyulu, A.V. and Srinivas, A. 2019. Components of organic farming. In: Gopinath, K.A. and Ramanjaneyulu, A.V. (eds) Organic Farming, Divya Publishing House, A Division of Astral International Pvt. Ltd, New Delhi, pp 35-60.

Mishra, J.S. and Singh, V.P. 2019. Weed management in organic crop production. In: Gopinath, K.A. and Ramanjaneyulu, A.V. (eds) Organic Farming, Divya Publishing House, A Division of Astral International Pvt. Ltd, New Delhi, pp 137-150.

Nataraj, K.C., Malleshwari Sadhineni, S.N., Gopinath, K.A. and Srinivasa Rao, Ch. 2019. Utilization of locally available resources for nutrient management in organic farming. In: Gopinath, K.A. and Ramanjaneyulu, A.V. (eds) Organic Farming, Divya Publishing House, A Division of Astral International Pvt. Ltd, New Delhi, pp 89-112.

Neelima, T.L., Ramanjaneyulu, A.V., Kumar, S.R., Gopinath, K.A. and Venkata Ramana, M.V. 2019. Vermicompost – A viable source in organic farming. In: Organic Farming, (Eds.). Gopinath, K.A. and Ramanjaneyulu, A.V. Divya Publishing House, A Division of Astral International Pvt. Ltd, New Delhi, pp 61-88.

Priyadarshini, M., Ramanjaneyulu, A.V., Chaitanya, T., Santosh Kumar, G. and Yakadri, M. 2022. Organic Products - Manufacturing and Marketing Companies in India. Chronicle of Bioresource Management 6(4): 095-103.

Naik, M., Umesh Hudedamani, Gopinath, K.A., Ramanjaneyulu, A.V., Ravishankar, N., Singh, V.K., Venkateshwarlu, G. and Srinivasa Rao, Ch. 2023. Organic Farming in India: Production Constraints and Policy Recommendations. ICAR-National Academy of Agricultural Research Management, Hyderabad, India, pp 30.

Shivanna, B. 2019. Management of insect pests in organic crop production. In: Gopinath, K.A. and Ramanjaneyulu, A.V. (eds) Organic Farming, Divya Publishing House, A Division of Astral International Pvt. Ltd, New Delhi, pp 151-174.

Trinadh, C.H. 2014. Prakrithi Vyavasayam – Palekar vidhanam Natural farming – Palekar method. Vachakam Publications, Vijayawada, Andhra Pradesh, India, pp 1-83.

15

Organic Certification System in India

Y. Yogamayi[1], Bir Singh Negi[2], R. Ravi Teja[3] and M. Koteshwari[3]

[1]*National Institute of Agriculture Extension Management (MANAGE), Rajendranagar, Hyderabad-500 030, Telangana, India*

[2]*Former Advisor Organic, APEDA*

[3]*ICAR-National Academy of Agricultural Research Management Hyderabad-500 0030, Telangana, India*

Abstract

Organic food sales increased steadily from the late 20th century. Greater environmental awareness, coupled with concerns over the health impacts of pesticide residues and consumption of genetically modified (GMO) crops, fostered the growth of the organic sector. Realizing the benefits of organic food, especially after COVID-19, consumers are also demanding organically grown food and fiber and are willing to pay premium price. But to tap the market, there is an inevitable need for certification. Recently launched Food Safety and Standards (Organic Food) Regulation, 2017 mandates that all food products sold as organic must be certified and comply the basic requirements of the Act. The organic food offered or promoted for sale shall also comply with all the applicable provisions of either National Programme for Organic Production (NPOP) or Participatory Guarantee System for India (PGS-India). Organic certification is a process certification. The requirement to confirm that the product is organic include labelling on the package and traceability for both domestic and export needs to be understood. For import and export of organic produce, transaction certificate issued by an Accredited Certification Body is required. These requirements are essential for fresh and processed food items. In this chapter, we tried to eloborate the process and procedures for organic certification.

Keywords: *Organic farming, certification, food Safety and Standards*

Organic food sales increased steadily from the late 20th century. Greater environmental awareness, coupled with concerns over the health impacts of pesticide residues and consumption of genetically modified (GMO) crops, fostered the growth of the organic sector. Realizing the benefits of organic food, especially after COVID-19, consumers are also demanding organically grown food and fiber and are willing to pay premium prices. But to tap the market there is an inevitable need for certification. Recently launched Food Safety and Standards (Organic Food) Regulation, 2017 mandates that all food products sold as organic must be certified and must comply the basic requirements of the Act.

1. Food Safety and Standards (Organic Food) Regulation 2017

Food Safety and Standard Authority of India (FSSAI) has notified ***Food Safety and Standards (Organic Food) Regulation, 2017 under FSS Act 2006***

A) Mandatory requirement: No person shall manufacture, pack, sell, offer for sale, market or otherwise distribute or import any organic food unless they comply with the requirements laid down under these regulations.

B) Applicability of the systems

1. The organic food offered or promoted for sale shall also comply with all the applicable provisions of one of the following systems, namely:
   1. National Programme for Organic Production (NPOP);
   2. Participatory Guarantee System for India (PGS-India).

   Any other system or standards as may be notified by the Food Authority from time to time.
2. The Organic food which is marketed through direct sales by the small original producer or producer organization, as determined by the Food Authority from time to time, to the end consumer shall be exempted from the provisions of the systems referred above.
3. Aggregators or Intermediaries who collect organic food from small original producer or producer organization and sell it to the end consumer directly, are exempted from the provisions of the systems referred in sub-regulation (1) and they shall maintain records of traceability and comply with any of the provisions of the systems mentioned in sub-regulation (1) and organic food shall not carry Food Safety and Standard Authority of India's organic logo.
4. In-conversion products under PGS-India may be labelled as 'PGS-Green' and may also be labelled as 'In-conversion to organic' and the conversion products under NPOP may be labelled as 'In-conversion to organic' and shall mention the year of conversion and such in-conversion products shall not carry Food Safety and Standard Authority of India's organic logo.

C) Labelling (Food Safety and Standards (Organic Food) Regulation 2017 under FSS Act 2006)

1. Labelling on the package of organic food shall convey full and accurate information on the organic status of the product. Such product may carry a certification or quality assurance mark of one of the systems mentioned at (1) above in addition to the Food Safety and Standard Authority of India's organic logo.
2. All organic foods shall comply with the packaging and labelling requirements specified under the Food Safety and Standards (Packaging and Labelling) Regulations, 2011 in addition to the labelling requirements under one of the applicable systems mentioned at (1) above.
3. Traceability - shall be established up to the producer level as applicable under the systems mentioned at (1) above and it shall include any other requirements prescribed by the food authority to maintain the organic integrity of the food product.
4. Requirement to comply with the provision of the other regulations made under the Act - without prejudice to the provisions of these regulations, all organic food shall comply with the relevant provisions, as applicable, under the Food Safety and Standards (Food Product Standards and Food Additives) Regulations, 2011. The organic food shall also comply with relevant provisions, as applicable under the Food Safety and Standards (Contaminants, Toxins and Residues) Regulations, 2011 except for residues of insecticides for which the maximum limits shall be 5% of the maximum limits prescribed or Level of Quantification (LoQ) whichever is higher.
5. The seller of organic food either exclusively or as part of his retail merchandise shall display such food in a manner distinguishable from the display of non-organic food.

D) Import of organic food

1. Organic food imports under bilateral or multilateral agreements on the basis of equivalence of standards between National Programme for Organic Production (NPOP) and the organic standards of the respective exporting countries shall not be required to be re-certified on import to India subject to their compliance with the provisions of the act, the rules and regulation made there under.
2. The organic food consignments referred to in sub-regulation (1) shall be accompanied by a transaction certificate issued by an accredited certification body covered under the terms of the equivalence agreement.

E) Labelling requirements for Export

All products labelled as organic shall be required to be certified under NPOP and shall bear logo of Jaivik Bharat (FSSAI organic food logo) along with the concerned certification programme logos.

2. Organic Certification

Organic certification is a "process certification" (not a product certification) intended for producers of organic food and other organic agricultural products. In general, any business directly involved in organic food production can be certified, including seed suppliers, farmers, food processors, retailers and restaurants. The requirements vary from country to country, and generally involve a set of production standards for growing, storage, processing, packaging and shipping.

3. Why Organic Certification?

- Organic certification is intended to assure quality and prevent fraud
- For organic producers, certification identifies suppliers of products approved for use in certified operations
- For consumers, "certified organic" serves as a product assurance, similar to "low fat", "100% whole wheat", or "no artificial preservatives"
- Certification is essentially aimed at regulating and facilitating the sale of organic products to consumers for promoting customer trust
- It ensures that the stake holders present in the certification programme are undergoing verification activities by third party certification body or peer appriasal system
- It brings homogeneity in the practices followed during organic food production ensuring the quality of the food

4. Existing Organic Certification Systems in India

- India is among the first few developing countries to have developed and launched a credible third-party process certification system i.e National Programme on Organic Production (NPOP) launched during 2001 for export of organic commodities served as sole quality assurance system. This NPOP certification is being handled by APEDA under Ministry of Commerce and Industry, GOI
- Later keeping in view of the prohibitively high cost of third certification and complexity in documentation, a farmer group centric certification system was launched during 2012 under National Project on Organic Farming (NPOF) by the Department of Agriculture, Cooperation

& Farmers Welfare, Ministry of Agriculture & Farmers Welfare, Government of India. i.e. PGS-India (Participatory Guarantee System of India) is a quality assurance initiative that is locally relevant, emphasize the participation of stakeholders, including producers and consumers and operate outside the frame of third party certification.

5. National Programme for Organic Production:

(Ref: National Programme for Organic Production Standards ***Seventh Edition: November 2014 by APEDA, Department of Commerce, Ministry of Commerce and Industry New Delhi)***

5.1. Crop production: (Individual farmer)

a) **Crop production plan:** The producer is required to develop an organic crop production plan including crops to be grown, (field wise), description of practices and procedures, list of inputs used, source of organic planting material (seeds and seedlings), monitoring, management and record keeping procedures.

b) **Conversion requirements and duration:** A mandatory conversion period of 24 months for plants/crops other than perennials and 36 months for perennial plants shall be followed. However, a reduction in conversion period up to 12 months can be granted if the standards are being met from last several years with enough documentary/practicing evidences

c) **Landscape:** Organic farming shall contribute beneficially to the ecosystem. The certification programme shall set standards/procedures for a minimum percentage of the farm area to facilitate biodiversity and nature conservation.

d) **Choice of crops and varieties:** All seeds and plant material shall be certified organic. When certified organic seed and plant materials are not available, chemically untreated conventional seed and plant material shall be used. The use of genetically engineered seeds, transgenic plants or plant material is prohibited.

e) **Diversity in crop production & management plan:** Where appropriate, the organic farms shall be required to maintain sufficient diversity in a manner that considers pressure from insects, weeds, diseases and other pests, while maintaining or increasing soil, organic matter, fertility, microbial activity and general soil health.

f) **Nutrient management:** Biodegradable material of microbial, plant or animal origin produced on organic farms shall form the basis of the nutrient management. Non-synthetic mineral fertilizers and brought-in bio fertilizers (biological origin) shall be regarded as supplementary. Mineral fertilizers shall only be used in a supplementary role to carbon-based materials and shall be applied in their natural composition and shall not be rendered more soluble by chemical treatment. Fertilization management should minimize nutrient losses and the desired pH levels shall be maintained in the soil by the producer. List of permitted, restricted and prohibited inputs for nutrient management are given as per NPOP standards (Table 1).

g) **Pest, disease and weed management:** Emphasis is placed on the use of a balanced fertilizing programme, use of crops and varieties well-adapted to the environment, fertile soils of high biological activity, adapted rotations, intercropping, green manures, etc. Products used for pest, disease and weed management, prepared at the farm from local plants, animals and microorganisms, shall be allowed. The use of synthetic herbicides, fungicides, growth regulators, synthetic dyes insecticides and other pesticides are prohibited. List of permitted, restricted and prohibited inputs for pest, disease and weed management are given as per NPOP standards (Table 2).

h) **Contamination control:** All relevant measures shall be taken to minimize contamination from outside and within the farm. The buffer zone should be sufficient in size to prevent the possibility of unintended contact of prohibited substances applied to adjacent conventional land areas/farms.

i) **Soil and water conservation:** Relevant measures shall be taken to prevent erosion, salination of soil, excessive and improper use of water and the pollution of ground and surface water. Clearing of land through the means of burning organic matter e.g. slash-and-burn, straw burning shall be restricted to the minimum. The clearing of primary forest is prohibited.

j) **Collection of non-cultivated material of plant origin/ forest produces:** The collection of wild plants and parts thereof, grown naturally and in forest shall be certified as organic. Organic operators should collect products only from within the boundaries of the clearly defined wild collection area and the collection area shall be at appropriate distance from the conventional farming, pollution and contamination.

5.2. Grower Group Certification: Growers groups are organized group of farmers/producers who intend to produce organic products/engage in organic processes in accordance with the National Standards of Organic Production (NSOP) with following requirements:

a) Shall apply to grower groups, farmers' cooperatives, contract production and small-scale processing units.
b) The producers in the group must apply similar production systems and the farms should be in geographical proximity
c) The grower group shall consist of minimum 25 and maximum 500 farmers.
d) Individual farms with land holding of 4 ha (10 acres) and above can also be a part of the group but the total area of such farms shall be less than 50% of the total area of the group (To be inspected separately every year by the accredited certification body).
e) The grower group shall be based on the Internal Control System (ICS).
f) The ICS will have a registered legal identity and have a constitution of the organization and shall be presented by an organizational chart with specified roles and responsibilities
g) Internal Control System (ICS) manual containing policies and procedures shoulkd be developed.
h) Identification of farmers and creation of awareness about Grower Group Certification
i) Qualified/experienced personnel for maintaining the Internal Control System
j) Internal Standards shall be prepared in local language by the ICS Manager
k) If the farmers are illiterate, the Internal Standards shall contain illustrations in the text for better understanding
l) The ICS personnel shall not have any conflict of interest that might hinder the work. All possible conflicts shall be declared in a written statement.
m) The group will market the products under a single entity. For trading the products from the group of producers, the ICS shall draw up relevant procedures
n) At least two internal inspections of the group (one in growing season of each crop for 100% farmers) shall be carried out by the internal inspector and will be documented.
o) In case of severe non-compliance, the results will be reported immediately to the ICS manager and all measures will be taken according to the internal sanction procedures.
p) In case of non-compliances, the ICS shall take corrective or mitigating measures.

q) The ICS manager will have a defined procedure to approve or impose sanction on the farmers in the group. All internal farm checklists are screened by internal approval staff with special focus on the critical control points of risk/difficult cases.

r) A competent person will train each internal inspector annually. The date of the training, list of participants and content of the training of all ICS staff needs to be documented.

s) The ICS manager will organize regular training to the farmers in the group

t) ICS shall develop clear buying procedures to ensure genuineness of the products from the group

u) Yields will be estimated for each crop for individual farmer in the group by the ICS and counter-checked with the estimates during external inspection by the accredited certification body

v) Procedures for implementation of sanctions will be defined in case of noncompliance.

w) Sanctions have to be documented (list of farmers issued sanctions, documentation of identified non-conformities in the files).

x) Farmers who have used prohibited inputs on their farms must undergo again the full conversion period (if they remain in the group). In such cases, it has to be checked whether the farmers have already delivered produce and whether this (now no longer certified) produce has been mingling with other produce, if this has been the case, the accredited certification body needs to be notified immediately and the mingled produce kept separate until further instructions.

y) The purchase or the warehouse manager during the handling of produce shall check the document to ensure the compliance with the NPOP standards.

z) Identification of the product at all stages of product flow during transition.

aa) Segregation of organic products from in conversion products.

ab) Fumigation of containers, irradiation/ionization etc. are prohibited.

ac) The location in the warehouse during storage must be labeled as `organic' or 'in-conversion'

5.3. Food processing and handling

a) **Organic food processing and handling plan:** The operator must develop an organic production and handling plan. This plan shall include description of practices and procedures pertaining to production, monitoring, management, separation measures, preventive measures for mixing and co-mingling, list of each substances/inputs used during production, storage and handling, record keeping system and decontamination, cleaning or disinfections practices.

b) **Pest control:**

- Pest avoidance by good manufacturing practices such as general cleanliness and hygiene.
- Recommended treatments are physical barriers, sound, ultra-sound, light and UV-light, traps (incl. pheromone traps and static bait traps), temperature control, controlled atmosphere and diatomaceous earth
- Treatments with pest regulating agents must thus be regarded as the last resort
- A plan for pest prevention and pest control should be developed
- For pest management and control the following measures shall be used in order of priority:
- Preventive methods such as disruption, elimination of habitat and access to facilities
- Mechanical, physical and biological methods
- Pesticidal substances contained in the appendices of the national standards
- Other substances used in traps
- Irradiation is prohibited.
- There shall never be direct or indirect contact between organic and prohibited substances.
- Persistent or carcinogenic pesticides and disinfectants are not permitted.

c) **Ingredients:**

- 100% of the ingredients used in processing shall be organic except where an organic ingredient is not available in sufficient quality or quantity and non organic ingredients may be used to a minimum extent only in case of essential technological need or for particular nutritional purpose which shall not be genetically engineered.
- Water and salt may be used & minerals (including trace elements), vitamins and similar isolated ingredients shall not be used.
- The certification programme may grant exceptions where use is legally required or where severe dietary or nutritional deficiency can be demonstrated.
- Ethylene gas is permitted for ripening

d) Processing methods

- The vital quality of an organic ingredient shall be maintained throughout each step of its processing methods and shall be chosen to limit the number and quantity of additives and processing aids.
- The approved kinds of processes include mechanical and physical, biological, smoking, extraction, precipitation & filtration.
- Extraction shall be either with water, ethanol, plant and animal oils, vinegar, carbon dioxide, nitrogen or carboxylic acids.
- Filtration substances shall not be made of asbestos and irradiation is not allowed.

e) Packaging

- Biodegradable, recyclable, reusable systems and eco-friendly packaging materials shall be used wherever possible.

f) Labeling

- Labeling shall convey clear and accurate information on the organic status of the product.
- When full standards requirements are fulfilled, products shall be sold as "produce of organic agriculture" or a similar description.
- The label for conversion products shall be clearly distinguishable from the label for organic products by mentioning the year of conversion.
- The name and address of the person or company legally responsible for the production or processing of the product shall be mentioned on the label.
- Product labels should list processing procedures which influence the product properties in a way not immediately obvious. All components of additives and processing aids shall be declared.
- Additional product information shall be made available on request.
- Ingredients or products derived from wild production shall be declared as such.
- Single ingredient products may be labelled as "Organic" when all standard requirements have been met.
- Multi ingredient products where not all ingredients including additives are of organic origin may be labelled in the following way (raw material weight):
- 95% of the ingredients are of certified organic origin- **"certified organic"** along with logo of the certification programme.

- 70% - 95% of the ingredients are of certified organic origin- "made with organic ingredients" with clear statement of the proportion of the organic ingredients. An indication that the product is covered by the certification programme should be used, close to the indication of proportion of organic ingredients.
- < 70% of the ingredients are of certified organic origin- "an ingredient is organic" may appear in the ingredients list. Such product may not be called "organic".
- Added water and salt shall not be included in the percentage calculations of organic ingredients.
- All raw materials of a multi-ingredient product shall be listed on the product label in order of their weight percentage. It shall be apparent which raw materials are of organic certified origin and which are not. All additives shall be listed with their full name.
- If herbs and/or spices constitute less than 2% of the total weight of the product, they may be listed as "spices" or "herbs" without stating the percentage.
- Organic products shall not be labelled as GE (genetic engineering) or GM (genetic modification) free in order to avoid potentially misleading claims about the end product. Any reference to genetic engineering on product labels shall be limited to the production method.
- Only CB approved labels are to be used by the operators

g) Storage & transport

- Controlled atmosphere, cooling, freezing, drying, humidity regulation are permitted.
- Organic products must be prevented from contamination and co-mingling with non-organic products and from contact with materials & substances not permitted for use in organic farming and handling.
- In units processing organic and non-organic, the organic products should be handled and stored separately.
- Storage areas and transport containers for organic product should be cleaned using methods and materials permitted in organic production.

5.4. Certification procedure in brief

- Application is made to the certification agency in the prescribed format with necessary farm and process details
- Screening of application by certification agency and if necessary further details/clarification sought

- Cost estimate comprising of certification charge, inspection charge, travel cost, reporting cost, laboratory charges etc is sent for acceptance
- Acceptance of cost by the grower/producer
- Signing of agreement between grower/producer and certification agency
- Certification agency seeks cropping/production/cultivation /processing plan and supply a copy of the standards to the grower/producer to follow
- Certification agency raises an invoice and asks the producer to release 50% of the certification cost in advance
- Grower/producer pays the fee
- Inspection schedule is worked out
- Inspection is carried out at one or more than one occasion
- If required unannounced inspection can also be done. In case of doubt inspection team can also draw plant/soil/raw material/input/product sample for laboratory analysis.
- Inspection report/(s) submitted to the certification committee
- Certification agency asks for final payment
- Final payment is made
- Certification is granted
- Grower/producer releases the stock for sale with certification mark

6. Participatory Guarantee System

PGS-India certification services are facilitated through an institutional network comprising of National Advisory Committee, National Executive Committee, PGS Secretariat, Zonal and Regional Councils and Local groups. Although majority of the institutional services are facilitated by the PGS secretariat free of cost, authorization of RCs, physical inspection of LGs by RCs, certification endorsement of LGs by RCs, physical inspection and grant of certification of individual farmers, processing units or large area certification are paid services and necessary fee shall be paid by the usersas per the decision of NEC as revised from time to time and/ or as agreed between RCs and LGs and other operators. (*Ref: PGS India Guidelines, PGS-India Operational Manual, National Centre of Organic Farming PGS-INDIA Secretariat Department of Agriculture Cooperation and Farmers Welfare Ministry of Agriculture and Farmers Welfare Government of India Published Year : 2021 Month : June)*

6.1. Individual producer

- In cases where an individual farmer (or a group of farmers less than 5) is interested in PGS-India certification and there is no group in the vicinity and no additional community members are ready to form a group, the individual producers can directly apply to the regional council for registration.

- Make an application to available regional council with signed application form,pledge and a signed declaration stating their current inability to form a local group.
- Invite other PGS-India group members to endorse the application after physical visit to farm. In case if no PGS-India group is accessible then request the Regional Council to physically verify the farm.
- Obtain PGS-India operational guidelines and standards from near by group or from RC.
- Request the RC to approve the registration and obtain user ID and password for PGS-India website.
- Request the Peer appraisers of nearby group to physically inspect the farm and fill peer appraisal form. Filled peer appraisal farm to be submitted to Regional Council. In case if no group is close by then request RC to do physical inspection and grant certification.
- Individual producer registration is an interim arrangement and the producer must initiate efforts to bring in other members from the village community to form the group in due course (maximum 2 years) and transform individual status to group status. In case if an individual farmer is unable to form a group even after 2 years, then Regional Council will attach the farmers with the nearest group.

6.2. Local Group (LG)

- Local group is the main functional and decision-making body under Participatory Guarantee System. It is a local group of farmers that live in the same village or close by villages and interact regularly with each other. Participation of consumers or representatives of traders or retailers in the group and its functioning should be encouraged as it strengthens the integrity and trust.
- A local group should comprise of minimum 5 members belonging to the same village or close by villages with continuous territory. Regional Councils can decide on maximum number of farmers per group keeping local situations in mind.
- Adequate participation of women farmers shall be ensured.
- At least a few (25%) members of the group shall be well versed with the PGS organic guarantee systems or certification system and National

Standards of Organic Production (NPOP) or have undergone training on PGS guarantee system organized by Regional Councils, Zonal Council or PGS Secretariat or have been part of the core team of other functional PGS group.

- All the members in the group have signed the PGS pledge and group agreement to adhere to the group's specific vision, participatory approach and collective responsibility.
- Although, there is no restriction on the size of holding of any individual farmer but in any case, the holding of one single member should not exceed 50% of the total land under the group.
- Under PGS organic guarantee system normally parallel production and part conversion is not allowed, therefore it is necessary that all group members need to bring their entire farm with livestock under organic management as per the PGS standards. However, Regional Councils in some cases may give exception and may also allow conversion in phases.
- Have access to PGS documents and preferably have access to computer and internet (optional)
- Registered with the concerned Regional Council and have obtained necessary user ID and password to upload data on PGS website.
- In case if farmer group is unable to operate on-line system of data up loading then the services can be availed from Regional Council or of any other facilitating agency or local NGO or service providers etc.
- In case an individual farmer or a group of farmers with less than 5 members is/are proposed to join existing LG, as advised by the RC, then to accept those farmers as members of the LG.
- LG need to register only once, as there is no concept of renewal in PGS-India, till the LG keeps on doing certification activities, uploading peer appraisal summary sheets on season to season basis on PGS-India website. Non- submission of peer appraisal summary sheet continuously for two seasons or 12 months shall result in suspension of the group. Revival of certification process for such groups will restart from the date of submission of fresh peer appraisal and fresh certification process will start with PGS-Green-1 status.
- Organize farmers in the group and each member individually sign PGS organic pledge and group agreement.
- Provide copies of PGS standards, operational manual and appraisal forms to all the members in the local language. If farmers are illiterate

then they need to be explained details and standards orally and through pictorial representations.

- Prepare necessary field documents with farm history. Each group shall maintain such documents in a group file comprising of application form, signed pledge, signed agreement, farm maps showing locations with GPS coordinates in respect of each member and last one year history on input usage and management practices.
- Elect group leader and core team of peer reviewers (minimum 3 in 5 members group). There is no upper limit. It will be an optimal situation if all members of the group can participate in peer reviews, as this contributes to capacity building and information exchange between farmers, and reduces conflicts of interests.
- Participate in the activities of any other registered PGS group to understand the functioning of the PGS Group.
- Implement standard requirements on the farms of all the group members and obtain endorsement from the other registered group. This endorsement is needed only once at the time of registration.
- Register group on PGS-India website and obtain registration approval from the Regional Council.
- Contact nearby PGS group for endorsement
- In case no PGS registered group is there in the vicinity for endorsement then State Agencies (State Agriculture Department District Officer) may be requested to verify the requirement and submit necessary verification report to RC. Else request RC to do verification and grant registration approval. RCOFs can also be requested for verification and endorsements of Local Groups for their recognition by Regional Council. RCs can endorse the groups only after physical verification/ inspection.
- In case of groups constituted under some government programme (such as PKVY) endorsement of the group shall be done by the authorised district officer/scheme in-charge of the implementing state government department.
- Organize time to time meetings and maintain attendance register. Participation of members in these meetings is a mandatory activity and is an indication of dedication of the member to the cause of group's guarantee scheme. There should be at least 2-4 times a year (2 for perennial crop group and 4 times a year for annual crop group) compulsory meetings at key times of the year depending on the season, the crops, etc. One/two for peer appraisal planning and one/two for decision making.

- Every member needs to attend at least 50% of the meetings in a year and sign in attendance register.
- Advise each other and share information to improve the capacity of the group as a whole.
- Organize regular training courses by inviting practicing organic farmers from other groups, RC members or experts of other state govt and non-govt agencies.
- Chalk out peer appraisal strategy and ensure timely appraisal of each farm at least twice a year. Peer reviewers will ensure to complete Peer review appraisal form, sign and submit to the group leader. Each farmer is to be appraised by at least a two member team. Inclusion of consumer's representative increases the credibility and trust.
- Inspection of peer reviewer's farms to be done by another peer reviewer group. To increase credibility and trust the group may have any number of peer reviewers.
- All peer appraisal sheets in respect of each group member needs to be maintained in hard copy or digitally by the local group for future supervision activity. These are to be made available in the public domain and provided to RC or statutory authority upon demand during physical supervision.
- At appropriate times the group decides which farmers are to be certified. Separate out farmers which are yet to comply with the certification requirements. List out defaulters and impose sanctions.
- Organize final decision meeting, explain the peer appraisal results to all the members. Collectively declare the group as conforming to PGS standards (in case of small groups, up to 10 members). If the group is large then elect a sub- group or certification committee, comprising of 5 or more members, which may review the results and decide upon the certification. Approval of majority group members is required only in the cases of negative decision (denial of certification or decertification). Full member body can also serve as an appeal body against the decision of the certification committee.
- At appropriate time prepare peer appraisal summary sheet with list of farmers declared certified with details of crops and expected quantity of produce.
- Upload peer appraisal summary sheet for the entire group on the PGS-India website along with necessary group decision and send signed hard copy to RC through post.
- On being approved by RC on-line, RC shall issue the certificate.

Certification process by Local Group (LG)

Step 1

- Form a group comprising of minimum 5 farmers (preferably belonging to one village or close by villages with contiguous territory).
- To avoid loss of group status in the event of few farmers are sanctioned or leave the group, it is advisable to have group of about 7-10 farmers
- Collect registration and farm history sheet from all the members.
- Obtain copies of PGS-India Standards and PGS-India operational documents from nearest RC and distribute to all the members. These documents can also be downloaded from PGS-India website.
- Convene a meeting of all the members to sign the pledge.
- All members need to amend/ modify their practices to comply with the requirement of PGS-India standards and ensure continued compliance.
- Prepare local group operational manual detailing requirement of documents to be maintained by farmers, peer appraisal methodology and checkpoints to be assessed, based on the broad guidelines of PGS-India norms (available on PGS-India website)
- Ensure that members those who have not brought their entire land holding/ operations under PGS-India programme (split production) maintain clear and distinct separation of organic units with non-organic units including farms separated by buffer zones, separate handling and thrashing facilities, separate storages etc. Such farmers also need to demonstrate their contamination control abilities to ensure the integrity of organic operations.
- Register the group on-line on PGS Website and generate user ID and password.
- Request RC to grant registration.

Step 2

- Follow the PGS guidelines on group meetings, key field trainings and knowledge sharing.
- Ensure that all members abide by the standards in their production process
- Keep vigil on other members' farms and if some non-compliances are noticed then inform group members during group meetings
- Organize group meetings from time to time and maintain necessary records.

- Ensure timely organization of training programmes in collaboration with RC.
- Invite practicing organic farmers and other local experts for problem solving in management issues such as in nutrient management, pest management etc during key trainings and group meetings.
- If member farmers are using or propose to use off-farm inputs then verify their organic status, discuss in group meetings and endorse or prohibit their use. Use of such inputs without group's approval shall be treated as non- compliance.
- LG need to maintain following documents for verification by RC/ ZC or PGS Secretariat:
- Application form of each member with last one year's farm history
- Copy of PGS-India pledge of each member
- Copy of agreement with RC
- Copy of individual peer appraisal sheets for last four seasons (24 months) till the physical inspection by RC
- Attendance and proceeding register for group meetings
- Attendance register for trainings
- Copy of peer appraisal summary sheet (season-wise as submitted to RC)

Step 3

- Chalk out peer appraisal schedule and constitute peer appraisal groups. Each group should have minimum 3 peer appraisers. Depending upon the number of farmers there can be any number of appraisal teams comprising of three or more members. At least one member in the appraisal team must be literate and well versed in filling the appraisal forms.
- Reciprocal review between two-member group farms is not allowed (*i.e.*, A reviews the B and B reviews A).
- Invite other group peer reviewers or representatives of consumers/ traders or local state agriculture department officer as invited member of the peer group (but their participation is not mandatory). This may increase the trust and credibility of the group guarantee and uphold transparency.
- Complete peer review of all the farms at least once in each season. Ensure that all farms have been reviewed objectively.
- Discuss review reports in the meetings and decide upon the organic status of each farm one by one.

- Segregate farmers who have fulfilled all the requirements and consider them for grant of certificate
- Discuss about the non-complying farmers and depending upon the nature and gravity of non-compliances issue sanctions.
- Denial of certification or exclusion of member from the group to be endorsed by the whole group and should be informed to the defaulting member

Step 4

- Appraisal paperwork is checked for completeness and a local group summary worksheet is prepared.
- Group or certification committee decides on certification and declares certification status of each group member.
- Enter all details on-line in PGS-India website and send signed copy of the appraisal summary worksheet to the RC. Alternatively send all details in hard copy along with signed copy of summary sheet to RC for uploading the data into PGS-India website.
- On receipt of approval from RC, scope certificate can be generated by group leader on-line and distributed to group members after his signature.
- Only the farmers which have completed full conversion period without any major or serious non-compliance be declared as "PGS-Organic". Farmers which have one or more major non-compliance or are under conversion period will be declared as "PGS-Green". No member can avail PGS-Green status for more than 3 years in case of annual crops and more than 4 years in case of plantation/ permanent crops. In cases where after the completion of 3 or 4 years period a member is unable to qualify as PGS-Organic the same member will be sanctioned and excluded from the group.
- In case of major non-compliance, except on use of prohibited substances, the status of PGS-Organic can be downgraded to PGS-Green for one or two seasons. On continued default the member can be excluded from group. All cases of prohibited substance use such as chemical fertilizers, pesticides, hormones etc will be treated as serious non-compliance and such members' certification status be downgraded to ZERO hour and they need to be either excluded or to registered as new member and shall have to undergo full conversion period.
- Only the crops sown after Registration approval of the group by RC shall qualify for "PGS-Green".

- Considering previous crops seasons data prior to date of registration approval by RC will be treated as major non-compliance and in all such cases the certification status of the group will be downgraded to registration.
- Scope certificates are issued with estimated yields as approved during peer appraisals.
- On harvest of crops, group needs to upload actual yields. Uploading of actual yields is necessary to obtain Transaction Certificates (TC).
- On approval of yields by RC, group can sell its produce as PGS certified produce and can generate Transaction Certificate on-line. TCs can be issued for each and every farmer member separately.
- PGS certified products can be sold with PGS-India logo as per prescribed terms and conditions for use of logo.

Step 5

- In case if RC returns the decision with reasons of return, the LG need to undertake corrective actions and resubmit the revised decision in 15 days.
- In case of certificate denial by RC, if not satisfied LG can appeal to their respective zonal council with intimation of PGS Secretariat for decision revie.

6.3. Off-farm processing and handling units/facilities away from pgs-groups

To ensure collection, aggregation, processing and handling of PGS certified organicproduce from one or more PGS groups for collective processing and trading, PGS- India programme provides for specific provisions for approval of stand-alone facilitiesaway from the PGS groups. As such activities are not participatory, the approval system shall be based on conformity assessment system.

- Only the Regional Councils authorized for "Off-farm processing and handling Approval" can register such units and approve their facilities for organic processing after necessary conformity assessment.
- Applicant processing and trading unit shall be legally registered body with necessary licenses and approval from FSSAI and/ or such other licensing authorities as required by law of the land.
- Have necessary infrastructure for intended processing and handling.
- Fully aware and conversant with PGS-India standards and certification requirements and have developed operational manual, formats and

checklists for entire operation, flow of material through the process and capable of maintaining uninterrupted audit trail starting from receipt of raw material(s), processing and handling operation, packing and finally up to sale to non-PGS operator or consumer.

- Processing and handling agencies shall have to grant full access to documentation, recipes, raw material and processed products inventory and sale and purchase records to Regional Council. Withholding of information, documentation and physical access to facilities may result in denial of certification.
- PGS secretariat will monitor such units on periodic basis

6.4. Certification process for large contogous traditional/default organic area

- Village council/ gram panchayat shall define the area to be brought under certification and prepare maps.
- Prepare farmer-wise documentation, i.e. name and family details of farmer, land details, livestock details, individual farmers land location in the map. All farmers will be taken into account and the entire defined area to be covered.
- Create village-wise group keeping one village one group concept
- Obtain PGS-India application form and PGS pledge from all the farmers
- Ensure that there is no synthetic input use history in the entire defined area at least for last 3 years and also that no licenses have been issued for sale of synthetic inputs.
- Request the State Government to issue necessary documents of the area being synthetic input usage free from last 3 years
- Make an application to the Regional Council with copy of all the above papers for registration. Applications to be made village-wise/ gram panchayat-wise
- Submit applications from all gram panchayats and villages from the defined area. Only one RC to be selected for one area.
- Undertake first peer appraisal through local farmers peer committee in each village and fill consolidated peer appraisal form.
- Submit all details to the concerned regional council and zonal council
- Upload required data into PGS-India portal.
- Facilitate PGS-NEC appointed committee for inspection and verification of the traditional organic status and for recommendation on conversion status.

- On receipt of organic certification, the village can undertake marketing of certified organic produce under one umbrella and under one brand name.
- Transaction certificates for sale can be generated from PGS portal under the common village name. Certificates and TCs in large area certification will be issued village group-wise. Individual farmers and processing units shall not qualify under large area certification.
- For renewal of certification every year at least one peer appraisal is done involving all the farmers and consolidated peer appraisal sheet, village-wise is prepared and submitted to the Regional Council.

Table 1. List of permitted, restricted and prohibited inputs for Nutrient Management

Inputs	**Condition for use**
Matter produced on an organic farm unit	
Farmyard & poultry manure, slurry, cow urine	Permitted
Crop residues and green manure	Permitted
Straw and other mulches	Permitted
Matter produced outside the organic farm unit	
Blood meal, meat meal, bone meal and feather meal without preservatives	Restricted
Compost made from any carbon based residues (animal excrement including poultry)	Restricted
Farmyard manure, slurry, cow urine (preferably after control fermentation and/or appropriate dilution) "factory" farming sources not permitted	Restricted
Fish and fish products without preservatives	Restricted
Guano	Restricted
Human excrement	Prohibited
By-products from the food and textile industries of biodegradable material of microbial, plant or animal origin without any synthetic additives	Restricted
Peat without synthetic additives	Prohibited for soil
	Conditioning
Sawdust, wood shavings, wood provided it comes from untreated wood	Permitted
Seaweed and seaweed products obtained by physical processes extraction with water or aqueous acid and/or alkaline solution	Restricted
Sewage sludge and urban composts from separated sources which are monitored for contamination	Restricted
Straw	Restricted
Vermicasts	Restricted
Animal charcoal	Restricted

Compost and spent mushroom and vermiculate substances	Restricted
Compost from organic household reference	Restricted
Compost from plant residues	Permitted
By products from oil palm, coconut and cocoa (including empty	Restricted
fruit bunch, palm oil mill effluent (pome), cocoa peat and empty cocoa pods)	
By products of industries processing ingredients from organic agriculture	Restricted
Minerals	
Basic slag	Restricted
Calcareous and magnesium rock	Restricted
Calcified seaweed	Permitted
Calcium chloride	Permitted
Calcium carbonate of natural origin (chalk, limestone, gypsum	Permitted
and phosphate chalk)	
Mineral potassium with low chlorine content (e.g. sulphate of potash, kainite, sylvinite, patenkali)	Restricted
Natural phosphates (e.g. Rock phosphate)	Restricted
Pulverised rock	Restricted
Sodium chloride	Permitted
Trace elements (Boron, ferrous, manganese, molybdenum, zinc)	Restricted
Wood ash from untreated wood	Restricted
Potassium sulphate	Restricted
Magnesium sulphate (Epson salt)	Permitted
Gypsum (Calcium sulphate)	Permitted
Silage and silage extract	Permitted excluding
	Ammonium silage
Aluminum calcium phosphate	Restricted
Sulphur	Restricted
Stone meal	Restricted
Clay (bentonite, perlite, zeolite)	Permitted
Microbiological preparations	
Bacterial preparations (bio fertilizers)	Permitted
Biodynamic preparations	Permitted
Plant preparations and botanical extracts	Permitted
Vermiculate	Permitted
Peat	Permitted

Tabel 2. List of permitted, restricted and prohibited inputs for pest, disease and weed management

Inputs	Condition for use
Substances from plant and animal origin	
Azadirachta indica (neem preparations)	Permitted
Neem oil	Restricted
Preparation of rotenone from *Derris elliptica Lonchocarpus, Thephrosia species*	Restricted
Gelatine	Permitted
Propolis	Restricted
Plant based extracts – garlic, pongamia etc.	Permitted
Preparation on basis of pyrethrins extracted from *Chrysanthemum cinerariaefolium*, containing possibly a synergist *Pyrethrum cinerafolium*	Restricted
Preparation from *Quassia amara*	Restricted
Release of parasite predators of insect pests	Restricted
Preparation from *Ryania species*	Restricted
Inputs	Condition for use
Tobacco tea	Prohibited
Lecithin	Restricted
Casein	Permitted
Sea weeds, sea weed meal, sea weed extracts, sea salt and salty water	Restricted
Extract from mushroom (shitake fungus)	Permitted
Extract from *Chlorella*	Permitted
Fermented product from *Aspergillus*	Restricted
Natural acids (vinegar)	Restricted
Chloride of lime/soda	Restricted
Clay (e.g. bentonite, perlite, vermiculite, zeolite)	Permitted
Copper salts / inorganic salts (Bordeaux mix, copper hydroxide, copper oxychloride) used as a fungicide depending upon the crop and under the supervision of accredited certification Body	Restricted
Mineral powders eg : stone meal	Prohibited
Diatomaceous earth	Restricted
Light mineral oils	Restricted
Permanganate of potash	Restricted
Lime sulphur (calcium polysulphide)	Restricted
Silicates, clay (bentonite)	Restricted
Sodium bicarbonate	Restricted

Inputs	Condition for use
Sulphur (as a fungicide, acaricide, repellant)	Restricted
Microorganism used for biological pest control	
Viral preparation (eg. Granulosis virus, Nuclear Polyhedrosis Virus etc.	Permitted
Fungal preparations (*Trichoderma spp.*)	Permitted
Bacterial preparations (*Bacillus spp)*	Permitted
Parasites, predators and sterilized insects	Permitted
Others	
Carbon dioxide and nitrogen gas	Restricted
Soft soap (potassium soap)	Permitted
Ethyl alcohol	Prohibited
Homeopathic and ayurvedic preparations	Permitted
Herbal and biodynamic preparations	Permitted
Traps	
Physical methods (chromatic traps, mechanical traps, sticky traps and pheromones)	Permitted

7. Conclusion

The NPOP and PGS systems of certification provide details of processes for production, conversion requirements, nutrient & pest management, contamination, landscape, choices of crop varieties, production & management plan, ingredients, processing and handling, packaging, labelling, storage & transport etc. for individual farmer and group certification. Various steps involved in certification process by individual and local groups have also been dealt in the paper. It also dealt with certification process for large contiguous Traditional/ Default organic area. Details of inputs permitted, restricted and prohibited have also been illustrated in paper. The paper will be helpful to various stakeholders *viz,* farmers, processors, traders, exporters, importers, consumers, academia's etc. at large in understanding various nuances in certification of organic produce.

8. Way Forward

Major focus on scope categories such as livestock, aquaculture, mushrooms, Honey, Textile and cosmetics will add to the certification of allied agriculture products. Strengthening exporters with export infrastructure and enhancing quality assessment capacities in labs, massive promotion of India Organic brands and doubling participation in international fairs will make Indian brands stand high in the other countries. Doubling the BSMs and RBSMs in India and other countries will promote the potential organic production.

Developing synergies between government entities for development of exportable commodity specific concentrated clusters in value chain mode in all the States with adequate financial assistance, encouraging clusters for NPOP/PGS certification under existing PKVY scheme and to make sure all clusters to have linkages with capacity building and market facilitation programmes of APEDA will facilitate community based organic production and marketing activities for higher profits. Identification of nodal department for promotion of agricultural exports with dedicated desk for organic products shall provide instant problem solving. Inclusion of organic agricultural exports in the state agricultural export policy will give a new swift to organic agriculture.

9. References

Food Safety and Standards (Organic Food) Regulation 2017 under FSS Act 2006

National Programme for Organic Production Standards (7th Edition): November 2014 by APEDA, Department of Commerce, Ministry of Commerce and Industry, New Delhi.

PGS India Guidelines, PGS-India Operational Manual, National Centre of Organic Farming PGS-INDIA Secretariat Department of Agriculture Cooperation and Farmers Welfare Ministry of Agriculture and Farmers Welfare Government of India. 2021.

16

Guidelines for Conducting Research on Organic Farming

K. Lakshman[1], B.G. Shivakumar[2], B. Gangaiah[3], A.R. Sharma[4] Jayanta Layek[5], Sushmita Munda[6], V. Karunakaran[7], G. Prasad Babu[8], N. Ramesh Naik[1], D. Akshay[9] and E. Vidya Maduri[10]

[1]Agricultural Research Station, Acharya N.G. Ranga Agricultural University Ananthapuram-515 001, Andhra Pradesh, India

[2]ICAR-Indian Grassland and Fodder Research Institute, Southern Regional Research Station, Dharwad-580 005, Karnataka, India

[3]ICAR-Indian Institute of Millets Research, Hyderabad-500 030 Telangana, India

[4]Rani Lakshmibai Central Agricultural University, Jhansi-284 003, Uttar Pradesh India

[5]ICAR-Indian Institute of Agricultural Biotechnology, Ranchi-834 010, Jharkhand India

[6]ICAR-National Rice Research Institute, Cuttack-753 006, Odisha India

[7]Agricultural College and Research Institute, Eachangkottai Thanjavur Tamil Nadu India

[8]ICAR-KVK, Kandukur, Prakasam-523 105, Andhra Pradesh, India

[9]Agricultural College, Bapatla, Acharya N.G. Ranga Agricultural University Guntur-522 101, Andhra Pradesh, India

[10]ICAR-Indian Agricultural Reseaech Institute, New Delhi-110 012

Abstract

Organic farming research is complex and intertwined with various aspects of agro-climatic conditions, edaphic factors, organic sources of nutrients and pesticides, cropping systems and cropping patterns. This research should encompass all these issues for achieving meaningful findings for practical utility. Although the broad contours of agricultural research are similar, the background information on various natural resources, agricultural practices, socio-economic factors of farming community, the scale and magnitude of research areas and the duration make organic farming research truly more complex. Further the refinement of organic farming research findings need a greater understanding of the whole spectrum of agrarian scenario of a given geographical area. In spite being more complex and confounded the organic farming research needs a greater attention for evolving more eco-friendly agricultural practices for safeguarding the fragile natural environment and yet produces healthy and chemical-free food commodities for the consumers. This chapter throws light on research issues, methodologies, and associated issues in organic farming research.

Keywords: *Organic farming, organic nutrients, organic pesticides, research issues, research methodologies, socio-economic interrelationships*

1. Introduction

Organic farming is gaining a lot of importance in recent years owing to raising awareness among the consumers of the agricultural commodities about the health concerns of the indiscriminate use of chemical fertilizers and pesticides (Anonymous 2015; Lakhsman *et al.,* 2024). The increasing demand for organically produced agricultural commodities and the premier prices have attracted the farmers' interest and motivated them to switch over to organic production systems. The shifting to organic farming by the farmers needs careful consideration of several issues starting from the suitability of soils, selection of crops, adoption of package of practices, processing, storage and marketing. There are several finer aspects which the farmers have to gloss over before they decide to change their production system to organic farming. Although the broad contours are known to all the farmers, there is a need to look into the research findings in the areas of farmers' interest for changing over to organic farming. The research evidence on the practical viability in a newer area will boost the confidence of the farming community for undertaking the organic farming. It further helps in systematic planning by the policy makers. The traditional research carried out with the modern agricultural practices may not be appropriate for organic farming. There are fundamental differences in the approaches followed in the organic farming research where the emphasis is on using all the natural resources in a complementary way so that there is hardly any need for synthetic inputs in the form of fertilizers and pesticides. It calls for systematic inventorization of resources available and additional organic inputs needed to make organic farming successful. It begins with assessing the suitability of soils, agro-climatic conditions, selection of crops, adoption of management practices, harvesting, processing, storage and marketing of organically-produced food commodities. The practical feasibility, economic viability and social acceptability of organic farming in newer areas need to be established through sound organic farming research before it is taken to the farming community.

2. Research Issues for Organic Farming

The organic farming research is more comprehensive and needs to be carried out with background information on various issues. A brief insight into the most important issues necessary for formulating organic farming research have been enumerated below:

2.1. Agro-climatic conditions

India was practicing a truly organic farming since time immemorial up to independence in 1947 (Chandrashekar, 2010). This farming system was well defined. The selection of crops and production practices were in tune with agro-climatic conditions such as the rains received and soil type in given location. However, after the independence due to urgency in augmenting food production, chemical farming with the use of chemical fertilizers and pesticides and building irrigation projects led to abandonment of the traditional farming. In a hurry to enhance food production, the clear delineation that existed earlier with reference to crops cultivated, cropping pattern and production technology followed was given up and the crops unheard earlier in a given location, modern production practices with newer high-yielding varieties and use of chemicals began in a big way. This resulted in growing of crops not appropriate in a given area owing to availability of irrigation facilities and chemical fertilizers diminished the importance of agro-climate based farming systems. Since the organic production system is based on synchronize utilization of natural resources and complementary use of available farm resources, there is a need for evolving systems based on the season, rainfall, beginning and end of rainy season, temperature lows and highs, wind velocity and humidity for meaningful and synergistic use of natural resources. Thus, the formulation of agricultural practices in general and organic farming in particular need to be reconsidered. The suitability of the existing crops should be assessed based on the rainfall pattern, quantum of rainfall, duration of agricultural season and sustainability in long-run.

The Agro-Climatic Regional (AGR) Planning aims at more scientific utilization of natural and manmade resources available in a given location. India is divided into 15 Agro –Climatic regions delineated on the basis of agro-climatic factors such as soil type, rainfall, temperature and water resources. An overall development profile of each region is formulated through an optimal mix of land stock management, crop production, animal husbandry, aquaculture, horticulture, forestry and agro-processing activities (Sandhu *et al.*, 2016).

2.2. Edaphic factors

Soil is the most important natural resource that determines the success or failure of agriculture. There are a wide variety of soils in India, such as alluvial, red, black, mountain, desert, saline and alkaline, laterite and peat soil (Figure 1). The soil physical, chemical and biological properties help in selection of crops and production technologies (Diekmann, 2013). The understanding of soil is crucial for agricultural planning, land-use management, environmental assessments, and natural resource planning. It aids in optimizing land productivity and implementing sustainable agricultural practices. In general soils having levelled topography, inherent high soil fertility, free from soil borne pests and diseases, and good water drainage facility are considered ideal for organic farming. Selection of crops based on the soil type will result in complementarity among the resources. There has been reasonable understanding about the soil type and suitable crops for cultivation (Table 1). Although there is diversity in crop distribution due to irrigation facilities and high yielding varieties of crops, there is a need to identify crops and varieties based on suitability of soils prevalent in a given area for organic farming research.

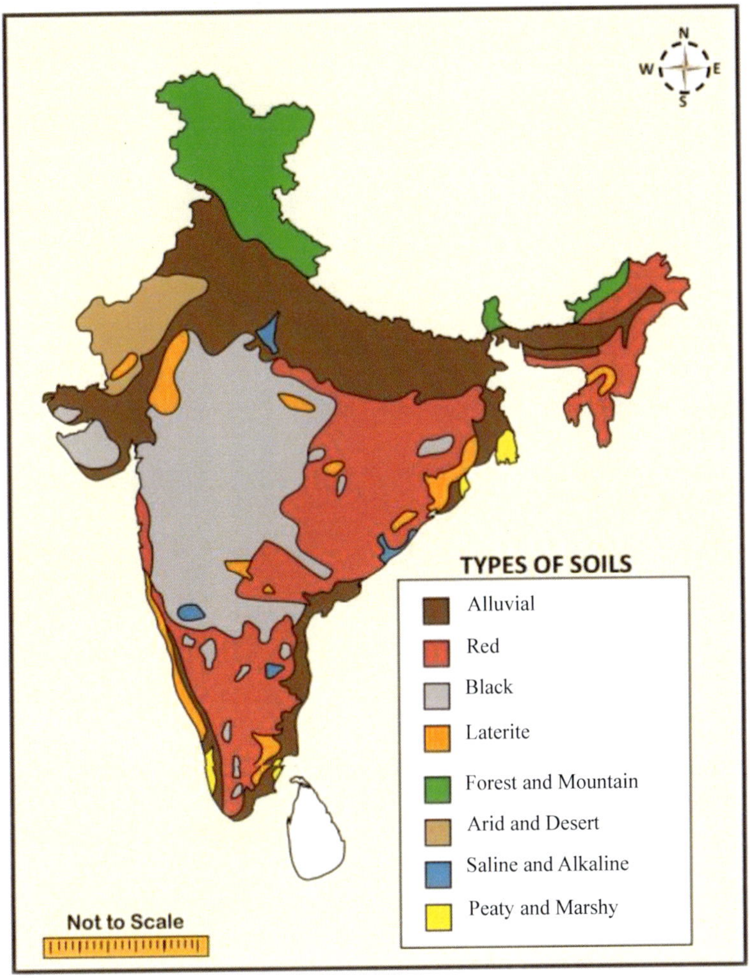

Fig. 1. Soil types in India

Table 1. Soil types and suitable crops

Soil type	Distribution	Characteristics	Crops suitable
Alluvial soils	River valley of Ganges, Brahmaputra; Plains of Uttar Pradesh, Punjab, Haryana, West Bengal, Bihar	Rich in nutrients viz., Potassium, phosphorus, Calcium	Rice, wheat, sugarcane, mustard, chickpea, jute etc.
Black soils	Maharashtra, Madhya Pradesh, Gujarat, Telangana, Karnataka, Andhra Pradesh	Rich in iron, calcium, magnesium, aluminum, etc.	Cotton, sorghum, pigeonpea, bajra, chickpea, safflower, sunflower,

Red soils	Karnataka, Kerala, Tamil Nadu, Central India,	More sandy, less clayey, poor water holding capacity, rich in iron, less soil organic matter	Ragi, groundnut, sugarcane, rice, potato, millets, tobacco, greengram, blackgram, horsegram, field bean
Laterite soils	Coastal areas of Gujarat, Maharashtra, Goa, Karnataka, Kerala,	Rich in iron and alluminium, acidic in nature, poor water holding capacity, very hard and brick like	Cashew, coconut, tapioca, rubber, coffee
Forest and mountain soils	Jammu and Kashmir, Himachal Pradesh, Uttarakhand, Western and Eastern Ghats	Rich in humus, acidic in soil reaction, deficit in potassium, phosphorus and calcium,	Apple, oranges, coffee, tea, rice, maize, potato, cardamom, black pepper
Arid and desert soils	Rajasthan, Parts of Gujarat, Punjab and Haryana	Alkaline in reaction, poor organic matter, rich in calcium	Bajra, barely, mothbean
Saline and alkali soils	Karnataka, Telangana, Andhra Pradesh, Bihar, Uttar Pradesh, Punjab, Haryana, Rajasthan, Maharashtra	High in salts viz, sodium, magnesium, calcium	Only salinity tolerant varieties of crops
Peaty and marshy soils	Coastal areas of Kerala, Odisha, Tamil Nadu, West Bengal, Bihar, Uttarakhand	High soluble salts, organic matter, poor in potassium, phosphate	Paddy, jute

2.3. Crops and cropping systems

Indeed the selection of crops and cropping systems has been of late governed by the irrigation facilities and nutrient management. With the advent of high efficacy pesticides and availability of crops and varieties, there has been spread of non-traditional crops in different regions of the country. The organic farming systems based on the non-traditional crops in a given locality, may not be sustainable in long-run and farmers may not realize the expected monetary returns from adoption of such practices. Crops too vary in their response to organic farming particularly in transition period. The grain yield of wheat was markedly lower under organic management (36-65% and 23-54% less in the first and second year of transition, respectively) than with the conventional treatment (Gopinath *et al.*, 2008). Similarly, the yields of bell pepper and french bean under organic management were significantly lower (25.2-45.9% and 29.5-46.2%, respectively) than under conventional management during a 2-year conversion period (Gopinath *et al.*, 2011). It has been reported that some crops and varieties are more suitable for organic farming (Ravishankar *et al.,* 2021). Hence only such crops and varieties should be given emphasis for greater impact in terms of higher yield. Recently a news item made the rounds in Karnataka highlighting the success story of a farmer growing apple near Bengaluru (Anonymous, 2024). However, there was a swift and prompt rejoinder from scientists expressing apprehensions about viability of such cultivation in long-run. Hence organic farming research should also take in to consideration, the traditional crops and cropping systems for long-term adoptability, sustainability and profitability. Suitable crops and cropping systems for organic farming in different states have been identified and suggested (Table 2 and 3).

Table 2. State-wise major crops grown under organic farming in India (Avasthe *et al.*, 2016)

State	Crops
Andhra Pradesh	Cotton, maize, pulses, oilseeds, fruits, vegetables
Arunachal Pradesh	Maize/sorghum, pulses, oilseeds, tea/coffee, herbs/medicinal plants
Assam	Tea/coffee, fruits, vegetables
Chhattisgarh	Rice, wheat, vegetables
Delhi	Wheat, vegetables
Goa	Fruits, vegetables
Gujarat	Cotton, pulses, oilseeds, vegetables
Haryana	Basmati rice, wheat, maize, vegetables
Himachal Pradesh	Wheat, fruits, vegetables
Jammu & Kashmir	Spices, fruits, vegetables
Karnataka	Cotton, rainfed wheat, maize, sorghum, pulses
Kerala	Spices, vegetables, herbs
Manipur	Spices, vegetables, herbs
Maharashtra	Cotton, rice, wheat, pulses, oilseeds, spices, vegetables
Madhya Pradesh	Soybean, wheat, vegetables
Meghalaya	Spices, vegetables
Punjab	Basmati rice, wheat, vegetables
Sikkim	Maize, rice, buckwheat, finger millet, vegetables, spices, herbs
Rajasthan	Cotton, wheat, seed spices, vegetables
Tamil Nadu	Tea, herbs, spices
Uttar Pradesh	Rice, wheat, maize, vegetables
Uttarakhand	Basmati rice, vegetables, maize, sorghum, herbs, spices
West Bengal	Tea, vegetables

Table 3. State-wise cropping systems under organic farming in India (Ravishankar *et al.*, 2017)

State	Cropping systems
Chhattisgarh	Soybean-chickpea; Soybean-onion; Rice-chickpea
Himachal Pradesh	Maize-garlic; Cauliflower-pea-tomato; Coriander-pea-tomato
Jharkhand	Rice (Basmati type)-wheat; Rice (Basmati type)-lentil; Rice (Basmati type)-linseed; Rice (Basmati type)-potato
Karnataka	Soybean-wheat; Groundnut-sorghum; Chilli+cotton; Maize-chickpea
Kerala	Turmeric-fallow; Ginger-fallow; Black pepper; Elephant foot yam + green manure cowpea; Green manure cowpea-yams + green manure cowpea; Taro + green manure cowpea
Maharashtra	Rice-groundnut; Rice-dolichos bean; Rice-cucumber; Rice-red pumpkin
Madhya Pradesh	Soybean-wheat; Soybean-mustard; Soybean-chickpea; Soybean-isabgol/linseed; Green manure (Sunhemp)-scented rice-durum wheat; Scented rice-chickpea-maize (Fodder); Scented rice-berseem (Fodder+Seed); Scented rice-vegetable pea-sorghum (Fodder)
Meghalaya	Rice-carrot (Raised beds in lowland); Rice-tomato (Raised beds in lowland); Maize+soybean-french bean (Upland)
Punjab	Maize-potato-summer greengram; Turmeric-onion; Basmati rice-wheat-green manure (Sesbania); Maize-durum wheat-cowpea (Fodder); Maize-berseem-bajra (Fodder system); Maize-berseem-maize+cowpea (Fodder System)
Sikkim	Maize (green cobs)-black gram (Pahenlo dal)-buckwheat; Maize+beans- vegetable pea; Maize+beans-rajmash; Maize+beans-toria; Rice-vegetable pea-maize (green cobs); Rice-fenugreek (Leafy vegetable)-baby corn; Rice-sunflower-dhaincha (Green manuring); Rice-vegetable pea; Vegetable cropping sequence for low cost plastic tunnels; Vegetable cropping sequence for low cost plastic rain shelter; Organic fruit production techniques
Rajasthan	Blackgram-fenugreek; Clusterbean-chickpea; Sesame-cumin; Groundnut-mustard; Groundnut-wheat; Castor-greengram; Maize-wheat (Durum); Soybean-coriander
Tamil Nadu	Cotton-maize-green manure (sesbania); Chillies-sunflower-green manure (Sesbania); Beetroot-maize-green manure (sesbania)
Uttar Pradesh	Basmati rice–wheat-Sesbania green manure; Coarse rice–barley+mustard–greengram; Maize (Grain)–potato–okra; Maize (Green cobs)–mustard+radish-Sesbania green manure
Uttarakhand	Basmati rice- wheat-Sesbania; Basmati rice-lentil-Sesbania; Basmati rice-vegetable pea-Sesbania; Basmati rice-Brassica napus –Sesbania; Basmati rice-chickpea–Sesbania (Under biodynamic practices)

2.4. Organic sources of nutrients

The organic farming thrives on availability of organic sources of nutrients. The agri- enterprises like animal husbandry, horticulture, crop residues and other rural wastes constitute major source of organic nutrient for organic farming. The prospects of organic farming could be brighter only if organic farming is practiced in areas where there is abundant scope for availability of organic sources of nutrients. Hence, the organic farming research should consider the availability and also reliability of such sources in future. The areas where the animal husbandry is flourishing owing to various reasons, may constitute a bright spot for organic farming. Likewise areas where the availability of fodder is abundant indicate the suitability of such areas for initiating organic farming research. The animal husbandry and fodder availability forms a significant basis for organic farming on sustainable basis (Figure 2).

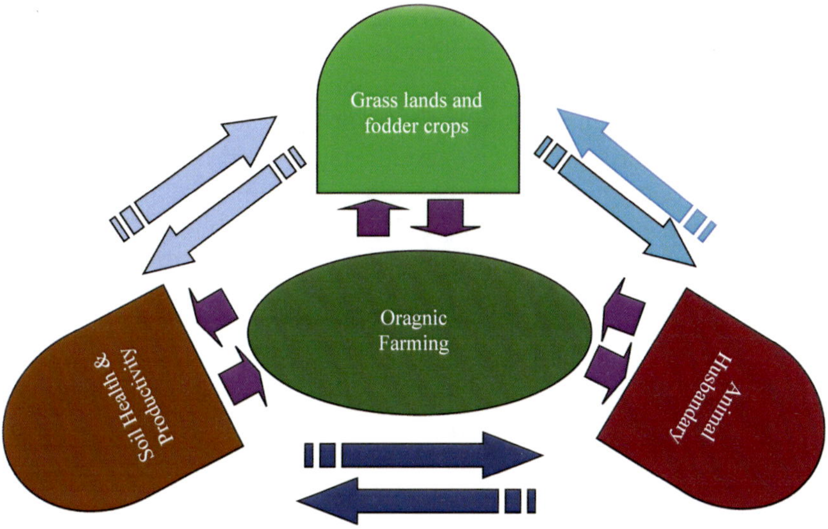

Fig. 2. Interrelationship between animal husbandary, grass lands and fodder crops

2.5. Incidence of pests, diseases and weeds

Plant protection from incidence of insect-pests, diseases, weeds under organic farming throws a major challenge. The endemic areas for various pests need to be avoided for organic farming, else the final practices emanating from such research, may be less acceptable or quite expensive. Thus, a thorough analysis of such incidences in a given locality should be undertaken. The alternative view could be to address such issues through organic ways before applied research is initiated for developing viable organic farming practices.

2.6. Plant protection practices

A large number of plant protection measures have been developed both organic and inorganic in nature. The organic measures mostly include pesticides of natural origin. The neem-based pesticides have been in vogue in many areas following organic farming. However, the use of chemical pesticides has been on rise in many areas owing to non-availability of such organic formulations. Hence a fact check for the availability of organic plant protection chemicals constitutes a major area before initiating organic farming research. When farmers are already practicing intensive agriculture practices under irrigated farming and using chemical fertilizers, pesticides; such areas should be low priority areas for initiating organic farming research afresh.

2.7. Level of mechanization

There has been an increasing degree of mechanization of farming in many areas owing to non-availability of manual labourers, urgency of the agricultural operations like timely sowing, inter-cultivation during rainy season with a small window of rain-free period, plant protection measures. The organic farming practices too should take into consideration to amalgamate such mechanization practices complementarily without compromising on the principles of organic farming. Further research on organic farming may be initiated to bring about harmony of various activities under mechanized farming with appropriate modifications to suit newer organic farming niches.

2.8. Storage, transportation and marketing facilities available

The chemical farming in the process has established protocols for storage, transportation and marketing involving usage of chemicals to protect and conserve the agricultural commodities for longer period of time. Thus, there is a need to take precautions of developing contamination-free storage, transportation and marketing facilities for organic produce. These operations should be devoid of using chemicals and cost effective. Further, organic produce owing to higher prices may require longer storage, transportation and establishment of newer markets. The organic farming research should also consider these aspects carefully.

2.9. Farmers' flair

The kind of farming practices also depend on the farmers' interest and concern for the chemical-free environment. Farmers who have been attracted towards organic farming for want of chemical free produce, freeing the ecosystem from the hazards of chemicals, tend to easily adopt the organic farming practices that are going to emanate from the proposed organic farming research. However, all the farmers may not be so keen owing to difficulties associated with adoption of organic farming. For example, the crop yields may be reduced in the initial years of switch over to organic production, there may not be preferential market for organic produce, non-availability of organic sources of nutrients, pesticides etc. These issues should be properly considered before planning organic farming research in newer areas.

2.10. Socio-economic conditions

Ultimately it is the socio-economic conditions prevailing in a given area which determine whether the farming community is going to adopt the newer practices of organic farming or not. The level of education, economic status, land holding pattern, complementary enterprises, viz. animal husbandry, cropping systems, cropping pattern, agrarian developments like irrigation facilities, marketing facilities etc. They need to be studied to formulate newer organic farming research programme so that later on the findings will be easily accepted by the farming community. There could be favourable attitude towards organic farming with connections between age, education background, farm size, organic farming benefits and social factors (Patidar and Patidar, 2015).

3. Methodologies for organic farming research

The organic farming research procedures differ significantly as compared to chemical farming research programmes. The experimental techniques adopted in the organic farming research at times may not yield meaningful results. The organic farming research is more a probe into nature and harness the nature for enhancing the production without any detrimental impact on it. Here the emphasis is on the synergistic use of locally available resources and avoiding external inputs of synthetic nature. In organic farming research, the response in terms of quantitative and qualitative output may not be as conspicuous as in chemical farming. The response may be slow, non-uniform and non-repetitive at times unlike in chemical farming. Hence, to arrive at plausible research findings we will have to conduct organic farming research with a lot of care and precautions. To make sure, the responses are exclusively due to treatments, the additive factors playing the role in between should be minimized. It can be achieved by careful consideration of the following:

3.1. Experimental area

The organic farming experiments tend to be vitiated by the spatial variabilities present in the soil owing to previous cultivation practices. It is necessary to first overcome the heterogeneous soil fertility status. Likewise, the occurrence of noxious weeds may also affect the performance of organic treatments. If organic farming experiments are conducted in small piece of land such heterogeneity may confound the effect of treatments. Hence to overcome such differences, the experimental plots should be reasonably larger. As such there is no fixed size for organic farming experiments. The size varies on the topography of the experimental land, fertility gradient, uniformity in the physico-chemical properties and number of treatments and replications proposed to be tested. However, it is agreed that the experimental treatment plots should be as large as possible to mask the smaller variabilities existing in the plot and even out such effects to attribute the differences in the treatments exclusively to the treatments only. There are again large experimental areas ranging from 100-1000 m^2 plots in organic farming experiments.

The important aspects in organic farming research include testing of organic sources of nutrients and pesticides. There are several organic sources of nutrients viz., farm yard manure, green manure, bio-fertilizers, vermicompost, crop residues etc. These sources tend to have slower release of nutrients and hence longer period of effect. However, when compared with chemical fertilizers they seem to be less effective. Therefore, these sources should be compared over a longer period of time to arrive at better comparison. The treatments also should be arranged in such way that the comparisons among different sources are well balanced and meaningful. The organic sources of nutrients vis-à-vis absolute control, conventional farmers' practices or integrated nutrient sources may elicit better understanding of organic sources. Further the nutrient use efficiency, residual effect and nutrient dynamics studies for longer period of time will provide greater understanding of the impact of these organic sources. The observations on impact on quality aspects of the agricultural produce and other synergistic effect on soil micro-flora, micro-fauna and micronutrient availability status gives a better measure of impact of organic sources of nutrients. Likewise, the sources of pesticides viz., neem based formulations, bio-pesticides etc. too require a comprehensive study over a period time. They should also be studied in comparison to no pest control measure or farmers' practice before comparing with the chemical measures. The observations on such comparative studies should be based on larger number samples over a longer period of time. Further, besides the impact of treatments on the crop yield, the impact on quality aspects of crop produce and soil nutrient supplying capacity and physical, chemical and biological properties also should be taken into consideration. The incidence of pests and their dynamics over a period of time need to be monitored for evaluating the benefits of organic farming. Maintenance of soil health, fertility, productivity; effect on environment and ecosystem services are important aspects of organic farming. The holistic studies will help in realizing the long term benefits of organic farming.

3.2. Duration of experimentation

The duration of organic farming experiments varies depending upon the conversion period from chemical farming to uniformity achieved following some standard practices. Once required degree of uniformity is achieved in the experimental area, the experimental period depends on the expected period of influence of the treatments till the results could be attributed to the treatments imposed. Further, there will be variability in the duration of the experiments depending upon the nature of treatments. The fertility treatments need to be studied at least for a period of 3 years once the conversion period is completed. On the other hand, the plant protection trials may be for 2-3 years subject to the conditions that the intensity of the pests is uniform over the experimental period and the results are consistent. Further, the duration of experiments also varies with the nature of the treatments imposed and factors affecting the expected outcome.

3.3. Confirmation of the research findings

It is important to ensure that the results are tested with same set of treatments at different places in the same agro-climatic conditions so as to confirm the research findings before they are finalized and a package is developed for their transmission to the farming community. The confirmation of the research finding may be done preferably through multi-location and long-term experiments with due diligence for reliable research findings. Mid-term review of research finding may also be conducted to rule out miscellaneous factors influencing the treatment effects. Further, proper relations between treatments and results may be worked out for validating the findings.

3.4. Economic analysis

The most important aspect of organic farming research is the economic feasibility of organic farming systems. Therefore, the organic farming research should invariably emphasize on the economic aspects. There is a common notion that the chemical farming is very intensive and needs higher investment. However, the organic farming is not quite inexpensive. It also involves reasonable expenditure. The organic sources of nutrients and pesticides too involve lot of investment. Further, the decline in yield in organic farming means lesser gross returns. In that scenario, the monetary aspects of any new organic farming system should be studied in greater details. Although the major emphasis is on quality of the produce and environmental safety in organic farming, it cannot be an attractive proposition if it fails to deliver handsome returns to the practitioner. Development of high yielding sustainable organic farming systems along with good and remunerative market price is necessary to attract farmers towards these systems (Rani *et al.,* 2023).

3.5. Extension

Popularization of identified organic farming systems is essential. However, we have not been able to convince and wean away farmers from conventional chemical farming systems to organic farming systems in substantial way. The reasons may be the yield loss during conversion period, lesser productivity vis-à-vis the chemical farming, non-availability of organic sources of nutrients and pesticides, and lack of premium price for organically-produced commodities. There comes the need of innovative extension activities for demonstrating and popularizing the organic farming systems in the vicinity of the farming community to convince and make them adopt the organic farming.

4. Associated issues for organic farming research

The organic farming research activities cannot happen in isolation in research farms. They need to be comprehensive and holistic taking on board the various issues associated with the organic farming. Unless and until these issues are duly addressed, the success of organic farming may not be satisfactory and farming community may not be forthcoming to adopt the research findings more enthusiastically and wholeheartedly.

4.1. Availability of organic sources of nutrients

The organic farming proposes to meet the nutrient demand of the crops from the organic sources only, viz. farm yard manure, vermicompost, crop residues or biological sources like bio-fertilizers, and complementary use of leguminous crops (Lakshman *et al.,* 2024). The availability of organic manures in adequate amounts and at costs affordable by the farmers is a major problem. Further, there may be uncertainties with regard to their sufficient availability and efficacy (Kumar *et al.,* 2022). Their availability on farmers own field and in the near vicinity of the farmers make more sense. The transportation of organic manures from far off places may not be economically viable option. Further a system of production and supply of these sources need to be ensured. The emphasis should also be laid on their quality aspects making them rich sources of plant nutrients and free from chemicals particularly heavy elements when urban wastes are also used in organic manure preparation.

4.2. Availability of organic sources of pesticides

There are several organic pesticides available in the market. However, their authenticity and effectiveness need to be verified. The plant-based organic sources like neem formulations need to be free from adulterants and their continued availability in the face of escalating demands in the market should be another concern. The microbial consortia or fungal consortia could be better options for using as pesticides. However, farmers cannot have their own production units and lack knowhow about them. The organic pesticides need to be further developed for different crops and practices of their appropriate usage for high efficacy are the issues that need to be addressed in organic farming research mode only (Lakshman *et al.,* 2024).

4.3. Facilities for production, storage and distribution

Additional facilities for production of organic inputs, their scientific storage and distribution for the clientele farmers practicing organic farming should be encouraged (Elayaraja and Vijai, 2020). Farmers may not be able to involve and take care of these activities. The organic farming research should also consider these aspects and entrepreneurs should be propped up in the process and the availability of quality organic inputs particularly sources of nutrients and pesticides should be given adequate attention. Then once a organic farming practice is perfected, there won't be any delay in its adoption for want of good organic sources of nutrients and pesticides.

4.4. Market proximity

In the present marketing set-up, there is no separate facility for selling organic produce. Further, there are no empirical ways of identification of organically produced commodities. It is only process certification. Therefore, organic farmers' interest should be safeguarded by facilitating reliable market infrastructure for organic produces. Although the upper strata of society are ready to pay additional premium for the organic commodities, they are placed farm away from the organic production centers. Thus, there is a need for developing markets in the near vicinity of the production centers (Manjula and Indira Devi, 2021; Masoodi *et al.,* 2022; Kaur, 2023). Further, proper transportation, processing and storage for enhancing the shelf-life of organic products should be created.

4.5. Long-term sustainability

The organic farming research should also focus on the long-term viability and sustainability. The dynamics of nutrient availability for higher organic production, novel plant protection measures devoid of chemicals, ensuring no contamination of hazardous chemicals in the organic production systems inadvertently during irrigation, intercultural operations, harvesting, processing and storage should be addressed. Further policy issues for ensuring organic production zones away from chemical farming areas and vice versa should be looked into.

5. Conclusion

There is a need for organic farming system research for developing appropriate area-specific technologies. In the absence of well researched information, the full potential of organic production system cannot be realized. At times the principles of organic farming practices will have to be compromised owing to lack of information. The organic farming research should be developed with thorough understanding of agro-climatic conditions, soil types, crop needs, cropping systems, cropping pattern, utilization of complementary enterprises to good effect in a given area. It is possible only through well-structured research systems taking into consideration various facets of organic farming. The research methodologies too to be overhauled for ensuring the appropriate measures for better and accurate understanding of impact of various organic treatments to achieve higher quantity and quality of the produce.

6. Way forward

The increasing awareness about the hazards of chemical farming on the soil sustainability, ecosystem safety and quality of the agricultural produce has created a new hope in organic farming. However, due to the limited knowledge regarding various aspects of organic farming, the adoption is slow. There is a need for systematic research in organic farming encompassing various agro-climatic regions, soil types and socio-economic conditions. The task is enormous and requires a very comprehensive and holistic approach. It may be very tedious and time consuming. Yet looking to the emerging needs for organic commodities, the challenges of organic farming research should be given top priority for safe guarding the precious natural resources and health of the consumers.

7. References

Anonymous. 2015. Base Paper on Organic Farming. Indian Council of Agricultural Research, New Delhi.

Anonymous. 2024. Hoskote apples anyone? Farmer spins a success story. Deccan Herald, 30.30.2024. Bengaluru Edition.

Avasthe, R.K., Raghavendra Singh and Subash Babu. 2016. Organic pulses production in India: perspectives and opportunities. Indian Journal of Agronomy 61 (Special issue): 93-101.

Chandrashekar, H.M. 2010. Changing scenario of organic farming in India: An overview. International NGO Journal 5 (1): 34-39.

Diekmann, F. 2013. Organic farming – A research guide. Reference and User Services Quarterly 52(3): 197-204.

Elayaraja, M. and Vijai, C. 2020. Organic farming in India: Benefits and Challenges. European Journal of Molecular & Clinical Medicine 7(11): 3021-3029.

Gopinath, K.A., Supradip Saha and Mina, B.L., 2011. Effects of organic amendments on productivity and profitability of bell pepper-french bean-garden pea system and on soil properties during transition to organic production. Communications in Soil Science and Plant Analysis 42(21): 2572–2585.

Gopinath, K.A., Supradip Saha, Mina, B.L., Kundu, S., Harit Pande and Gupta, H.S. 2008. Influence of organic amendments on growth, yield and quality of wheat and on soil properties during transition to organic production. Nutrient Cycling in Agroecosystems 82: 51–60.

Kaur, H. 2023. Challenges and prospects for organic farming: A case study from India. Gyan Management Journal 17(2): 75-79.

Kumar, R., Rana, N., Kaur, M., Bhowmik, S., Kumar, M., Negi, A., Singh, S. and Janaki Raman, P. 2022. Organic farming status in India: A review. The Pharma Innovation Journal 11(12): 2964-2671.

Lakshman, K., Veerendra, M., Nataraj, K.C., Naseerudin, R., Ramesh Naik, N., Deva, S. and Reddy, B.S. 2024. Eco-friendly products and practices – A key to successful organic farming. Chronicles of Bioresource Management 8(1): 17-23.

Manjula, M. and Indira Devi, P. 2021. Organic farming in India: Catalysts that can help in transition. Ecology, Economy and Society–the INSEE 4 (1):21–29.

Masoodi, U.H., Bashir, K. and Malik, A.A. 2022. Organic farming in vegetables- The Indian scenario. Biological Forum – An International Journal 14(3): 564-572.

Patidar, S. and Patidar, H. 2015. A study of perception of farmers towards organic farming. International Journal of Application or Innovation in Engineering and Management. 4(3): 269-277.

Rani, M., Kaushik, P., Bhayana, S. and Kapoor, S. 2023. Impact of organic farming on soil health and nutritional quality of crops. Journal of the Saudi Society of Agricultural Sciences 22:560-569.

Ravishankar, N. Ansari, M.A., Panwar, A.S., Aulakh, C.S. Sharma, S.K., Suganthy, M., Suja, G. and Jaganathan, D. 2021. Organic farming research in India: Potential technologies and way forward. Indian Journal of Agronomy 66 (5th IAC Special issue): 142-162.

Ravishankar, N., Panwar, A.S. Kamta Prasad, Vipin Kumar and Bhaskar, S. 2017. Organic farming crop production guide, network project on organic farming, ICAR-Indian Institute of Farming Systems Research, Modipuram, Meerut-250 110, Uttar Pradesh, India, p. 586.

Sandhu, J. S., Gautam, U.S., Ghosh, P. K., Dubey, S. K., Atar Singh, Kumar, R.V. and Singh, S.V. 2016. Agro-climatic Region Centered Research and Development Planning (Central Plateau and Hill region), ICAR, New Delhi, p. 25.

Index

P

Q